Global Issues Series

General Editor: Jim Whitman

This exciting series encompasses three principal themes: the interaction of human and natural systems; cooperation and conflict; and the enactment of values. The series as a whole places an emphasis on the examination of complex systems and causal relations in political decision-making; problems of knowledge; authority, control and accountability in issues of scale; and the reconciliation of conflicting values and competing claims. Throughout the series the concentration is on an integration of existing disciplines towards the clarification of political possibility as well as impending crises.

Titles include:

Berhanykun Andemicael and John Mathiason
ELIMINATING WEAPONS OF MASS DESTRUCTION
Prospects for Effective International Verification

Robert Boardman
GOVERNANCE OF EARTH SYSTEMS
Science and Its Uses

Mike Bourne
ARMING CONFLICT
The Proliferation of Small Arms

John N. Clarke and Geoffrey R. Edwards (*editors*)
GLOBAL GOVERNANCE IN THE TWENTY-FIRST CENTURY

Malcolm Dando
NEUROSCIENCE AND THE FUTURE OF CHEMICAL-BIOLOGICAL WEAPONS

Neil Davison
"NON-LETHAL" WEAPONS

Nicole Deitelhoff and Klaus Dieter Wolf (*editors*)
CORPORATE SECURITY RESPONSIBILITY?
Corporate Governance Contributions to Peace and Security in Zones of Conflict

Toni Erskine (*editors*)
CAN INSTITUTIONS HAVE RESPONSIBILITIES?
Collective Moral Agency and International Relations

Moira Feil
GLOBAL GOVERNANCE AND CORPORATE RESPONSIBILITY IN CONFLICT ZONES

Annegret Flohr, Lothar Rieth, Sandra Schwindenhammer and Klaus Dieter Wolf
THE ROLE OF BUSINESS IN GLOBAL GOVERNANCE
Corporations as Norm-Entrepreneurs

Annegret Flohr
SELF-REGULATION AND LEGALIZATION
Making Global Rules for Banks and Corporations

Beth K. Greener
THE NEW INTERNATIONAL POLICING

David Karp and Kurt Mills (*editors*)
HUMAN RIGHTS PROTECTION IN GLOBAL POLITICS
The Responsibilities of States and Non-State Actors

Alexander Kelle, Kathryn Nixdorff and Malcolm Dando
CONTROLLING BIOCHEMICAL WEAPONS
Adapting Multilateral Arms Control for the 21st Century

Kelley Lee
HEALTH IMPACTS OF GLOBALIZATION (*editor*)
Towards Global Governance

Kelley Lee
GLOBALIZATION AND HEALTH
An Introduction

Catherine Lu
JUST AND UNJUST INTERVENTIONS IN WORLD POLITICS
Public and Private

Robert L. Ostergard Jr. (*editor*)
HIV, AIDS AND THE THREAT TO NATIONAL AND INTERNATIONAL SECURITY

Graham S. Pearson
THE UNSCOM SAGA
Chemical and Biological Weapons Non-Proliferation

Graham S. Pearson
THE SEARCH FOR IRAQ'S WEAPONS OF MASS DESTRUCTION
Inspection, Verification and Non-Proliferation

Nick Ritchie
A NUCLEAR WEAPONS-FREE WORLD?
Britain, Trident and the Challenges Ahead

Julian Schofield
STRATEGIC NUCLEAR SHARING

David Scott
"THE CHINESE CENTURY"?
The Challenge to Global Order

Andrew Taylor
STATE FAILURE

Marco Verweij and Michael Thompson (*editors*)
CLUMSY SOLUTIONS FOR A COMPLEX WORLD
Governance, Politics and Plural Perceptions

Marco Verweij
CLUMSY SOLUTIONS FOR A WICKED WORLD
How to Improve Global Governance

Global Issues Series
Series Standing Order ISBN 978-0-333-79483-8
(*outside North America only*)

You can receive future titles in this series as they are published by placing a standing order. Please contact your bookseller or, in case of difficulty, write to us at the address below with your name and address, the title of the series and the ISBN quoted above.

Customer Services Department, Macmillan Distribution Ltd, Houndmills, Basingstoke, Hampshire RG21 6XS, England

Chemical Control

Regulation of Incapacitating Chemical Agent Weapons, Riot Control Agents and their Means of Delivery

Michael Crowley
University of Bradford, UK

© Michael Crowley 2016
Foreword © Alastair Hay 2016

All rights reserved. No reproduction, copy or transmission of this publication may be made without written permission.

No portion of this publication may be reproduced, copied or transmitted save with written permission or in accordance with the provisions of the Copyright, Designs and Patents Act 1988, or under the terms of any licence permitting limited copying issued by the Copyright Licensing Agency, Saffron House, 6–10 Kirby Street, London EC1N 8TS.

Any person who does any unauthorized act in relation to this publication may be liable to criminal prosecution and civil claims for damages.

The author has asserted his right to be identified as the author of this work in accordance with the Copyright, Designs and Patents Act 1988.

First published 2016 by
PALGRAVE MACMILLAN

Palgrave Macmillan in the UK is an imprint of Macmillan Publishers Limited, registered in England, company number 785998, of Houndmills, Basingstoke, Hampshire RG21 6XS.

Palgrave Macmillan in the US is a division of St Martin's Press LLC, 175 Fifth Avenue, New York, NY 10010.

Palgrave Macmillan is the global academic imprint of the above companies and has companies and representatives throughout the world.

Palgrave® and Macmillan® are registered trademarks in the United States, the United Kingdom, Europe and other countries.

ISBN 978–1–137–46713–3

This book is printed on paper suitable for recycling and made from fully managed and sustained forest sources. Logging, pulping and manufacturing processes are expected to conform to the environmental regulations of the country of origin.

A catalogue record for this book is available from the British Library.

A catalog record for this book is available from the Library of Congress.

Contents

List of Figures and Tables	vi
Foreword by Alastair Hay	vii
Acknowledgements	xi
List of Abbreviations	xii

1	Introduction	1
2	Incapacitating Chemical Agent Weapons	9
3	Riot Control Agents	39
4	Means of Delivering or Dispersing Riot Control Agents	88
5	Application of the Chemical Weapons Convention to ICA Weapons, Riot Control Agents and Related Means of Delivery	107
6	Arms Control and Disarmament Agreements Applicable to ICA Weapons and Riot Control Agents	134
7	International Humanitarian Law Applicable to ICA Weapons and Riot Control Agents	152
8	Human Rights Law Applicable to ICA Weapons and Riot Control Agents	166
9	International Criminal Law Applicable to ICA Weapons and Riot Control Agents	190
10	Mechanisms to Regulate the Transfer of ICA Weapons, Riot Control Agents and Related Means of Delivery	199
11	Application of the United Nations Drug Control Conventions to ICA Weapons	223
12	The Role of Civil Society in Combating the Misuse of Incapacitating Chemical Agents and Riot Control Agents	229
13	Conclusions and Recommendations	263
Notes		278
Index		372

Figures and Tables

Figures

2.1 Biochemical threat spectrum   10
2.2 Selected indicative drug classes and chemicals with potential
     utility as ICA weapons   11

Tables

3.1 Comparative estimated toxicity for humans of selected RCAs   40
3.2 Alleged inappropriate use of RCAs and their means of delivery
     by law enforcement officials from 1st January 2009 to 31st
     December 2013   51

Foreword

Several times over the last 25 years I have had to provide expert evidence for the courts on the effects of exposure to riot control agents. Fortunately, in all but one of these instances the person who was affected by the chemical agent survived and without serious ill effects. Tragically, in one instance the man died and the jury in that case ruled that death was due to a combination of exposure to the riot control agent CS and to positional asphyxia, the latter brought about by the way in which several policemen held him on the ground.

Finding the evidence necessary to make my reports for these cases would have been a much simpler affair if I had had access to this book of Michael Crowley's. This is the resource I would turn to if I wanted to get details about the toxic properties of individual agents and if I needed something as arcane as a discussion on their shelf life, because some police force was using up old stock.

But for me, the most disturbing material Michael Crowley has assembled is in his third chapter, on the inappropriate use of riot control agents. These agents have been deployed almost universally to assist police forces when crowds get out of control. They are an addition to the police's armoury, and meant to help avoid resort to more extreme measures such as batons or guns. In other words, these chemical agents are a sort of interface between doing nothing and possibly shooting to regain control of the streets. Because they are so irritant, at very low air concentrations, when anyone is exposed to them and their eyes start to smart and breathing hurts, the general reaction is to leave the scene and find some fresh air. This is certainly how I responded when inadvertently exposed to CS in Rome many years ago when I turned a corner and found myself in a cloud of the chemical, discharged into a group of demonstrators protesting about the need for abortion law reform.

Regrettably, many of those enforcing the law use riot control agents in ways that were never intended and, in so doing, endanger lives. Ninety-five countries are documented in Chapter 3 because they used riot control agents in ways that were a direct violation of human rights. The evidence for these abuses has been collected by many different organizations over the years, evidence which Crowley has abstracted, detailed and assembled in tabular form. Under headings which range from torture and ill-treatment, through suppression of assembly, indiscriminate and excessive use, deployment in enclosed spaces to serious injury or deaths, the statistics make for grim but absolutely essential reading. For the evidence is accumulating that some form of guidance is urgently needed on how riot control agents are used. The abuses are increasing. This is not what was meant to happen. But unless there is clear guidance on appropriate use of riot control agents anyone with a grievance who happens to be in a demonstration may find themselves opposed increasingly by law enforcement officers who feel they have a free rein.

This concern is amplified by the evidence that Michael assembles about delivery systems for munitions containing riot control agents or any other crowd control chemical. Why would a country wish to have a vehicle-mounted, automatic grenade launcher that can discharge 200 grenades in a minute? Just who will be the recipient of such a weapon? We will know at some point as the launcher is currently being promoted by an ordnance company in China. But China is not alone in developing these delivery systems that are really the preserve of the military. Detailed in Chapter 4 are a range of "wide area" RCA means of delivery including large calibre mortar munitions or other projectiles, some weighing many kilograms and which can be fired over many kilometres. Turkey, Russia, China and the United States appear to have wanted (or at least to have developed and in certain cases marketed) such devices.

The military should not be interested in deploying riot control agents as they are forbidden in armed conflict, as Crowley explains. If not to be used in wars, however, how does one explain these delivery systems that can drop huge quantities of highly irritant chemicals in precise locations in a matter of seconds? Are there other intentions behind such delivery systems? Are they meant for those regimes that prop themselves up through fear or other restrictive measures? Regrettably, because riot control agents are legal for law enforcement purposes, and law enforcement has not been adequately defined in the Chemical Weapons Convention (CWC) which prohibits chemical warfare, certain States may feel free to decide what they use for controlling crowds.

Another grey area in the CWC is incapacitants. Although these too have no universally agreed definition, in general they are regarded as substances that will cause a prolonged, but non-permanent disability, and they generally operate by affecting the central nervous system. A number of countries are very interested in these substances, which came to prominence in October 2002 when they were used to incapacitate heavily armed Chechen separatists who were holding 900 people hostage in a theatre in Moscow. The separatists had threatened to kill all the hostages and negotiations to end the standoff were still deadlocked after three days. Using the Moscow sewer system, Russian Spetsnaz special forces gained entry to the theatre where they released a highly volatile anaesthetizing agent through the ventilation system and into the seating area where everyone was being held.

To ensure rapid incapacitation of those seated furthest from the air vents, huge quantities of the agent were released. Sedation happened fairly rapidly for most. Thirty minutes after release of the incapacitant, which was later identified as derivatives of the anaesthetic opiate fentanyl, special forces stormed the building and killed all the separatists. But because this was carried out so secretly, none of those attending to the sedated hostages had any idea what had been used. The doctors who treated those affected were also in the dark. The huge quantities of the chemical used delivered many times the lethal dose to those seated close to the air vents. The outcome was inevitable. One hundred and twenty-five died from the effects of the opiate mixture which slows breathing, and, in high doses, stops it completely.

The fact that the majority of the hostages were saved was viewed by Russia as an overwhelming successful outcome to the siege. Many other governments thought likewise. Faced with a similar situation they too liked the idea of being able to resort to a chemical that would deliver a rapid knockout dose to hostage takers. This interest persists. Countries can use whatever incapacitant they choose if it is for law enforcement. And they do not have to declare what they have in their policing arsenal. This is a worrying development and requires international action. The real problem is in defining law enforcement and avoiding graduation into armed conflict where incapacitant use is strictly forbidden.

The search for the ideal incapacitant is on. This would be one possessing an enormous safety factor – the outcome of which is a dose that will only incapacitate rather than kill. But it is a fool's errand. There are no chemicals at present that come anywhere near providing a sufficient safety factor. Taken together with how incapacitants are likely to be deployed, there are unlikely to be situations where everyone will survive. Crowley covers this ground superbly in his text.

What is also impressive about this book is the way in which it has been assembled. It is designed in such a way that anyone wishing to find ways of applying some kind of regulatory mechanism to control use of arms will find invaluable. Crowley uses the term 'Holistic Arms Control'. He envisages that it will first involve an exploration of weapons and related technology that require control, followed by review of applicable laws and treaties and then synthesizing all into an approach that will facilitate regulation.

Crowley's key concerns are the incapacitants and riot control agents. He provides evidence throughout this book that these substances require agreement at the international level on what constitutes appropriate use. But what he also provides is the mechanism to help achieve this. By furnishing us with the properties of these chemicals, how they can be deployed, what treaties govern their use at present but also how laws on human rights and control of drugs as well as international humanitarian and criminal law can be brought to bear, he makes the task so much easier.

None of what Crowley would like to see happen will be achieved if no one acts. So his penultimate chapter is how civil society, you and I in other words, can work together to bring about change. Civil society is a broad church and there will be many interest groups. Some will focus on ethical principles, for example concerning those trying to find ideal incapacitants or even those doctors who treat victims of torture but do not speak out. But there will be others including lawyers and scholars as well as victims' groups who will also have a role to play in this movement. And a veritable movement is what is needed if we are to see proper regulation of chemicals which are meant to support the law but which can be so easily used to usurp it.

Alastair Hay, PhD, OBE
Professor of Environmental Toxicology
LICAMM Institute
School of Medicine
University of Leeds

Acknowledgements

First and foremost, I would like to thank Professor Malcolm Dando, from Bradford Disarmament Research Centre, for convincing me to research and write this book. He led me into this dark and tangled forest, encouraged me to stray from the narrow path, and to explore what lay beyond.

I would also like to thank all those from academia, government, non-governmental organizations and inter-governmental organizations who have been so generous in sharing their expertise and time with me. I very much appreciate your patience and forbearance as I asked my naïve and simple questions, wittingly or unwittingly opened up old and new controversies, and generally poked around in areas that were none of my business.

I would like to thank my past and present colleagues, especially those from Bradford University and the Omega Research Foundation, for all their invaluable advice and cooperation. In particular, I would like to acknowledge with gratitude access to Omega's company research, funded in part by the Joseph Rowntree Charitable Trust (JRCT) and the European Instrument for Democracy and Human Rights (EIDHR). In addition, I would also like to thank JRCT for their continuing support of the Bradford Non-lethal Weapons Research Project.

Lastly I would like to thank my wife, Jools, for all the encouragement, patience and support she has given me throughout the writing of this book.

This work is dedicated, with all my love, to Jools and to my two wonderful boys, Dan and Ben.

Abbreviations

AAD	Africa Aerospace and Defence (South Africa)
AFOSR	Air Force Office of Scientific Research (US)
AG	Australia Group
AI	Amnesty International
ANAO	associate national adhering authority
ARCAT	advanced RCA technology
ARDEC	Armament Research, Development and Engineering Center (US Army)
ASCIM DI	ASCIM Defense Industry
ATT	Arms Trade Treaty
BAA	Broad Agency Announcement
BDRC	Bradford Disarmament Research Centre
BiH	Bosnia and Herzegovina
BMA	British Medical Association
BNLWRP	Bradford Non-Lethal Weapons Research Project
BTWC	Biological and Toxin Weapons Convention
BWA	biological warfare agents
BZ	3-Quinuclidinyl benzilate
CAS	Chemical Abstracts Service
CBW	chemical and biological weapons
CESCR	Committee on Economic, Social and Cultural Rights (UN)
CHEMRAWN	Committee on Chemical Research Applied to World Needs (IUPAC)
CIDTP	cruel, inhuman or degrading treatment or punishment
CIPATE	China (Beijing) International Exhibition and Symposium on Police and Anti-Terrorism Technology and Equipment
CN	2-Chloroacetophenone
CNS	central nervous system
CNS	Centre for Non-Proliferation Studies
COIN	counter-insurgency operations
CPS	UN Convention on Psychotropic Substances
CPT	European Committee for the Prevention of Torture and Inhuman or Degrading Treatment or Punishment
CR	dibenz[b,f][1,4]oxazepine

List of Abbreviations xiii

CS	2-Chlorobenzylidenemalonitrile (*o*-Chlorobenzylidene malononitirile)
CSP	Conference of States Parties (OPCW)
CWC	Chemical Weapons Convention
CWCC	Chemical Weapons Convention Coalition
CWS	Chemical Warfare Service (US Army)
DM	Adamsite (Diphenylaminechlorarsine)
DoD	Department of Defense (US)
DPIC	Death Penalty Information Center
DRC	Democratic Republic of the Congo
DRDE	Defence Research & Development Establishment (India)
DRDO	Defence Research & Development Organization (India)
DSTL	Defence Science and Technology Laboratory (UK)
EC	European Commission (EU)
EC	Executive Council (OPCW)
ECBC	Edgewood Chemical Biological Center (US Army)
ECHR	European Convention on Human Rights
ECOWAS	Economic Community of West African States
ECtHR	European Court of Human Rights
ED	effective dose
EMR	educational module resource
EU	European Union
FARC	Fuerzas Armadas Revolucionarias de Colombia
FDI	Furkan Defense Industry
FFM	fact-finding mission (OPCW)
FIDH	International Federation for Human Rights
FSB	Federal Security Service (Russian Federation)
GPC	General Purpose Criterion
HAC	holistic arms control
HPCR	Humanitarian Policy and Conflict Research
HPCSA	Health Professions Council of South Africa
HRW	Human Rights Watch
HSP	Harvard Sussex Program on Chemical and Biological Weapons
IACHR	Inter-American Commission on Human Rights
ICA	incapacitating chemical agent
ICC	International Criminal Court
ICCPR	International Covenant on Civil and Political Rights
ICESCR	International Covenant on Economic, Social and Cultural Rights

ICoC	International Code of Conduct for Private Security Service Providers
ICRC	International Committee of the Red Cross
ICTR	International Criminal Tribunal for Rwanda
ICTY	International Criminal Tribunal for the former Yugoslavia
IDEF	International Defense Industry Fair (Turkey)
IDEX	International Defence Exhibition and Conference (United Arab Emirates)
IFSEC	International Fire and Security Exhibition and Conference (South Africa)
IHL	international humanitarian law
IHRL	international human rights law
IIBR	Israel Institute for Biological Research
INCB	International Narcotics Control Board
INLDT	Institute for Non-Lethal Defense Technologies
ISS	Institute for Security Studies
ISU	Implementation Support Unit (BTWC)
IUPAC	International Union of Pure and Applied Chemistry
JAG	Judge Advocate General (US)
JNLWD	Joint Non-Lethal Weapons Directorate (US)
JNLWP	Joint Non-Lethal Weapons Program (US)
KFOR	Kosovo Force
LD	lethal dose
LECTAC	Law Enforcement and Corrections Technology Advisory Council (US)
LSD	Lysergic acid diethylamide
LTTE	Liberation Tigers of Tamil Eelam (Tamil Tigers)
MAARS	Modular Advanced Armed Robotic System
MINUSTAH	United Nations Stabilization Mission in Haiti
MKEK	Makina ve Kimya Endüstrisi Endütrisi
MoD	Ministry of Defence (UK)
MOOTW	military operations other than war
MOUT	military operations in urban terrain
MSP	Meeting of States Parties
NAO	national adhering authority
NATO	North Atlantic Treaty Organization
NEER	Non-Lethal Environmental Evaluation and Remediation Center
NGO	non-governmental organization
NIH	National Institutes of Health (US)
NIJ	National Institute of Justice (US)
NKE	non-kinetic energy

NLW	non-lethal weapons
NORINCO	North Industries Corporation
NRC	National Research Council (US)
NSABB	National Science Advisory Board for Biosecurity (US)
OAS	Organization of American States
OC	oleoresin capsicum
OECD	Organisation for Economic Co-operation and Development
OIE	World Organisation for Animal Health
OMCT	World Organisation Against Torture
OPCW	Organisation for the Prohibition of Chemical Weapons
OR	opiate receptors
ORF	Omega Research Foundation
OSCE	Organization for Security and Cooperation in Europe
PAVA	Pelargonic acid vanillylamide
PCC	Professional Conduct Committee (HPCSA)
PHR	Physicians for Human Rights
PJV	Porgera Joint Venture
PLA	People's Liberation Army (China)
PMC	private military company
POW	prisoner of war
PS	chloropicrin
PSC	private security company
PSU	Pennsylvania State University
QNA	QinetiQ North America, Inc.
R&D	research and development
RCA	riot control agent
RELEX	Foreign Relations Counsellors Working Group (European Union)
RENAMO	Mozambican National Resistance
RPG	rocket-propelled grenade
RTO	Research & Technology Organization (NATO)
SAB	Scientific Advisory Board (OPCW)
SAR	structure–activity relationship
SCND	UN Single Convention on Narcotic Drugs
SEB	staphylococcal enterotoxin B
SFOR	Stabilization Force
SIRUS	superfluous injury or unnecessary suffering
STOA	Scientific and Technology Options Assessment (European Parliament)

TRC	Truth and Reconciliation Commission (South Africa)
TS	Technical Secretariat (OPCW)
UAV	unmanned aerial vehicle
UN	United Nations
UNBP	United Nations Basic Principles on the Use of Force and Firearms by Law Enforcement Officials
UNCoC	United Nations Code of Conduct for Law Enforcement Officials
UNESCO	United Nations Educational, Scientific and Cultural Organization
UNGA	United Nations General Assembly
UNITA	União Nacional para a Independência Total de Angola
UNMIL	United Nations Mission in Liberia
UNODA	United Nations Office for Disarmament Affairs
UNSC	United Nations Security Council
UNSCR	United Nations Security Council Resolution
UNSG	United Nations Secretary General
WA	Wassenaar Arrangement
WHO	World Health Organization
WMA	World Medical Association
WMD	weapons of mass destruction

1
Introduction

1.1. Introduction

On 10 December 2013, the Organisation for the Prohibition of Chemical Weapons (OPCW) received the Nobel Peace Prize for its ongoing activities overseeing the destruction of chemical weapons in Syria and for its continuing efforts supporting implementation of the Chemical Weapons Convention (CWC) and the global effort to eradicate all such weapons.

In his Nobel Lecture, the OPCW Director General, Ambassador Ahmet Üzümcü, described the remarkable success of the Chemical Weapons Convention and the attendant control regime:

> For sixteen years now, the OPCW has been overseeing the elimination of an entire category of weapons of mass destruction. Our task is to consign chemical weapons to history, forever. A task we have been carrying out with quiet determination, and no small measure of success. Under the terms of the Chemical Weapons Convention, the OPCW has so far verified the destruction of more than 80% of all declared chemical weapons. We have also implemented a wide range of measures to prevent such weapons from re-emerging. And with 190 states now members of this global ban, we are hastening the vision of a world free of chemical weapons to reality.[1]

The CWC stands as an important bulwark, preventing the use of chemical weapons against both military personnel and civilian populations, under any circumstances. Through the tireless work of the OPCW and its Member States, real advances have been and are continuing to be made to identify and destroy all existing arsenals of chemical weapons across the world. These accomplishments are critically important for safeguarding international peace and security and to be truly comprehensive and permanent must be fully supported by the international community in word and deed.

To date, however, the international community has been unwilling or unable to effectively address the real and growing dangers to human security posed by the failure of States to adequately regulate two classes of weapon employing chemical agents, namely incapacitating chemical agent (ICA) weapons and riot control agents (RCAs). This publication attempts to highlight the consequences of this ongoing failure, and through a holistic arms control (HAC) framework seeks to explore the full range of applicable regulatory mechanisms and international law that either constrains or prohibits the use of these weapons.

1.2. Incapacitating chemical agent weapons

At around 5 a.m. on 26 October 2002, Russian security forces, in their attempt to save 900 hostages held in a Moscow theatre by armed Chechen fighters, employed a secret ICA weapon believed to affect the central nervous system. Although the bulk of the hostages were freed, more than 120 of them were killed by the chemical agent and many more continue to suffer long-term health problems.

To this day, the Russian authorities refuse to publicly disclose the weapon they employed, nor will they provide any details of the nature and levels of ICA weapons they may have developed or stockpiled. Despite the official silence, there is evidence, documented by the author, of continued research in relevant fields by Russian scientists, including computer modelling of so-called "calmative" "gas flows" in enclosed spaces, as well as exploration of the interaction of potential ICAs with human receptor sites. And Russia is not alone; China and Israel have also developed ICA weapons targeting individuals, while a number of other States have conducted research that is potentially applicable to the study or development of such weapons.

ICA weapons can be described as weapons employing a disparate range of substances – potentially including pharmaceutical chemicals, bioregulators and toxins – that are purportedly intended to act on the body's core biochemical and physiological systems (such as the central nervous system) to cause prolonged but non-permanent disability. They include centrally acting agents producing loss of consciousness, sedation, hallucination, incoherence, disorientation, or paralysis. At inappropriate doses, however, death can result.

Proponents of ICA weapons have long promoted their development and use in certain extreme law enforcement scenarios, where there is a need to rapidly and completely incapacitate an individual or a group without causing permanent disability or fatality. They have also been raised as a possible tool in a variety of military operations, especially in locations where fighters and civilians are in close proximity or intermingled.

In contrast, a broad range of observers, including scientific and medical professionals, arms control organizations, international legal experts and human rights and humanitarian organizations, as well as a number of countries, have criticized ICA weapons, contending that their use presents potentially grave dangers to health and well-being.

Critics have expressed a variety of additional concerns, including the risk of "creeping legitimization" of ICA weapons as the norm against the weaponization of toxicity is eroded; the dangers of ICA weapons proliferation to both State and non-State actors; the employment of ICA weapons to facilitate torture and other human rights violations; the further misuse and militarization of the life sciences; the potential for States to use law enforcement ICA weapons development as a cover for covert offensive chemical weapons programmes; and the danger of creating a "slippery slope" that could lead in the end to chemical warfare. Disquiet about unregulated State research into ICA weapons is further exacerbated by the rapid advances in relevant science and technology, particularly genomics, synthetic biology, medical pharmacology and neuroscience.

1.3. Riot control agents and related means of delivery

Riot control agents (RCAs), commonly known as tear gases, are typically potent sensory chemical irritants normally with low lethality. They are regularly employed by law enforcement officials throughout the world for activities such as the dispersal of assemblies posing an imminent threat of serious injury, and the incapacitation of violent individuals. When utilized in accordance with manufacturers' instructions and in line with international human rights standards, RCAs can provide an important alternative to other applications of force more likely to result in injury or death, such as firearms. However, they are also open to misuse. And such misuse appears to be widespread and serious.

Research conducted by the author has uncovered reports from United Nations (UN) and regional human rights bodies and international non-governmental human rights organizations of human rights violations committed by law enforcement officials utilizing RCAs in at least 95 countries or territories from 1 January 2009 to 31 December 2013. RCAs have reportedly been used to suppress the right to assembly, in the excessive use of force, and to cause ill-treatment and torture. In some instances the misuse of RCAs, particularly in enclosed spaces, has reportedly resulted in serious injury or death. In other instances, RCAs have reportedly been employed together with firearms by police or security forces in massacres of unarmed people.

A further area of concern is the current development and marketing of a range of large calibre munitions and other delivery systems that can be utilized for dispersing significant amounts of RCA over wide areas or over extended distances. Research undertaken by the author has uncovered the development, testing, production, possession or promotion by State or commercial entities in at least 15 countries of some 40 "wide area" RCA means of delivery. These "wide area" RCA means of delivery have included large RCA "smoke" generators and irritant sprayers; multiple munition launchers; automatic grenade launchers; rocket propelled grenades; mortar munitions; large calibre aerial munitions; heliborne munition dispensers; cluster bombs; and RCA launchers fitted to unmanned aerial and ground vehicles. Certain forms of "wide area" RCA means of delivery may have restricted utility in large-scale law enforcement situations provided they meet and are used in strict conformity with human rights standards. Other forms of "wide area" RCA means of delivery that have been developed – such as artillery shells, aerial bombs, large calibre mortar shells and cluster munitions – are completely inappropriate for any form of law enforcement, having possible utility instead in large-scale human rights abuses or armed conflict. They should be considered to be chemical weapons and treated accordingly.

1.4. Holistic arms control (HAC)

For many years the governmental and non-governmental arms control communities have sought to develop strategies to combat the proliferation of chemical and biological weapons to State and non-State actors. Recognizing that reliance upon a single disarmament or arms control agreement alone would not guarantee success, scholars have explored a number of concepts, seeking to broaden the range of possible regulatory mechanisms. Utilizing and building upon such work, particularly the concepts of "preventative arms control",[2] and "webs of prevention (or protection)",[3] the author has developed a "holistic arms control" (HAC) framework for regulation, and has sought to apply it to ICA weapons, RCAs, and related means of delivery.

Although the proposed HAC analytical framework concentrates upon existing arms control and disarmament measures, it attempts to widen the range of applicable mechanisms for regulation, and also the nature of the actors involved in such regulatory measures.

Consequently, HAC can be thought of as a framework for analysis to aid the development of a comprehensive, layered and flexible approach to arms control that:

- Is tailored for and unique to the specific type of weapon or technology under consideration rather than for a broad grouping of weapons;
- Is not necessarily limited to a single existing arms control or disarmament treaty, but actively explores and seeks to incorporate States' existing responsibilities under the full range of relevant international law and applicable agreements;

- Potentially covers all stages of a weapon's existence, i.e., research, development, mass production, stockpiling, deployment, use, transfer and destruction;
- Seeks to clearly identify the types of permitted and prohibited weapons, acknowledges where existing ambiguity lies and highlights potential mechanisms for resolving such ambiguities;
- Seeks to clearly identify existing constraints upon the permitted use of weapons, i.e., legitimate targets (e.g. whether this would include armed combatants, terrorist organisations, criminals, civilians), legitimate types of operations (e.g. whether this would include law enforcement, military operations other than war, armed conflict) and how such operations should be conducted (rules of engagement); acknowledges where existing ambiguity lies and highlights potential mechanisms for resolving such ambiguities;
- Seeks to identify measures to facilitate effective national implementation as well as reporting, transparency, verification and enforcement mechanisms;
- Seeks to identify measures for regular "feedback", both "official" (e.g. annual State reporting) and "unofficial" (e.g. civil society briefings to State Parties or regime secretariats, media reports), and comprehensive review of the relevant control regime(s);
- Explores mechanisms to facilitate regime adaptation to ensure continued effectiveness and relevancy in response to developments including in:

  - the nature of the use/misuse of weapons in practice;
  - science and technology (so that the regime is able to regulate weapons and technologies that have not yet been invented);
  - international law, particularly that limiting types and use of weapons;

- Recognizing that States are the prime actors in existing regulatory regimes, allows for and encourages participation by the full range of relevant stakeholders.

HAC consists of the following three-stage process:

- *Stage one:* an examination of the nature of the weapons and technology to be controlled and exploration of the current and potential future scenarios of application, together with the attendant national and human security concerns of inappropriate use. During this stage, the potential relevance of advances in science and technology is assessed.
- *Stage two:* an analysis of the full range of potentially applicable international law, (arms control, disarmament and other) instruments and

attendant control regimes; highlighting strengths, weaknesses and ambiguities in these mechanisms. The potential roles of relevant non-State actors are also explored.
- *Stage three:* following an analysis of information derived from stages one and two, a comprehensive strategy is developed to strengthen existing mechanisms and/or introduce new mechanisms to facilitate effective regulation or prohibition of the weapon or weapons-related technology of concern.

The manner in which HAC is employed will be dependent upon the expertise, experience and resources that a researcher or research body can call upon, as well as the nature of the weapon or weapons-related technology being investigated. Consequently, given time and resource constraints, the application of HAC will necessarily be a compromise between:

- Breadth – range of potentially applicable instruments and control regimes, international law and other potential regulatory actors [mechanisms] examined;
- Depth – level to which each relevant mechanism is examined.

In order to achieve the optimal balance between breadth and depth, a two-step research process is followed. Firstly, an initial, brief, impressionistic, "wide area" overview of the existing regulatory terrain is undertaken. This survey should not examine potentially applicable mechanisms in great depth and should not be restricted to "traditional" arms control and disarmament mechanisms, but at a minimum should also include relevant international law, as well as other "outlier" mechanisms.

Following this scene-setting survey, the researcher will be able to determine which mechanisms show the greatest potential for regulating the weapon or technology of concern, and consequently where the main focus of subsequent in-depth research should be concentrated. The following factors are to be taken into account when determining the potential relevance and effectiveness of each regulatory mechanism during the subsequent in-depth research process and in the development of a HAC strategy:

- range and nature of items currently regulated;
- scope of State and non-State actors regulated;
- scope of activities regulated (whether the mechanism covers research, development, production, stockpiling, transfer, deployment and use of the weapon or just certain aspects of this cycle);
- number, nature and geographical spread of States adhering to the mechanism;
- authority and legal underpinning of the mechanism;

- stringency of the relevant obligations (i.e. absolute prohibition of activities, control of activities, advisory guidelines to be taken into account);
- clarity of definition of relevant obligations and corresponding areas of ambiguity;
- level and nature of national implementation required, and mechanisms established to monitor and facilitate such implementation;
- importance currently given (and/or potential responsiveness of mechanism) to regulating weapons or technology under review;
- ability of mechanism to respond to relevant advances in science and technology.

1.5. Chapter overview

The structure of this publication has been shaped by the three-stage process envisaged for an HAC analysis of ICA weapons and RCAs. Consequently, Chapters 2 and 3, corresponding to HAC stage one, provide an overview of the chemical agents that are the object of study, exploring their properties, the current or proposed employment scenarios and the potential implications of advances in relevant science and technology. Chapter 4 explores a second important aspect of weaponization, namely, the mechanisms developed to disperse the relevant chemical agents over a wide area or deliver them to the targets.

Chapters 5–12, comprising stage two of the HAC analysis, explore the range of control mechanisms that could potentially be applied to the regulation of these agents and related means of delivery. Following an initial overview survey, it became clear that the CWC has been the main focus of State activities in this area. Consequently, Chapter 5 provides a summary of the treaty followed by an analysis of its application to ICA weapons, RCAs and related means of delivery, exploring areas of ambiguity and contested interpretation amongst State Parties and the consequent implications for effective regulation.

Chapter 6 analyses the relevance of the legal prohibitions against chemical and biological weapons enshrined in the Geneva Protocol, the Biological and Toxin Weapons Convention (BTWC) and UN Security Council Resolution 1540. In addition, the potential utility of the UN Secretary General's investigatory mechanism is explored, and the use of this mechanism in Syria examined.

Chapter 7 examines obligations derived from international humanitarian law (both international customary law and treaty law) most notably the Geneva Conventions and Additional Protocols, and the consequent prohibitions on the use of toxic chemicals in armed conflict. Chapter 8 surveys potentially applicable human rights law (again, both international customary law and treaty law) concentrating upon that relating to the rights to life; to liberty and security; to freedom from torture and other cruel, inhuman or degrading treatment or punishment; to engage in "peaceful protest"; and to health; together with attendant obligations on the restraint of force.

Chapter 9 explores the potential application of the Single Convention on Narcotic Drugs and the UN Convention on Psychotropic Substances to the regulation of ICA weapons. Chapter 10 explores the potential applicability of international criminal law and judicial mechanisms (and specifically the International Criminal Court) to cases involving the serious misuse of ICA weapons and RCAs.

Chapter 11 examines the potential application of control regimes which prohibit or regulate the transfer of: chemical and biological weapons (and so-called "weapons of mass destruction"); conventional arms, dual use goods and related (para)military equipment; and security equipment utilized in torture and ill-treatment. In addition, the chapter also explores the utility of legally binding arms embargoes introduced by either the UN or certain regional organisations to address threats to international peace and security arising from armed aggression and, increasingly, gross human rights abuses or breaches of international humanitarian law.

Chapter 12 examines the potential roles that can be played in regulation by civil society, including: societal monitoring and verification activities; development of a "culture of responsibility" amongst the scientific and medical communities built upon strong normative and ethical standards; and the role of civil society organisations in developing and advocating mechanisms to strengthen relevant control regimes, particularly the CWC.

Finally, Chapter 13 summarizes the findings of the previous chapters and, utilizing their findings, seeks to develop appropriate strategies for the regulation of ICA weapons, RCAs and their related means of delivery.

2
Incapacitating Chemical Agent Weapons

2.1. Introduction

As part of the first stage of the HAC analytic process, this chapter examines the properties of incapacitating chemical agents (ICAs), and explores proposed scenarios for their application as weapons, highlighting attendant arms control and human security concerns about the development and proliferation of such ICA weapons. The chapter briefly surveys historical weaponization programmes, and examines contemporary reports of research, development, possession or use of ICA weapons. It also examines advances in relevant science and technology, and explores the potential implications for arms control.

2.2. Describing incapacitating chemical agents

The ongoing revolution in the life sciences that has proceeded particularly since the 1980s has resulted in the boundary between chemistry and biology becoming increasingly blurred. This, in turn, has meant that the distinction between certain chemical and biological weapons has become less useful.[1] Rather than thinking of chemical and biological weapons threats as distinct, some analysts, including Aas,[2] Dando,[3] Davison,[4] Hemsley[5] and Pearson,[6] have argued that it is more useful to conceptualize such agents as lying along a continuous biochemical threat spectrum from the classical chemical agents on one extreme (i.e. nerve, blood and blister agents), through mid-spectrum agents and on to biological agents (including traditional and genetically modified biological agents) (Figure 2.1).

This chapter will focus upon those mid-spectrum agents[7] – which include pharmaceutical chemicals, bioregulators and toxins – that have been considered as potential ICAs weapons. Although certain States,[8] and pluri-lateral organizations such as NATO,[9] have sought to characterize ICAs, there is currently no internationally accepted definition for these chemical agents.

Indeed, certain leading scientific experts and international organizations believe that such a technical definition is not possible. A January 2012

10 *Chemical Control*

Classical chemical weapons	Industrial pharmaceutical chemicals	Bioregulators and peptides	Toxins	Genetically modified biological weapons	Traditional biological weapons
Blood agents	Fentanyl	Neurotransmitters	Staphylococcal enterotoxin B	Modified bacteria and viruses	Bacteria
Blister agents	Carfentanil	Hormones			Viruses
Choking agents	Remifentanil	Cytokines	Botulinum toxin		Rickettsia
Nerve agents	Etorphine		Ricin		
	Dexmedetomidine		Saxitoxin		Anthrax
	Midazolam				Plague
					Tularemia
← Chemical Weapons Convention (CWC) →					
		← Biological and Toxin Weapons Convention (BTWC) →			
				← Infect →	
← Poison →					

Figure 2.1 Biochemical threat spectrum[10]

report of an expert meeting organized by Spiez Laboratory concluded that: "…because there is no clear-cut line between (non-lethal) ICA and more lethal chemical warfare agents, a scientifically meaningful definition cannot easily be made. One can describe several toxicological effects that could be used to "incapacitate", but in principle there is no way to draw a line between ICAs and lethal agents."[11]

Whilst recognizing the contested nature of this discourse, a provisional working description of ICAs based upon the 2012 Royal Society definition will be employed in this publication. Consequently, ICAs will be considered as: substances whose purported intended purpose is to cause prolonged but non-permanent disability; they include centrally acting agents producing loss of consciousness, sedation, hallucination, incoherence, paralysis, disorientation or other such effects.[12]

For the purposes of this publication, an "ICA weapon" will be considered to comprise an ICA and/or associated means of delivery, developed with the purported intention of temporarily incapacitating but not killing a target. Although a theoretically "ideal" candidate agent for an ICA weapon would have a very high safety margin,[13] in practice known candidate agents typically possess very low safety margins, so the effects of ICA weapons are in fact variable and can include death.[14] ICA weapons are distinct from RCAs, which act mostly on the peripheral nervous system to produce rapid sensory irritation of the eyes, mucus membranes and skin, and whose effects disappear shortly after termination of exposure.

There are a wide variety of substances that could potentially be employed as candidate agents for ICA weapons. A study, published in 2000, by the Applied Research Laboratory and the College of Medicine at Pennsylvania State University (PSU), identified a range of drug classes which the researchers considered as having potential utility as ICA weapons, including: anaesthetic agents, skeletal muscle relaxants, opioid analgesics, anxiolytics, antipsychotics, antidepressants and sedative-hypnotic agents.[15] (See Figure 2.2 for a selected summary of the study.) Many of these substances

Selected drug classes	Examples	Site of action
Benzodiazepines	Diazepam	GABA receptors
	Midazolam	
	Etizolam	
Alpha$_2$ adrenergic receptor agonists	Dexmedetomidine	Alpha$_2$-adrenergic receptors
Dopamine D3 receptor agonists	Pramipexole	D3 receptors
	CI-1007	
Selective serotonin reuptake inhibitors	Fluoxetine	5-HT transporter
	WO-09500194	
Serotonin 5-HT$_{1A}$ receptor agonists	Buspirone	5-HT$_{1A}$ receptor
	Lesopitron	
	MCK-242	
Opioid receptors and Mu agonists	Morphine	Mu opioid reception
	Carfentanil	
Neurolept anaesthetics	Propofol	GABA receptors
	Droperidol and fentanyl combination	DA, NE and GABA receptors
	Phencyclidines	Opioid receptors
Cholecystokinin B receptor antagonists	CI-988	CCK-B receptor
	CI-1015	

Figure 2.2 Selected indicative drug classes and chemicals with potential utility as ICA weapons[16]

have been legitimately used by the medical or veterinary professions as tranquilizing or anesthetizing agents.[17]

2.3. Potential dangers and proposed utility

Proponents of ICA weapons have promoted their development and use in certain law enforcement scenarios such as hostage-taking situations (see Russian Federation case study in Section 2.3.1), where there is a need to rapidly and completely incapacitate single individuals or a group without causing death or permanent disability. ICA weapons have also been raised as a possible tool in a variety of military operations, especially in situations where combatants and non-combatants are mixed.[18] Such perceptions of utility were noted in the 2011 report of a high-level expert panel convened by the Director General of the OPCW, which stated:

> [D]istinctions between law enforcement, counter-terrorism, counter-insurgency and low-intensity warfare may get blurred, and certain types of chemical weapons such as ICAs may appear to offer tactical solutions to operational scenarios where civilians and combatants cannot easily be separated or distinguished.[19]

In contrast, a broad range of observers, including scientific and medical professionals, arms control organizations, international legal experts, human rights monitors and humanitarian organizations, as well as a number of States, have voiced their disquiet about the development and utilization of ICA weapons. Amongst the issues raised have been:

(i) *Creeping legitimization and the erosion of the norm against weaponization of toxicity:* Perry Robinson has described how "Today's regime against chemical/biological-warfare armament... derives its reach and strength from that fundamental norm of State behaviour that eschews fighting with poison or infectious disease. Fragment the norm, as by asserting that this or that form of toxicity is not really a part of it, and the foundation of the regime may be weakened."[20]

Perry Robinson has consequently argued that previous attempts by certain States, particularly the United States, to legitimize the development and use of ICA weapons have threatened to fragment the norm. Describing the US "Advanced RCA Technology" (ARCAT) development projects of the 1990s, which included work on the fentanyls and "other such intensely toxic chemicals", Perry Robinson has stated that: "The process that can be seen here is a surreptitious equation of toxicity with lethal toxicity. In this attempt to loosen the CWC constraint on the weaponization of other forms of toxicity we have started to see a creeping legitimization of non-WMD CBW [chemical and biological weapons that are not weapons of mass destruction]..."²¹ Perry Robinson has argued that this "creeping legitimization" presents the greatest danger to the existing prohibitions on chemical and biological weapons and to the re-emergence of chemical and biological warfare.²²

Similarly, Swiss Government expert Mogl has contended that the "almost universal willingness for CWC compliance" will be undermined if ICAs become a "designer" weapons agent employed (in "Special Operations") only by developed countries. He has argued that: "Any ICA (incapacitating chemical agent), even an agent with very high safety ratio, is effective due to its toxic properties. In the absence of clear standards for the use of such chemicals, it will be difficult to defend the prohibition of some toxic agents, whilst allowing or tolerating others."²³

(ii) *Proliferation and legitimization by States:* A range of researchers and organizations have highlighted the implications of State proliferation of ICA weapons. Dr Alan Pearson has noted that "... efforts to develop incapacitating biochemical weapons may well gather steam as more nations become intrigued by them and, observing the efforts of Russia and the United States, become convinced not only that effective and acceptably 'non-lethal' incapacitating agents can be found, but that their use will be legitimized".²⁴ In September 2012, the International Committee of the Red Cross (ICRC) warned that continued development and use of ICA weapons for law enforcement "is likely to present broad and unpredictable risks for security, including inevitable proliferation".²⁵ In such a proliferation scenario, "acquisition of weapons by specialised police units or special forces, and even by military forces in international operations such as peacekeeping, could be expected". Consequently, "use of these weapons, or demand for such use, may range from limited domestic law enforcement scenarios to wider military operations in which the boundaries between law enforcement and conduct of hostilities in armed conflict can become blurred".²⁶ The ICRC has noted that proliferation would "likely occur among different forces within countries and among a growing circle of countries".²⁷ Furthermore, the ICRC has argued that this spread would be unpredictable and unlikely to be uniform, with different countries potentially developing "different toxic chemicals with different effects as weapons for use in a variety of circumstances".²⁸

(iii) *Proliferation to, and misuse by, non-State actors:* Analysts have highlighted the potential utility of ICA weapons for a range of non-State actors, including criminals, terrorists, paramilitary organizations and armed factions in failing or failed States, many of whom would not feel as constrained as States by international law and concerns about lethality.²⁹ The future use of ICA weapons by private military or security companies is another related area of concern, given the inadequate regulation of such entities to date.[30]

(iv) *Use as a lethal force multiplier:* While being promoted by their advocates as a "less lethal" weapon, ICA weapons may in future be used by military and/or law enforcement agencies, not as an alternative to lethal force, but as a means to make lethal force more deadly. For example, as discussed in Section 2.3.1, following their employment of an ICA weapon to end the October 2002 Moscow theatre siege, Russian security forces reportedly shot all those Chechen hostage takers who had been rendered unconscious by the ICA, in apparent violation of international human rights law.[31]

(v) *Facilitation of torture and other human rights violations:* Human rights organizations have documented the reported use of ICAs for "repressive psychiatry" and/or as "truth serums" to facilitate interrogation.[32] As well as potentially being utilized for the torture and ill-treatment of individuals, ICA weapons could also facilitate repression of groups by, for example, allowing the capture, en masse, of large numbers of people participating in peaceful demonstrations. In addition, human rights organizations and anti-death penalty campaigners have documented the employment of certain ICAs by a number of countries in judicially sanctioned executions through so-called "lethal injections".[33]

(vi) *Malign applications of the life sciences:* Analysts including Wheelis and Dando,[34] Perry Robinson,[35] Dando,[36] and the ICRC[37] have warned that the continuing utilization of the life sciences in the development of ICA weapons could potentially open the way to more malign objectives. In some scenarios, the widespread repression of groups or entire populations has been raised. The British Medical Association described this danger in its 2007 report: "Using existing drugs as weapons means knowingly moving towards the top of a 'slippery slope' at the bottom of which is the spectre of 'militarization' of biology, this could include intentional manipulation of peoples' emotions, memories, immune responses or even fertility."[38]

(vii) *Camouflage for lethal chemical weapons programme:* Perry Robinson has highlighted how States could exploit the CWC's limited transparency mechanisms, potentially applicable to ICAs and other toxic chemicals designated for use in law enforcement, to hide illicit activities. He has argued that:

> If a CWC State Party were challenged to explain why it was conducting development, production or stockpiling of toxic chemicals that it had not declared to the OPCW, it could assert, rightly or wrongly, that the activity was nothing to do with chemical weapons, but was for the non-prohibited purpose of law enforcement... A great loophole thus exists within the CWC's international verification system, endangering confidence in the Treaty.[39]

2.3.1. Case study: Use of an ICA weapon by the Russian Federation

Concerns about the development and employment of ICA weapons were heightened following the use of presumed derivatives of fentanyl by Russian Security Forces to free 900 hostages held by heavily armed Chechen separatists in the Dubrovka Theatre in Moscow, in October 2002.[40] The Russian security forces pumped the aerosolized ICA into the theatre, putting the hostages and some of the hostage takers into a "deep sleep". Approximately 30 minutes later, members of the Russian Spetsnaz Special Forces stormed the theatre and killed all of the hostage takers,[41] including those unconscious from the ICA. According to an October 2003 statement by the press department of the Moscow city Prosecutor's Office, 125 hostages died from the effects of the ICA, some of them while in hospital; an additional five hostages were reportedly killed by the hostage takers.[42] In addition, it has been reported that an undetermined, but large, additional number of hostages suffered long-term damage, or died prematurely in the years after the siege.[43]

Treatment of the hostages who had been poisoned was delayed and compromised by the refusal of the Russian authorities to state publicly what type of ICA had been used in the theatre for four days after the siege had ended.[44] On 30th October 2002, the Russian Health Minister, Yuri Shevchenko, identified the incapacitating agent as "a mixture of derivative substances of the fast action opiate Fentanyl".[45] Mr Shevchenko further stated that: "I officially declare: chemical substances which might have fallen under the jurisdiction of the international convention on banning chemical weapons were not used during the special operation."[46] However, the Minister refused to be more precise about the chemicals used, even on 11th December 2002, when faced with a parliamentary question. He said it was a "State secret".[47]

In 2012, a paper by Riches and colleagues detailed the results of trace analysis undertaken by researchers from the United Kingdom's Defence Science and Technology Laboratory (DSTL) at Porton Down of extracts of clothing and urine from survivors of the Moscow theatre siege. The paper indicated that the ICA weapon comprised a mixture of two anaesthetics, carfentanil and remifentanil.[48] At the time of writing, the Russian authorities have not publicly responded to this paper.

Whilst some commentators have characterized the Russian Federation use of an ICA weapon as a success, attributing the failure of the Chechen group to kill large numbers of their hostages to the effects of the ICA, others disagree. For example, Wheelis has maintained that:

> The agent rendered most of the hostages and hostage takers unconscious (although not all), but it took at least several minutes to do so. Hostages had time to make cell phone calls, and the male hostage takers had time to exit the theatre into the surrounding corridors, avoiding incapacitation. There was ample time for the female hostage takers to detonate their bombs; why they didn't remains unexplained, but it was

certainly not because the drug incapacitated them too quickly for them to act. In this respect, the drug was a failure.[49]

In a subsequent interview with the author, Wheelis argued that the failure of the hostage takers to detonate their bombs "certainly was not due directly to incapacitation. [It] may have been a command failure (the women with the detonators were left alone in the theater while the commanders left to battle the FSB in the hallways), or the bombs may have been dummies or defective".[50] He contended that the use of an ICA weapon: "was a stupid gamble that could easily have led to disaster, and almost certainly would if repeated in similar situation".[51]

2.4. Disputed feasibility of developing "acceptable" ICA weapons

Although proponents of ICA weapons promote the potential benefits of their use as "non-lethal" or "less lethal" weapons,[52] many in the medical and scientific communities have questioned the feasibility of developing weapons employing ICAs that do not kill or seriously injure a significant proportion of the target population. In 2003, Klotz, Furmanski, and Wheelis developed a predictive model illustrating "why seemingly non-lethal incapacitating agents may be quite lethal in actual use".[53] In their conclusion they stated:

> We have shown, at least within the approximations of our simple (but generous) two receptor equilibrium model, that even with a therapeutic index of 1,000 (above any known anaesthetic or sedative agent), a chemical agent used as an incapacitating weapon can be expected to cause about 10% fatalities.[54]

Furthermore, as Pearson has noted, even such predictive modelling will potentially underestimate fatalities when an ICA weapon is used in real-life situations where there is uncontrollable variability "both in terms of exposure (uneven concentration and exposure time) and within the target population (age, size, gender, health status and individual susceptibility)".[55] As a result of such considerations, the British Medical Association, in 2007, stated:

> The agent whereby people could be incapacitated without risk of death in a tactical situation does not exist and is unlikely to in the foreseeable future. In such a situation, it is and will continue to be almost impossible to deliver the right agent to the right people in the right dose without exposing the wrong people, or delivering the wrong dose.[56]

This position has subsequently been reiterated by a range of respected scientific organizations. In 2012, for example, a study conducted by the Royal Society concluded that:

> It is not technically feasible to develop an absolutely safe incapacitating chemical agent and delivery system combination because of inherent variables such as the size, health and age of the target population, secondary injury (e.g. airway obstruction), and the requirement for medical aftercare.[57]

However, there is a danger, highlighted by Pearson and others, that "increased interest in incapacitants will generate pressures that lead to the use and proliferation of weapons that are deemed 'good enough'. In other words, if and when 'success' comes, it may be due more to a redefinition of acceptability than to advances in science and technology".[58] Pearson has contended that if this were to occur then "the institutionalization, conventionalization, and marketization of the new chemical weapons...may well lead to an ever-expanding definition of acceptability, ever-broader range of uses, and a more powerful array of new and improved agents".[59]

2.5. Historical ICA weapons research and development programmes

From the late 1940s onwards, military, security or police entities and related State policy-making bodies of certain countries have explored the potential utility of ICA weapons. States that reportedly conducted research and attempted development of ICA weapons or acquired such weapons at some stage prior to the signing of the CWC in 1993 included: Albania, China, Iraq, Israel, (Apartheid) South Africa, the Soviet Union, the United Kingdom, the United States of America and Yugoslavia.[60]

2.5.1. Illustrative case study: US research and development of ICA weapons

Details of the historical ICA weapons research and development (R&D) programmes that have been made public are partial and of varying reliability. However, information released by the United States into its past programmes show that the range of chemicals that were under consideration, as potential ICAs with weapons utility, was extensive. For example, the 1997 United States Army textbook, *Medical Aspects of Chemical and Biological Warfare*, stated that:

> Virtually every imaginable chemical technique for producing military incapacitation has been tried at some time. Between 1953 and 1973, at the predecessor laboratories to what is now the US Army Medical Research

Institute of Chemical Defense, many of these were discussed and, when deemed feasible, systematically tested. Chemicals whose predominant effects were in the central nervous system were of primary interest and received the most intensive study...[61]

The authors went on to suggest that almost all such agents could be put into one of four classes: stimulants, depressants, psychedelics and deliriants. Stimulants include, for example, amphetamines and cocaine, depressants include barbiturates, and psychedelics include LSD (Lysergic acid diethylamide). Delirium, "an incapacitating syndrome, involving confusion, hallucinosis, disorganized speech and behavior", can be produced by a wide variety of drugs. But as the text pointed out: "From this large number of possibilities, chemical compounds in a single subgroup – the 'anticholinergics' – are regarded as most likely to be used as military incapacitating agents."[62]

The US studies that were carried out on such deliriant agents have been described in some detail,[63] and one, BZ (3-quinuclidinyl benzilate), was weaponized by the United States. Approximately 60,000 kilograms (130,000 pounds) of BZ were manufactured and the agent was weaponized in two munitions – the 175-lb M44 generator cluster and the 750-lb M43 cluster bomb – that entered the US arsenal in 1964.[64] Although the US military proposed initiating use of BZ along with CS in the Vietnam War,[65] there were no confirmed reports that BZ has ever been utilized by US forces in armed conflict.[66]

From 1955 to 1975, the US military and contractors from the US pharmaceutical industry and elsewhere conducted research into a wide range of other ICAs with potential weapons utility,[67] many of which were tested on US military volunteers.[68] In the late 1970s, the US Army conducted some advanced development, including work on a pilot plant for production of the glycolate EA 3834A and a filling facility for an XM96 66mm ICA rocket warhead, although these munitions do not appear to have entered the US arsenal.[69] From the late 1970s onwards, there is no evidence of further advanced large-scale ICA weapons development. BZ was removed from the US arsenal; the stockpile of BZ was eventually destroyed by incineration between 1988 and 1990, and the BZ filing plant subsequently destroyed in 1999.[70]

Although the US ICA weapons programme concentrated upon pharmaceutical chemicals, there are indications that toxins were also explored as potential ICA weapons, specifically Staphylococcal enterotoxin B (SEB), which is produced by *Staphylococcus aureus*. SEB (also known as Agent PG and UC) is a super-potent incapacitating agent whose inhalation ED50 in man is estimated to be at least three orders of magnitude smaller than the corresponding LD50 of nerve agents such as sarin or Agent VX.[71] High-dose, microgram-level exposures to SEB will result in fatalities, whilst inhalation exposure to nanogram or lower levels may be severely incapacitating.[72] SEB represents many (perhaps hundreds) of related biologically active super-antigens that are readily isolated and manipulated by recombinant DNA techniques. All of these super-antigens are presumed to have a similar mode of biological action, but very little data are available for confirmation.[73]

During the 1960s, SEB was studied extensively as an ICA weapon in the US offensive programme. In 1968, during DTC Test 68–50, dry-agent spray-tanks filled with Agent PG mounted on American *Phantom* strike aircraft were released over caged monkeys and other animals at sea off Eniwetok Atoll in the Marshall Islands. The test data reportedly indicated a 30% casualty rate over an area of 2,400 km^2 per weapon.[74] As far as is known, PG was never employed as a weapon, and all US R&D of such offensive toxin weapons ended with a moratorium announced by President Nixon on 14th February 1970.[75]

2.6. Analysis of contemporary dual-use research potentially applicable to ICA weapons and reported development or use of ICA weapons

There are indications from open source information that research potentially applicable to ICA weapons development continued after the coming into force of the CWC in 1997 and may still be taking place in certain States, although the range of agent types under active consideration may have narrowed. As the Royal Society *Brain Waves* study report stated:

> Many different forms of incapacitation were investigated during the Cold War, but with increasing emphasis on rapid action and short duration of effects, contemporary interest has tended to focus on sedative-hypnotic agents that reduce alertness and, as the dose increases, produce sedation, sleep, anaesthesia and death.[76]

Consequently, a range of studies have explored dual-use research concerning a range of pharmaceutical chemicals including opioids, benzodiazepines, alpha 2 adrenoceptor agonists and neuroleptic anaesthetic agents, and discussed their potential application as ICA weapons.[77]

In 2008, the International Union of Pure and Applied Chemistry (IUPAC) noted that:

> Many of the chemicals that are being synthesized and screened as part of the drug discovery efforts... will have incapacitating properties that could make them suitable as so-called "nonlethal" agents .... Efforts are reportedly underway in some States Parties to develop weapons with nonlethal properties for use in law enforcement situations. But such weapons may also be thought to have utility in counter-terrorism or urban warfare situations.[78]

In 2008, a National Research Council (NRC) report on *Emerging Cognitive Neuroscience and Related Technologies*[79] highlighted several areas of contemporary and possible future R&D applicable to ICA weaponization, including "medical pharmacology, with particular attention to more potent fentanyl derivatives and inhalation anaesthetics".[80] The report noted that: "existing pharmacological agents could be used in a nefarious way. An example would be currently used agents, such as alpha blockers, that would work quickly to drop blood pressure if delivered in high doses. In addition, anti-cholinergic agents could cause molecular changes that lead to temporary blindness".[81]

In its 2010 report the ICRC stated that: "There is clearly an ongoing attraction to 'incapacitating chemical agents' but it is not easy to determine the extent to which this has moved along the spectrum from academia and industrial circles into the law enforcement, security and military apparatuses of States."[82]

A senior US State Department official interviewed by the author stated that "there are concerns that some others might be looking at incapacitants".[83] However, the official believed that the situation is more complex than it first appears, as it is "very hard to know" the intention behind such research.[84] "Once you have a very prominent event like [the Moscow siege] most chemical defence establishments would be remiss if they didn't at least do a little bit of work to figure out how to protect themselves. Going into the old conundrum 'of whether working on something which could be justified for defensive purposes signifies something more?'... there you start to look at the broader international behaviour of the State."[85]

In the light of such ongoing concerns, a survey of contemporary dual-use chemical and life science research of potential relevance to the study or development of ICAs and associated means of delivery was conducted by the author in collaboration with Professor Malcolm Dando,[86] in two stages.[87] In stage one, an initial survey of relevant open source literature was undertaken, including scientific and medical databases and publications detailing research activities in relevant disciplines; Government documents pertaining to past State ICA weapons development programmes; technology monitoring or evaluation reports from relevant bodies such as IUPAC and the Scientific Advisory Board (SAB) of the OPCW. The information obtained was reviewed against a range of indicators of potential concern so as to narrow the focus of subsequent in-depth research conducted in stage two to a discrete number of illustrative country case studies. In addition to documenting a range of potentially applicable research, the survey also collected and analysed information concerning reported development, possession or use of ICA weapons in certain States.

2.6.1. China

From at least the mid-1990s, there appears to have been development, production and promotion in China of a weapon incorporating an ICA for use in law enforcement operations against individuals. In 1995, marketing materials distributed internationally by the State-owned China North Industries Corporation (NORINCO) promoted the "BBQ-901 anaesthetic gun system". This weapon discharges a projectile, with an effective range of 40 m, which on impact injects a liquid ICA into the target.[88] From 1996 to 2006, an essentially identical entry on the BBQ-901 appeared in *Jane's Police and Security Equipment*, reporting that: "Depending upon the particular anaesthetic specified, the victim will be rendered unconscious within 1 to 3 minutes, this time obviously varying with individuals and with the placement of the projectile." Furthermore "the effects" reportedly would "wear off after 3 or 4 more minutes, giving sufficient time to place the victim in restraint".[89] According to NORINCO, the BBQ-901 was intended "for SWAT units and other special usage... It can be used for reconnaissance and capture of criminals in a concealed place. It is also used as a riot control weapon to subdue the ruffians and maintain public order."[90]

The essentially similar, if not identical, "BBQ-901 narcosis gun" was promoted by a second Chinese State-owned company, State 9616 Plant, at the Asia Pacific China Police Expo in 2004, and subsequently at the same venue in 2006.[91] According to the company's brochure, the BBQ-901 is a "self-researched weapon", and State 9616 Plant was subsequently awarded the "State Second-Class Science and Technology Advancement Diploma" for its manufacture.[92] The BBQ-901 "is used to scout and capture snugly for patrol. It is an advanced weapon for obtaining important intelligence or completing other special mission."[93] The marketing materials stated that the weapon's "characteristic[s]" included "silence, high precision, quick narcotism, eximious reliability, quick revival", and that it was "a pioneer in the middle of police and military narcosis-gun".[94]

In March 2012, *Defence Asia Review* reported that a "recent public display" by the Hong Kong garrison of the People's Liberation Army (PLA) included "a BBQ-901 tranquiliser gun (a pistol-type air gun fitted with a folding stock)".[95] In 2011, the author of this article, Gordon Arthur, had photographed a previous display of the BBQ-901 by the PLA in Hong Kong.[96]

China has clearly developed weapons employing ICAs for law enforcement purposes, which are in the possession of the PLA. Such weapons appear to be restricted to those targeting individuals. As of August 2015, there is no further information publicly available regarding the stockpiling or employment of this weapon within China, nor of international transfers.

To date, China has made no statement clarifying whether any Chinese research entity has conducted or is conducting research activities related to the development of ICA weapons targeting groups of individuals, and if so, for what purposes. However, there are indications of potential interest in this area.

In July 2005, *Military Review*, a US Army journal, contained a speculative article by two Chinese analysts – Guo Ji-wei, Director of the Department of Medical Affairs, Southwest Hospital, Chongqing, and Yang Xue-sen, a biotechnology lecturer and writer – in which they stated: "In the field of military affairs, modern biotechnology maintains a rapid pace of development and plays an important role in medical protection. However, it is gradually revealing a character of aggression as well. Therefore, it is of increasing military value."[97]

The authors further claimed that "... war through the command of biotechnology..." will "...ultimately, lead to success through ultramicro, nonlethal and reversible effects".[98]

In 2011, a paper by Qi, Cheng, Zuo, Li and Fan, all from the Institute of Chemical Defence, examined the degradation pathways of fentanyl and its analogues.[99] The authors noted that:

[T]hese kinds of compounds can also be utilised as incapacitants in countering terrorism. In October 2002, the analogues of fentanyl were reported to be successfully used in the Accident [sic] of rescuing hostages in Russia. In recent years, the analgesic and anesthetic medicines have gained attention in the world over. The dealing methods of these compounds are of great importance to criminalistics and countering terrorism.[100]

2.6.2. Czech Republic

In 2000, the Czech military funded a research project entitled "Analgesic-sedative and anesthetic agents used for emergency conditions – sedatives" (MO 03021100007),[101] which was led by Dr Fusek from the Czech Army's Purkyne Military Medical Academy in Hradec Kralove. The full details of this research have not been made public. However, one of Dr Fusek's colleagues, Dr Schreiberová, then an anaesthesiologist at University Hospital, Hradec Kralove, has recorded that: "In 2000 she started cooperation with the Institute for Clinical and Experimental Medicine in Prague (Ass. Prof. Hess) and Medical Military Academy in Hradec Kralove (Prof. Fusek)."[102] According to Dr Schreiberová, "the themes of these studies were anaesthesia and analgesia under specific conditions in disaster medicine and *the potential use of anaesthetic agents as non lethal weapons*" (emphasis added).[103]

Fusek, Hess and Schreiberová subsequently authored a paper, presented at the 3rd Ettlingen European Symposium on Non-Lethal Weapons in early May 2005, describing their investigations over several years relating to pharmaceutical chemicals that could be employed, in their words, as "pharmacological non-lethal weapons".[104] The authors reported administering rhesus monkeys with various "pharmacological cocktails" in order to determine which combinations and doses resulted in "fully reversible immobilization".[105] The paper also described how "Fully reversible analgesic sedation was... tested in man", utilizing the triple combination of dexmedetomidine, midazolam and fentanyl given to patients undergoing surgery, and a second combination of dexmedetomidine, midazolam and ketamine, which was tested on ten nurses.[106]

The researchers also investigated a number of alternative means of agent delivery, including via inhalation administration, which was initially tested on rats;[107] subsequent inhalation experiments utilizing human "volunteers", which included children, were also reported.[108] The researchers explored trans-buccal and sub-lingual administration in cats and rhesus monkeys; and conjunctival, nasal, sub-lingual and trans-dermal administration in rabbits.[109] Although apparently not tested in these studies, the researchers highlighted the potential application of paintball technology as a possible delivery mechanism.[110]

A shortened version of the Ettlingen paper, now entitled "Ultrapotent Opioids as Non-Lethal Weapons", was presented at an international meeting held under the auspices of NATO's Research Technology Organization (RTO) at the end of May 2005.[111] The meeting was part of an RTO Task Group on Prophylaxis and Therapy against Chemical Agents. In their paper for the meeting, Hess, Schreiberová and Fusek stated:

> There is a possibility of pharmacological control of an individual behaving aggressively. The demonstration, that this is not mere science fiction, we were able to see in October 2002 during a terrorist attack at Dubrovka Theatre in Moscow, Russia. The anti-terrorist commando employed, against Chechnya terrorists, fentanyl in aerosol or its derivative to render them harmless.[112]

Although Hess, Schreiberová and Fusek noted that "At present, their use contradicts the conventions on the use of chemical weapons" and "The issue also involves numerous legal aspects",[113] the paper summarized the authors' attempts to investigate chemical agents that could potentially be utilized in what they termed "non-lethal weapons" (NLW). The authors concluded that "many agents used in everyday practice in anesthesiology can be employed as pharmacological non-lethal weapons. *An anesthetist familiar with the pharmacokinetics and pharmacodynamics of these agents is thus familiar with this use. As a result, he or she can play a role in combating terrorism*" (emphasis added).[114]

In October 2005, Dr Schreiberová described the group's findings in a presentation – entitled "Pharmacological Non-Lethal Weapons" – to the Jane's Less Lethal Weapons 2005 Conference.[115] A copy of the associated PowerPoint presentation provided details of the research methodology employed, and included images of the application of the agents to a rabbit, macaque monkeys and human subjects.[116] One slide is entitled "Inhalation and nasal analgosedation in children" and has an image of a child in a hospital bed.[117]

In 2007, in a paper authored by Hess, Schreiberová, Fusek, Málek and Votava, which was presented at the 4th Ettlingen European Symposium on Non-Lethal Weapons, the researchers described how they "decided to test new combinations [of drugs] for suppression or complete abolition of aggressive behaviour" in macaque monkeys.[118] The researchers stated that: "All tested combinations resulted in macaques in reduction or complete loss of aggressiveness. Optimal combinations was naphtylmedetomidine + dextrorotatory isomer of ketamine + hyaluronidase. The onset of effect was rapid and we achieved complete manipulability of the animal with low motoric sedation."[119] Furthermore, the researchers argued that: "the results can be used to pacify aggressive people during medical treatment (mental disease), terrorist attacks and *during [sic] production of new pharmacological nonlethal weapons*" (emphasis added).[120]

The papers given at the NATO RTO Task Group meeting, the Jane's Less Lethal Weapons Conference and the two Ettlingen European Non-Lethal Weapons Symposia appeared to present the research findings in terms of their potential applicability to the development of what the authors called "pharmacological non-lethal weapons". Other papers by Czech researchers into the application of chemical agents for the treatment of aggressive states or to induce immobilization contained no explicit reference to their potential application as such so-called "non-lethal weapons".

In June 2010, Hess, Votava, Schreiberová, Málek and Horáček published a paper describing their studies inducing immobilization in orang-utans and chimpanzees, utilizing a naphthylmedetomidine–ketamine-hyaluronidase combination.[121] Subsequent papers by Votava, Hess, Schreiberová, Málek and Štein on "short term pharmacological immobilization in macaque monkeys",[122] and by Hess, Votava, Slíva, Málek, Kurzová and Štein, exploring the effects of ephedrine on "psychomotor recovery from anesthesia in macaque monkeys",[123] were published in 2011 and 2012, respectively. Although the results of these later studies were presented in terms of facilitating the relocation and painless medical examination of the animals, such research may also potentially be applicable to the study or development of weapons employing ICAs.

In July 2014 correspondence to the author and Professor Dando, the Czech CWC National Authority stated:

> The purpose of the research into toxicological properties of some incapacitating agents with code numbers MO 0302 11 00007, Grant Agency: Ministry of Defence, and Project Number: NR/8508-3/05 Grant Agency: Ministry of Health was purely medical such as for combating pain in incurable patients, easing anxiety during medical procedures or helping with treatment of non-cooperative or aggressive patients, as well as developing cocktail of anaesthetics usable in treating of high number of casualties during mass disasters. All these goals are fully legitimate,

and as they are not considered as research for protective purposes against chemical weapons, they are not reportable to the Technical Secretariat of the OPCW.[124]

The CWC National Authority categorically declared that: "There was no connection of the research with creation of any sort of weapons or devices which could be used for military or police purposes."[125]

With regard to publications by Czech researchers which appeared to frame their research findings in terms of the potential development of "pharmacological non-lethal weapons", the CWC National Authority stated:

> We interviewed principal authors of the projects and subsequent publications referring to incapacitation chemical weapons. We came to conclusion that their research programmes had justifiable medical goals, but *their reporting in public media exceeded actual results of the research thus creating a false impression of possible development of some sort of chemical weapons.*"[126] Publications referring to the possible use of opioids, anaesthetics and other toxic chemicals with incapacitating effects as chemical weapons [were] based solely on approximation of this legitimate research of the properties of these chemicals into the area [of] possible misuse.[127] [Emphasis added]

Furthermore, the CWC National Authority stated:

> Based on these findings, we organised several presentations for the Ministry of Health, the Ministry of Defence and the Ministry of the Interior with the aim to improve their knowledge of the [Chemical Weapons] Convention and the National Legislation for its implementation. We put emphasis on to the prohibition of development of any sort of chemical weapons, the area of riot control agents as well as their responsibility to report any development of a novel riot control agent to the Czech National Authority.[128]

2.6.3. India

A review of publicly available scientific papers indicates that researchers at the Defence Research & Development Establishment (DRDE) of the Defence Research & Development Organization (DRDO) have undertaken wide-ranging research into the synthesis, aerosolization and bio-efficacy of fentanyl and/or its analogues. In 2005, DRDE/DRDO researchers Gupta, Ganesan, Pande and Malhotra published a paper detailing "a straightforward one pot synthesis of fentanyl".[129] The authors stated that the method was "very simple and efficient"; the reaction taking place "under mild conditions and at room temperature";[130] and that this "method can also be used for the synthesis of fentanyl analogues".[131]

In 2008, a paper by Gupta, Ganesan, Gutch, Manral and Dubey described the application of thermogravimetry techniques for the estimation of vapour pressure and related thermodynamic properties of fentanyl.[132] The paper highlighted the importance of such data "for understanding and modelling the thermal aerosol formation of fentanyl which in turn is required for the development of its aerosol delivery system".[133] In 2009, a paper by Manral, Gupta, Suryanarayana, Ganesan and Malhotra detailed the group's investigations utilizing flash pyrolysis to explore the thermal behaviour of fentanyl at different temperatures,[134] and noted that the study "will be useful while developing technologies for thermal aerosol generation of fentanyl and related compounds".[135]

In 2009, a paper by Manral, Muniappan, Gupta, Ganesan, Malhotra and Vijayaraghavan documented their work exploring exposure to aerosolized fentanyl in mice.[136] The authors undertook the study "with a view to determine the effect of fentanyl aerosols on the breathing pattern of mice during and after exposure and to estimate a safety limit".[137] The researchers reported that "on exposure to fentanyl aerosol, a decrease in the respiratory rates of mice was observed, which recovered when exposure stopped. Mortality occurred on exposure to higher concentrations of fentanyl aerosols."[138] They concluded that "Although fentanyl aerosol did not cause any sensory and pulmonary irritation and since the RD50[139] and LC50[140] are very close, indicating a low safety margin, this type of sedative should not be used as an incapacitating agent."[141]

Work by DRDO/DRDE researchers related to the synthesis and bio-efficacy of fentanyl and its analogues continued in collaboration with researchers from other organizations. In 2010, a paper by Yadav, Chauhan, Ganesan, Gupta, Chauhan and Gokulan described their review of alternate methods for synthesizing fentanyl and their work to determine the structure–activity relationship (SAR) of fentanyl analogues.[142] Subsequently, a 2013 paper by Gupta, Yadav, Bhutia, Singh, Rao, Gujar, Ganesan and Bhattacharya described the synthesis of four fentanyl analogues.[143] The analogues were subsequently evaluated for their bio-efficacy. The study concluded that: "replacing the phenyl group of [the] phenethyl tail of fentanyl with different functional groups results in decreased toxicity of the molecules without sacrificing their potency. Thereby, enhanced therapeutic index could be achieved."[144]

Although a number of the published papers have highlighted the application of fentanyl for analgesia in a medical context, the specific purposes behind the DRDE/DRDO research activities were not specified, and the intended uses to which the resultant chemicals would be put remain unclear. Legitimate questions can be raised given the DRDO's stated role of working "... towards enhancing self-reliance in Defence Systems and [undertaking] design & development leading to production of world class weapon systems and equipment".[145] Furthermore, the organisation's website stated that DRDO is "working in various areas of military technology which include... armaments... advanced computing, simulation and *life sciences*" (emphasis added).[146]

Further insight into the intentions behind the DRDE/DRDO work on fentanyl and related pharmaceutical chemicals may be gained from analysing the biographical data available on DRDE personnel working on these or other DRDE/DRDO projects. For example, contributors to a 2007 paper included: "Dr Pradeep K. Gupta...[who] joined DRDE, Gwalior, in 2003. He is presently working as Scientist B in Synthetic Chemistry Division. *His area of work is synthesis and process development of non-lethal incapacitating agents*" (emphasis added).[147] Subsequently, the April 2010 edition of the *DRDO Newsletter*, the organization's "monthly house bulletin",[148] recorded that, among those who received a formal DRDO Award in 2009, was "Dr Pradeep K Gupta, Sc 'C', Defence Research & Development Establishment (DRDE), Gwalior", who was granted this award in recognition of his "significant contributions towards synthesis and optimisation of up-scaling fentanyl and other non-lethal incapacitating agents".[149]

On 22nd July 2014, in correspondence to the Bradford Non-Lethal Weapons Research Project (BNLWRP), the Indian CWC National Authority declared that: "India does not hold any stockpiles of weapons involving ICAs."[150] Furthermore, "India does not conduct research in order to develop weapons ICAs", nor does it "conduct research related to weaponisation of incapacitating agents" for defensive purposes such as the development of counter-measures.[151] In its response, the Indian CWC National Authority did confirm that "DRDE/DRDO undertakes research into fentanyl and its analogues", but that this was "only for purposes of characterization including its detection and protection aspects".[152] It is not known whether such activities have been reported to the OPCW as part of India's annual declaration of national programmes related to "protective purposes".[153]

2.6.4. Iran

Research scientists based in the Department of Chemistry at Imam Hossein University (IHU) have explored the structure–activity relationships of fentanyl and its analogues and have attempted to generate stable long-lasting aerosols of medetomidine and other potential ICAs; their work has been detailed in papers from 2007 till 2013.[154] The intentions behind this research and the potential uses to which the findings may be applied were unclear. Imam Hossein University (IHU) was co-founded in 1986 by Mohsen Reza'i, then Commander of the Sepāh e Pāsdārān (Army of Guardians), also known as the Iranian Revolutionary Guards (IRG).[155] The university has reportedly been run on military lines and used for training of IRG personnel.[156]

On 15th July 2014, in correspondence with the author, the Secretary of the Iranian CWC National Authority stated that the lead IHU researcher in this area was "interested in advance [of] academic and scientific chemical issues that [were] not prohibited by the Chemical Weapons Convention", that this "academic research [was] financed by [the] ministry of science and technology" and was "solely [for] scientific purposes".[157] The Secretary also explained that "IHU has held several training courses for its students and researchers to [make them] aware...with regard to the provisions of the CWC."[158]

2.6.5. Israel

Analysis of publicly available information indicates that Israel initiated a chemical weapons programme in the mid-1950s, which, according to Knip and Cohen, may have included work by the Israel Institute of Biological Research (IIBR) on chemical and toxin incapacitating agents.[159] Papers published by scientists working at IIBR during the 1960s till the end of the 1980s indicate research into a range of potential ICAs and/or related receptor sites.[160]

In its 2005 analysis of Israel's biological and chemical programmes, the Swedish Defence Research Institute concluded that: "The state previously developed offensive biological and chemical warfare capabilities. It has not been possible to conclude if these offensive programs still remain active today".[161] The report contended that:

> Israel has the scientific know-how and the industrial infrastructure to de novo produce and deploy militarily significant CBW [chemical and biological weapons] rapidly if so desired... In our view, the focus of the Israeli chemical and biological capacity today is to develop agents for small-scale covert use, i.e. a so-called 'dirty tricks' program.[162]

There has, to date, been one widely reported use of an ICA weapon by the Israeli security services (Mossad) in October 1997, in either a failed assassination attempt or a kidnapping operation that subsequently went awry.[163] According to a January 1998 *Jane's Intelligence Review* article: the chemical agent used was "believed to have been a synthetic opiate called Fentanyl which, absorbed through the skin and quickly metabolised, can kill within 48 hours and leaves no trace".[164] The Israeli intelligence team that conducted the operation reportedly included one physician, who "also carried an antidote known as Narcan or Naloxone in case something went wrong".[165] The target was Mr Khalid Mishal, a then-mid-ranking Hamas leader living in Jordan. The Israeli group reportedly followed Mr Mishal and attempted to deliver the fentanyl trans-dermally. Mr Mishal's driver, who witnessed the event, described the attack to the *New York Times*: "a man advanced toward Mr. Meshal and then lunged toward the area around his left ear. The Mossad agent's hand...was wrapped in a white bandage, with a small lead-colored protuberance in the palm."[166] The paper also interviewed Mr

Mishal, who stated: "I felt a loud noise in my ear... It was like a boom, like an electric shock. Then I had [a] shivering sensation in my body like an electric shock."[167] Mr Mishal was able to escape the attack, but, following the event, was reportedly seriously affected by the drug, and required significant medical attention. King Hussein of Jordan reportedly demanded that Israel provide an antidote. Mr Mishal subsequently made a full recovery.[168] On 5th October 1997, the Israeli Government publicly admitted responsibility for the attack, and on 7th October initiated an inquiry into the incident.[169]

It appears that the October 1997 operation may not have been an isolated event. *Jane's Intelligence Review* stated that "Israeli officials" interviewed by *Jane's* "indicated that Mossad has used Fentanyl in other operations, which they declined to describe, noting that it had a '100 per cent success rate'".[170] Similarly, in a *Time* magazine article, Israeli "government officials" reportedly stated that "the chosen method of assassination had been, until now, 'foolproof' " and that "the decision to act was taken based on the 100% success rate of this method, which left no fingerprints whatsoever. If they had done it the right way, no one would have noticed."[171] There have been no subsequent reports of use by the Israeli security or military forces of weapons employing fentanyl or other ICAs.

There is insufficient publicly available information to determine whether any Israeli entity is currently undertaking research into weapons employing ICAs, or whether Israel holds stockpiles of such weapons. Israel has made no clarificatory statement on this issue. There is limited information available indicating that the IIBR may be conducting work in potentially relevant dual-use fields, although the details of specific IIBR research projects are not available.[172]

2.6.6. Russian Federation

There are indications that, following the Moscow theatre incident, Russian researchers have continued work related to the potential employment of ICA weapons. In 2003, a paper by Klochikhin, Pirumov, Putilov and Selivanov, attempting to forecast future European "non-lethal" weapon application, was presented at the 2nd Ettlingen Symposium on Non-Lethal Weapons. In it, the authors stated: "Some experience of gas application in dramatic conditions of terrorists attack was gained in Moscow in 2002....The main problem is how to assess an impact of chemicals on a big crowd of civilians and terrorists between them in a concrete scenario and real conditions of application".[173] The authors noted that whilst "There has been significant success in the chemistry of calmatives... restriction of individual dosage is very important. There is still no perfect tranquillizing agent, but the problem of safety can be solved by the succeeding or simultaneous application of calmative and antidote. This can minimize potential fatality."[174]

A paper by Klochikhin, Lushnikov, Zagaynov, Putilov, Selivanov and Zatekvakhin, presented at the 3rd Ettlingen Symposium on Non-Lethal

Weapons in May 2005, described the computer modelling of a scenario in which aerosolized chemical "calmative" agent was introduced into a building where hostages were held captive. The paper stated that: "If the level of 95% efficiency is absolutely required to neutralize terrorists and to prevent mass destruction, there is no chance to eliminate hard consequences and fatalities. Calculations show that the majority of hostages can get serious poisoning and part of them – fatality. This is the cost of releasing if no other solutions [are] left."[175] The authors reported that: "One possible solution under discussion is to apply gaseous calmative agent and antidote together in the same composition or consequently after some delay. This is the way to control the value of impact and to decrease collateral damage."[176]

The researchers noted that "the real problem of chemical NLW is rather difficult. It requires serious efforts to develop reliable techniques and mathematic instruments for calculation of various scenarios...the full solution for such challenge demands the big intensive work of many scientific teams within several years."[177]

It appears that Russian researchers have continued work to develop computer models for the application of what they describe as "calmatives" against groups of individuals in enclosed spaces. In November 2009, Klochikhin and Selivanov presented a "Report on the 1st Phase of the Project 'Gas Flow'" to a meeting in London.[178] In the presentation, the authors described their work to develop computer code to generate 3-D simulations of "an effective scenario of calmative application", utilizing existing medical data on "calmatives" and physical data describing the nature of "gas" movement in enclosed spaces. The authors stated that the resultant computer code: "draws the gas; simulates gas transfer with air between rooms; calculates its concentrations in rooms; evaluates the calmative effects; shows the realistic simulation to define characters' status and gas concentration field to optimize the scenario gas effects".[179] No additional information about Project "Gas Flow" is currently in the public domain.

Further indications of potentially relevant Russian research were highlighted in the 2012 paper by Riches, Read, Black, Cooper and Timperley which stated that: "Scientific papers published by Russian military officers indicate an interest in fentanyls extending back 12 years: opioid receptor studies, fentanyl analysis and synthesis of fentanyl precursors."[180] Although this paper was not an official submission from the UK government to the OPCW, the study on which it was based was funded by the UK Ministry of Defence (MoD) and conducted by research scientists working at the UK DSTL. The paper was published with the permission of DSTL on behalf of the Controller of Her Majesty's Stationery Office. The author and Professor Dando subsequently identified further papers (published between 2005 and 2012) by certain Russian life scientists cited in the UK DSTL study, indicating additional research relating to opiate receptors (OR) and their interaction with OR ligands.[181]

To date (2015), the Russian Federation has provided no further details of the chemical or chemicals employed as an ICA weapon in the Moscow theatre siege, nor provided information as to whether stockpiles of weaponized ICAs are currently held in the Russian Federation. Furthermore, the Russian Federation has made no formal public statement clarifying whether research into the development and employment of weaponized ICAs is taking place, and if so, for what purposes.

2.6.7. United Kingdom

The UK government has released documents detailing the country's previous attempts to develop ICA weapons for military purposes from 1959 to 1972.[182] There is no evidence of a subsequent military ICA weapons development programme. There are indications of ICA research continuing into the 1980s, although the nature and purpose of such activities are not known.[183] In the early-to-mid-2000s, the UK government assessed the feasibility of introducing ICA weapons for certain law enforcement purposes,[184] but subsequently rejected this option.[185]

In 2013, the United Kingdom "unequivocally" declared that it "neither holds nor is developing any ICAs for law enforcement".[186] UK researchers based with the Defence Science and Technology Laboratory at Porton Down have conducted research into ICAs for "protective purposes",[187] as permitted under the CWC. The United Kingdom has provided some information on these activities to the OPCW in its annual Article X declarations, and also to the UK parliament.[188]

2.6.8. United States

The United States has a long history of research into ICA weapons dating back to the 1950s (as described in Section 2.5.1), which was halted in the late 1970s, with all ICA weapons stockpiles destroyed in the late 1980s and 1990s. The United States subsequently conducted research into ICA weapons for both military and law enforcement purposes, prior to and after the coming into force of the CWC, although there is no evidence of completed development or production of ICA weapons. In December 1999, the US Army's Edgewood Chemical Biological Center (ECBC) solicited research proposals for a three-phase project – CBD00-108 Chemical Immobilizing Agents for Non-Lethal Applications – the objective of which was to: "demonstrate the feasibility of a safe, reliable chemical immobilizing agent(s) for non-lethal (NL) applications in appropriate military missions and law enforcement situations".[189] Under Phase 1 of this programme, researchers would "conduct an analysis of promising new chemical immobilizing agents or combinations of agents", including "recent breakthroughs in the pharmacological classes such as anesthetics/analgesics, tranquilizers, hypnotics and neuromuscular blockers".[190] Researchers would "conduct a toxicological test program" with the "most promising" new immobilizing agents, to fill data gaps and consequently "establish the mode of immobilization, the effective dose(age) for immobilization, onset time and duration of effects, and safety ratio in the most appropriate animal species".[191]

Under Phase 2, input would be gathered from potential military and law enforcement users on the desired performance/operational characteristics; and the implications of the CWC for proposed scenarios of use would be determined. Following the selection of the "optimum scenario(s) of use", a series of "non-human primate and clinical tests" would be conducted "to establish safety and performance characteristics".[192]

Subsequently, an "appropriate delivery technique", for example, "an aerosol generator for dissemination for the inhalation route of entry, or a dart for injection in the intra-muscular route of entry" would be designed and demonstrated. Under Phase 3, dual-use applications of the technology were to be analysed.[193] The potential military uses cited included: "meeting U.S. and NATO objectives in peacekeeping missions; crowd control; embassy protection; rescue missions; and counter-terrorism".[194] The potential law enforcement uses highlighted included: "hostage and barricade situations; crowd control; close proximity encounters... to halt fleeing felons; and prison riots".[195] In June 2000, ECBC awarded a contract for Phase 1 of this project to OptiMetrics, Inc.[196] In November 2002, it was reported that Phase 1 had been completed.[197] It is not known whether Phase 2 or 3 were ever undertaken and if so, when or by whom.

In 2000, the Applied Research Laboratory and the College of Medicine at PSU published the results of its literature study and analysis of bio-medical research into a range of pharmaceutical agents, including "sedative-hypnotic agents, anesthetic agents, skeletal muscle relaxants, opioid analgesics, anxiolytics, antipsychotics, antidepressants and selected drugs of abuse", which attempted to "assess the potential use of calmatives as non-lethal techniques".[198] The report identified 10 classes of pharmaceutical agents and 32 representative agents or agent combinations as having a "high potential for further consideration as a non-lethal technique" (see Figure 2.2).[199]

In fiscal year 2001, the National Institute of Justice (NIJ) funded a three-phase project on NLW at the Institute for Non-Lethal Defense Technologies (INLDT) at PSU. Phase 2 of the project was to "... conduct an investigation of controlled exposure to calmative-based oleoresin capsicum".[200] Although publicly available information regarding this project is scarce, it apparently involved the combination of an ICA with the chemical irritant oleoresin capsicum (OC) in order to produce more profound effects.[201]

In 2003, the NRC issued a report reviewing prior and existing NLW research, examining relevant scientific and technological developments and recommending future areas of NLW research.[202] Whilst the report highlighted concerns regarding compliance with the CWC, the NRC panel recommended "increase[d] research in the field of human response to calmatives". They stated that: "Calmatives have potential as NLWs in many types of missions where calming of individuals or crowds is needed." The panel recommended that "The human effects of these compounds and their safety must have thorough evaluation under conditions simulating their mission uses."[203]

More recently, there has been uncertainty as to whether there was continuing interest by the US Armed Forces in dual-use research that could potentially have applicability in the study or development of ICA weapons. On 1st October 2009, the Air Force Office of Scientific Research (AFOSR) issued an initial solicitation for a $49 million multi-project research programme entitled "Advances in Bioscience for Airmen Performance" (BAA-09-02-RH).[204] Under this announcement, the 711th Human Performance Wing, Human Effectiveness Directorate solicited "white papers...for innovative science and technology projects to support advanced bioscience research". Specifically, the Biosciences and Performance Division was seeking "unique and innovative research concepts" that address its four "technical mission areas", which included "biobehavioral performance".[205]

Although the overarching goal of the "biobehavioral performance" mission area was to "develop bio-based methods and techniques to sustain and optimize airmen's cognitive performance", it included *inter alia*:

> Development of effective, reliable, and affordable alertness management, performance enhancing and emotional state modulation technologies. Includes non-medical neuroscience and biochemical pathway techniques. *Conversely, the chemical pathway area could include methods to degrade enemy performance and artificially overwhelm enemy cognitive capabilities.*[206]
>
> [Emphasis added]

In its original solicitation, the US Air Force anticipated "awarding 3–4 awards per year for this announcement"[207]; as of 2015, publicly available information indicates that contracts were awarded in 2012 and 2013 for projects apparently unrelated to ICAs.[208] The deadline for final submission of white papers for this research programme was 14th February 2014.

On 11th September 2014, following a request by the author and Professor Dando for further information on this project, a US Department of Defense (DoD) official stated that:

> The purpose of the program text for the bio behavioral performance technical area...was to be inclusive of all potential chemical pathways areas for study in order to sustain and optimize cognitive performance ... However, no grant was awarded for work under this technical area. Grants were awarded for work in other technical areas, but that work does not involve ICA research. The solicitation of and granting of any work under this project is compliant with the Chemical Weapons Convention...[209]

Whilst research into ICA weapons for military application appeared to end in the early to mid-2000s, there were indications that interest in developing ICA weapons for law enforcement continued. In April 2007, the NIJ convened a "community acceptance panel" to discuss the potential role of "calmative agents" in law enforcement.[210] The panel – which was comprised of experts from the scientific, toxicological and bio-ethical communities, as well as representatives from civil rights and human rights advocacy organizations, and the legal and law enforcement communities[211] – was tasked with "assessing the potential of developing new riot control agents,[212] such as chemical calmatives, as a viable addition or alternative to the law enforcement less lethal arsenal". It was envisaged that "such less lethal options would be delivered in situations and in a manner similar to pepper balls or OC, except the resulting effects would be designed to calm rather than irritate the target".[213]

According to an NIJ report of the meeting, the panel reached the "general consensus" that law enforcement officers need additional "less lethal" options, and that "pursuing new or updating existing research on the safety and viability of calmative agents was reasonable... It is important to note that the panel did not determine whether a tool could be developed, only that further research was an appropriate next step."[214] The NIJ subsequently awarded PSU $250,000 under grant 2007–DE–BX–K009,[215] to "explore the potential of operationalizing calmatives and to examine possible pharmaceuticals, technologies and legal issues".[216]

Further indications of interest in developing weapons utilizing ICAs for use in law enforcement have been enunciated by the Law Enforcement and Corrections Technology Advisory Council (LECTAC), a body composed of "senior leaders from law enforcement, corrections, courts, forensic science and other criminal justice agencies and professional organizations ... appointed by NIJ based on their records of distinguished service".[217] According to its publications, LECTAC was a "critical part of the National Institute of Justice's (NIJ) Research, Development, Test and Evaluation process" and provided "practitioner-based input on what technologies are most important and what technology gaps currently exist".[218]

The 2010 LECTEC report stated that the criminal justice community needed a "capability to inhibit metabolic functioning of individuals and groups (calmative agents) that is quick-acting, completely reversible and has no long-term physical or psychological effects", together with "a method of delivery that is capable of delivering at a variety of ranges to a target of one or many".[219] The LECTEC report further stated that the "required response" would be: "Immediate immobilization fully recoverable in two to 30 minutes", and this would entail "immediate and full impairment of physical function with full recovery, immediate disruption of ability to sense and interpret information with full recovery and immediate full compliance".[220] The LECTEC report noted that a "current project partially addressing this issue: 2007-DE-BX-K009" was under way, and this "study should yield sufficient data to focus and solicit manufacturer development effort".[221] It is not known whether and, if so, how the recommendations from LECTAC were taken forward by the NIJ.[222]

Despite a previous long-standing interest in research potentially related to ICA weapons development, in 2013, 2014 and 2015 senior US officials formally stated that no attempts to develop weapons employing such chemical agents currently take place.[223] For example, at the 19th Conference of States Parties to the CWC in December 2014, US Under Secretary for Arms Control and International Security, Ms Gottemoeller, stated that: "I want to reiterate that the United States is not developing, producing, stockpiling, or using incapacitating chemical agents."[224] It is not currently known whether the United States undertakes dual-use research related to ICAs for "protective purposes", and, if so, how and whether this is reported to the OPCW.

2.7. The implications of ongoing revolutionary scientific and technological developments

The survey of open source literature described in Sections 2.6.1–7 indicates that, since the mid-1990s, China, Israel and the Russian Federation have developed or acquired ICA weapons, whilst some additional States appear, at least, to have shown an interest in studying this area. Whilst such activities appear to have concentrated upon a limited range of existing pharmaceutical chemicals, future attempts at ICA weaponization may well explore a far wider range of potential agents.

Indeed, the search for candidate ICAs with potential weapons utility as well as suitable delivery mechanisms is likely to be informed by current and continuing advances in relevant science and technology, particularly genomics, synthetic biology, biotechnology, medical pharmacology, neuroscience and the understanding of human behaviour.[225] Whilst many of these advances have great potential to benefit mankind, for example in the development of more effective, safer medicines,[226] a range of scholars and scientific bodies have highlighted the potential implications of the misuse of such research for hostile purposes.

In 2005, Wheelis and Dando surveyed developments and future trends in neurobiology and concluded that there were indications that military interest was already directed towards the next generation of chemical agents affecting the brain and central nervous system:

> In addition to drugs causing calming or unconsciousness, compounds on the horizon with potential as military agents include noradrenaline antagonists such as propranolol to cause selective memory loss, cholecystokinin B agonists to cause panic attacks, and substance P agonists to induce depression. The question thus is not so much when

these capabilities will arise – because arise they certainly will – but what purposes will those with such capabilities pursue.[227]

In 2006, the US NRC produced a report entitled *Globalization, Biosecurity, and the Future of the Life Sciences*, which concluded that:

> Recent advances in understanding the mechanisms of action of bioregulatory compounds, signalling processes, and the regulation of human gene expression – combined with advances in chemistry, synthetic biology, nanotechnology, and other technologies – have opened up new and exceedingly challenging frontiers of concern.[228]

In 2007, McCreight highlighted the potential security implications (for the United States) of continued unregulated military neuroscience research. He observed that: "Many nation States have conducted both legitimate and military-related neuroscience research. There are no binding international norms or rules to govern legitimate research. There are no rules or mechanisms to regulate, halt or delay military research in neuroscience." These observations led him to conclude that "We have no protections or safeguards unless we take steps to insist on them."[229] He continued:

> Given more than 45 years of military investment in neuroscience thus far by several countries, despite limited results, we can expect some variety of weapons to emerge within 10 years...Unless measures are taken to halt existing military research into neuroscience we may face new categories of weapons before 2020 held by several nations both friendly and hostile.[230]

In a presentation to a 2010 ICRC expert meeting on ICAs, Trapp warned:

> The explosion of knowledge in neuroscience, bioregulators, receptor research, systems biology and related disciplines is likely to lead to the discovery, amongst others, of new physiologically-active compounds that can selectively interfere with certain regulatory functions in the brain or other organs, and presumably even modulate human behavior in a predictable manner. Some of these new compounds (or selective delivery methods) may well have a profile that could make them attractive as novel candidate chemical warfare agents.[231]

Advances in the discovery or synthetic production of potential incapacitating agents have occurred in parallel with developments in particle engineering and nanotechnology that could allow the delivery of biologically active chemicals to specific target organs or receptors. The implications of this were highlighted in the 2008 NRC report on *Emerging Cognitive Neuroscience and Related Technologies*, which warned that nanotechnologies could be used to overcome the blood–brain barrier and thereby "enable unparalleled access to the brain. Nanotechnologies can also exploit existing transport mechanisms to transmit substances into the brain in analogy with the Trojan horse".[232]

The 2008 NRC report also highlighted the potential threats resulting from developments in nanotechnologies or gas-phase techniques that allow dispersal of highly potent chemicals over wide areas. It noted that, at the present time, "pharmacological agents are not used as weapons of mass effect, because their large-scale deployment is impractical" as it was "currently impossible to get an effective dose to a combatant". However, the report stated that "technologies that could be available in the next 20 years would allow dispersal of agents in delivery vehicles that would be analogous to a pharmacological cluster bomb or a land mine".[233]

2.8. Conclusions

As part of the first stage of a holistic arms control analysis of ICA weapons, an examination was undertaken into their properties and effects. These ICAs can be considered as a diverse range of substances whose chemical action on specific biochemical processes and physiological systems, especially those affecting the higher regulatory activity of the central nervous system, produce a disabling condition, but at higher concentrations can potentially prove fatal.

Proponents of ICA weapons have promoted their development and use in certain law enforcement scenarios, particularly hostage situations; and as a possible tool in a variety of military operations, especially where combatants and non-combatants are mixed.

Opponents and sceptics have strongly contested the possibility of employing truly "less lethal" ICA weapons and have highlighted the grave dangers to the health of the targeted populations. They have raised concerns regarding the creeping legitimization of such agents with the erosion of the norm against the weaponization of toxicity; the risks of their proliferation to both State and non-State actors; their potential use as a lethal force multiplier; their applicability in the facilitation of torture and other human rights violations; and the militarization of the life sciences.

Over the last 50 years, although a number of States have sought to develop ICA weapons, to date only one State – the United States – has confirmed weaponizing them for military purposes (now discontinued). Publicly accessible information does indicate that China, Israel and the Russian Federation have acquired or developed ICA weapons, and that such weapons are either in the possession, or have been used by the law enforcement or security services of those countries since the coming into force of the CWC in 1997. The situation in other States is less certain: although there is evidence of relevant dual-use research in a number of countries, the full nature and purpose of such research in certain countries is often unclear, as are the intended applications to which it will be put. Currently, there is no evidence of concerted attempts by non-State actors, such as terrorist groups, to conduct research and development of ICAs weapons.[234]

Consequently, ICA weapons can be considered as an immature, limited and contested technology; however, one that could be radically affected by advances in science and technology and that may potentially proliferate and be misused by State and non-State actors.

3
Riot Control Agents

3.1. Introduction

As part of the first stage of the HAC analytic process, this chapter examines the properties of riot control agents (RCAs), explores contemporary research into these agents, examines the current scenarios of their application and highlights arms control and human security concerns pertaining to the misuse of such agents.

RCAs are potent sensory irritants normally of low lethality that produce dose and time-dependent acute site-specific toxicity. The most widely used are commonly known as "tear gases", and, when employed for military purposes, have often been called "harassing agents". Such chemicals interact pharmacologically with sensory nerve receptors associated with mucosal surfaces and the skin at the site of contamination, resulting in localized discomfort or pain with associated reflexes. Although intense lachrymation and sternutation are common reactions to exposure to RCAs, these compounds can elicit a diverse array of physiological effects.[1] Non-irritant chemicals have occasionally also been considered for riot control or other purposes.

RCAs are considered to have three common characteristics: rapid onset of effect(s) (seconds to a few minutes), relatively short duration of effects (typically 15–30 minutes) following cessation of exposure and decontamination and high safety ratio (i.e. large margins between the dosage of an RCA that is needed to cause tissue irritation or pain [effective dose – ED] and the dosage that could cause serious injury and in the worst case death [lethal dose – LD]).[2] RCAs can be divided into three broad types: lachrymators (irritants that cause tearing [watering of the eyes]), sternutators (substances that induce sneezing) and vomiting agents.[3] The comparative estimated toxicity for humans of various RCAs is summarized in Table 3.1.

Table 3.1 Comparative estimated toxicity for humans of selected RCAs[4]

Compound	Minimal Irritant Concentration (mg · m^{-3})	Estimated LCt$_{50}$ (mg · min m^{-3})	Estimated ICt$_{50}$ (mg · min m^{-3})
CN	0.3–1	8,500–25,000	20–50
CR	0.002	Greater than 100,000	Approximately 1
CS	0.004	25,000–150,000	5
DM	1–5	11,000–35,000	20–150
PS	2–9	2,000	–

3.2. Types of RCAs employed

3.2.1. Most commonly utilized RCAs

3.2.1.1. CN (2-chloroacetophenone)

CN was first synthesized in 1869 by Carl Graebe and was studied for military purposes during the latter part of World War I.[5] Not actually employed in the fighting, it was developed by the US Army Chemical Warfare Service (CWS) which formally adopted it in 1922.[6] Japan was reported as using CN against aboriginal Taiwanese in 1930.[7] CN was the RCA of choice for a number of States including the United States till the late 1950s, when it was replaced by the less harmful o-chlorobenzylidene malononitirile (CS).[8] CN is a solid and can be disseminated as smoke. It is also available in powder and liquid formulations. It has been utilized as the active ingredient of Mace sprays – used by police in the United States and other countries, and sold to the public for "self-defence". According to the 2013 Annual Report of the OPCW, 71 States have declared possession of CN for the purposes of riot control.[9] The clinical effects of CN are similar to those of CS, but it is more harmful and more likely to cause serious side effects. The harassing concentration of CN is two-and-a-half times that of CS, and laboratory studies rate its effects as equivalent to those of CS but threefold to tenfold more toxic.[10] Deaths from high concentrations of CN may occur and have been reported – normally when the agent has been used in enclosed spaces, such as prisons. Post-mortem examinations from such cases revealed oedema and congestion of the lungs, alveolar haemorrhage, necrosis of the mucosal lining of the lungs and bronchopneumonia.[11]

3.2.1.2. CS (2-Chlorobenzylidenemalonitrile, o-chlorobenzylidene malononitirile)

CS was first synthesized by Corson and Stoughton in the United States in 1928. According to Hu and colleagues, CS causes "a burning sensation in the eyes ... severe irritation of the respiratory tract, burning pain in the nose, sneezing, soreness and tightness of the chest. Even very light exposure can cause a rapid rise in blood pressure, and as this increases, gagging, nausea and vomiting occur."[12] CS did not become widely utilized as an RCA until well after World War II, when it was learned that the effect of CS was less harmful but more potent than that of CN.[13] CS was adopted for law enforcement purposes, initially by the United Kingdom, in the 1950s because it was more effective and readily usable than CN, since it can be dispersed by solution spraying, explosive dispersion or as smoke from a pyrotechnic mixture.[14] Despite some evidence of the detrimental effects of CS on human health, it remains the "tear gas" most commonly used by security forces.[15] The OPCW 2013 Annual Report records that 118 States have declared possession of CS for the purposes of riot control.[16] According to Hu and colleagues: "Its popularity among military and police forces stems partly from comparisons with the other tear gas agents, which suggests that CS is a more potent lachrymator and seems to cause less long-term injury, particularly with respect to the eye."[17] However, according to Salem, Gutting, Kluchinsky and colleagues, scientific investigations into the identification of CS-derived compounds and other thermal degradation products formed during the heat dispersion of CS have raised questions about the potential health risks associated with the use of high-temperature heat dispersion devices, particularly if used in enclosed spaces.[18]

According to US military manuals, CS has been "specially formulated for varied dissemination characteristics and/or effects as CSX, CS1 and CS2".[19] CSX, a form of CS developed for dissemination as a liquid rather than as a powder, stings and irritates the eyes, skin, nose, throat and lungs of exposed personnel.[20] CS1, comprising micro-pulverised CS powder mixed with silica, was formulated "to prolong persistency and increase effectiveness". As CS1 settles out of the air, it "readily contaminates terrain, vegetation, personnel and equipment". When disturbed, it re-aerosolises to cause respiratory and eye effects.[21] It can remain active for 14 days in an enclosed space such as "a dry tunnel or bunker", and about one week in the open air.[22] CS2 is a form of CS blended with silicone-treated silica aerogel, which enables it to repel water and prolongs its effectiveness for both immediate and surface contamination effects. When disturbed, CS2 re-aerosolizes to cause respiratory and eye irritation, up to two months after it has settled.[23]

In October 2007, the US government released a report summarizing research undertaken from May 2003 to September 2004 at Edgewood Chemical Biological Center,[24] with a view to improving the properties of CS and enhancing its effectiveness. The report described the synthesis – using microwave irradiation and novel catalysts – of 15 new CS analogues incorporating fluorine and fluorine-containing groups. The structures and physical properties and mass spectral characterization of this group of highly potent CS-type agents were described in the report. In their conclusions the researchers stated that: "The new CS-analogs are expected to be more potent than CS" and "the new compounds possess several distinct advantages over

CS".[25] No further information is publicly available as to whether subsequent R&D into these agents has taken place.

3.2.1.3. CR (dibenz[b,f][1,4]oxazepine)

CR was first synthesized by Higginbottom and Suschitzky in the United Kingdom in 1962.[26] Its effects are similar to those of CS, but it is about five times more potent. Limited data are available on CR but it appears to pose less immediate health risks than CS because it seems to have little effect on the lower airways or lungs. There do not appear to be persistent skin or eye effects.[27] It is dispersed in solution as a liquid or aerosol spray. As it is hydrolysed very slowly in water, it persists in the environment for long periods.[28] According to the OPCW 2013 Annual Report, 13 States have declared possession of CR for the purposes of riot control; however, the OPCW does not publicly identify these States.[29] One State, the United Kingdom, has publicly reported that CR has been stockpiled by its MoD alongside CS and other agents. Due to its high potency, it has been UK policy to retain it for use for counter-terrorist measures and not for normal policing activities.[30] CR was reportedly deployed to UK forces in Northern Ireland (in the Maze Prison) in 1974, and there were unconfirmed and contested reports of its use against rioting Republican prisoners in October of that year.[31] Although there were plans for the use of CR in trials in Hong Kong in January 1974, such trials were never carried out.[32]

3.2.1.4. Oleoresin capsicum (OC)

OC is a naturally occurring mixture of compounds extracted from more than 20 different species of the capsicum plant, which include chilli peppers, red peppers, jalapeno and paprika. Many of the physiological responses induced by OC are due to a family of compounds known as capsaicinoids. OC includes from 0.1% to 1% capsaicinoids by dry mass. The main capsaicinoid of interest as an RCA is capsaicin (trans-8-methyl-N-vanillyl-6-noneamide). The capsaicinoids content of OC is approximately 70% capsaicin, 20% dihydrocapsaicin, 7% norhydrocapsaicin, 1% homocapsaicin and 1% homodihydrocapsaicin – though its specific components can vary depending on the capsicum used. OC spray inflames the mucous membranes, causing coughing, gagging, choking, shortness of breath and an acute burning sensation on the skin and exposed areas.[33]

OC has found widespread use by law enforcement agencies for incapacitating dangerous individuals or as an RCA for crowds. It is also widely promoted by commercial companies to the general public for personal protection. According to the OPCW 2013 Annual Report, 27 States have declared possession of "capsaicinoids" for the purposes of riot control.[34] US forces deployed to Somalia carried "non-lethal" packages that included OC. Military police from several US army divisions, as well as several Marine Corps units, have used OC in the past, and, as of 2008, were investigating its capabilities.[35] There appears to be no international mechanism to ensure standardization in development and testing of OC products and, consequently, numerous formulations of OC have been developed and marketed (commonly referred to as pepper spray, pepper mace and pepper gas). Similarly, there is a wide diversity in the range of delivery and dispersal mechanisms, carriers and propellant systems employed.[36] Serious injuries and deaths have been associated with OC/pepper sprays.[37]

3.2.1.5. PAVA (Pelargonic acid vanillylamide)

PAVA pepper spray is a synthetic formulation of one active OC constituent, and is classified as an inflammatory, since it causes acute burning of the eyes, severe inflammation of the mucous membranes and upper respiratory tract, and produces coughing and gagging. It is less potent and causes less pain than natural capsaicin.[38] Various formulations of PAVA have been approved for use by police forces, for example, in Belgium, Germany, the Netherlands, Switzerland and the United Kingdom.[39]

3.2.2. Toxic chemicals previously employed but now considered inappropriate as RCAs

3.2.2.1. DM (Adamsite, diphenylaminechlorarsine)

First discovered in 1913 by German scientists through an expensive and difficult process, DM was more cheaply and simply developed in 1918 by Robert Adams in the United States. The United States produced DM by the end of World War I but did not use it; however, reports suggest that Italy may have employed it. In World War II all belligerent States produced DM, and smoke generators containing DM were developed.[40] For riot control purposes, because of its minimal effects on the eye, DM was mixed with CN, and this preparation was used by US troops during the Vietnam War.[41] DM is classified by the US military as a vomiting agent and a sternutator.[42] It is considered to be more toxic than many other RCAs.[43] Its acute effects in both laboratory animals and human volunteers following DM inhalation are "strikingly variable".[44] DM affects the upper respiratory tract, initially causing irritation of nasal and sinus mucosae. The irritation spreads into the throat, followed by coughing and choking, with eventual effects observed in the lower air passages and lungs.[45] There may be prolonged systemic effects, such as headache, chills, nausea, cramps, vomiting and diarrhoea. Several deaths following exposure to DM have been reported.[46] In December 2000, following a recommendation by the Scientific Advisory Board, the Executive Council of the OPCW concluded that DM was not suitable as an RCA.[47] Although DM is now considered obsolete as an RCA,[48] according to the OPCW 2007 Annual Report, two States were still declaring possession of DM for the purposes of riot control.[49] Although subsequent OPCW

Annual Reports do not specifically list DM in the relevant annexes, there is an undefined category of "other types" of chemical in which it may be incorporated.[50]

3.2.2.2. PS (chloropicrin)

PS was used extensively as a lachrymator, choking and vomiting agent by France, Germany, Russia and the United Kingdom during World War I.[51] Although previously used as a harassing agent, PS acts much like a pulmonary agent and is often classified as such.[52] Weaponized PS was primarily disseminated through wind dispersion in analogous fashion to chlorine, phosgene and mustard gas.[53] The human toxicity of PS following inhalation is primarily restricted to the small-to-medium bronchi, and death may result from pulmonary oedema, bronchopneumonia or *bronchiolitis obliterans*.[54] It is no longer used as an RCA and its employment as such would appear to be prohibited under the CWC as it is a Schedule 3 chemical.[55] However, according to the OPCW 2000 Annual Report, three States had declared possession of PS for the purposes of riot control;[56] this number dropped to two States in the following year,[57] and subsequently no State has specifically declared possession of PS for such purposes. In the OPCW 2002 Annual Report, one State declared possession of CNS (an RCA which contains 38.4% PS) for riot control purposes.[58] Although there is no record of any further declaration of PS or products containing PS in subsequent OPCW Annual Reports, there is an undefined category of "other types" of chemical in the relevant annexes where such agents may be incorporated.[59] The last reported case of a State's security forces utilizing PS occurred in Georgia in 1989.[60]

3.2.3. Malodorants

In addition to the "traditional" RCAs – such as the tear gases and chemical irritant "pepper" sprays described – there are certain chemical agents (notably malodorants) that some arms control experts believe can be classed as RCAs, but whose position, as of August 2015, has not been clarified by the CWC or by any of the policy-making organs of the OPCW.[61]

Although no internationally agreed definition currently exists, malodorants have been described by one commentator as "chemicals designed to target human olfactory receptors in order to provoke a physiological response, ranging from simple aversion, to – in more extreme cases – symptoms such as nausea and vomiting".[62]

Details of previous or contemporary State research and development of malodorants and related means of delivery are scarce. However some information concerning US and Israeli activities in this area has been made public. In its 1999 Annual Report, the US Joint Non-Lethal Weapons Program (JNLWP) reported that it was sponsoring a project that "investigates odorants and their effects on behaviour. It can be used for riot control, to clear facilities, to deny an area, or as a taggant."[63] In 2000–2001, information on research by the US Army's Edgewood Chemical Biological Centre into a range of candidate odours came to light,[64] whilst, in 2001, the Nonlethal Environmental Evaluation and Remediation Center (NEER) at Kansas State University reported on evaluations of two specific malodorant formulations for suitability as NLW.[65] During the early to mid-2000s, the US Army also explored the development of mechanisms that could potentially be employed for malodorant delivery, including an 81 mm mortar[66] and a 155 mm projectile.[67] There were indications that US bodies including the JNLWP subsequently continued to study malodorants and related delivery mechanisms, at least at the conceptual level;[68] and in June 2014, the US Office of Naval Research issued a Broad Agency Announcement (BAA), soliciting proposals for research into a range of "non-lethal" or "less lethal" technologies for fiscal year 2015, which included calls for research into malodorants.[69] In addition one US company, Security Devices International, Inc, has promoted a 40mm "non-lethal" malodorant Blunt Impact Projectile[70]; whilst a second US company, Mistral Inc., has promoted a malodorant and range of attendant delivery systems including a 60 oz. canister, 40 mm "less lethal" grenade and a skid sprayer with a 50 gallon tank.[71] To date, there have been no confirmed reports of the contemporary possession of malodorant weapons by any US military, security or police force.

In contrast to the United States where no malodorants have to date been employed in law enforcement operations, there are widespread reports that Israel has developed a weapon employing a malodorant agent and that this weapon, called the Skunk, has frequently been used by Israeli police and security forces. The Skunk was developed collaboratively by the Technological Development Department of the Israel Police and Ordortec Ltd, an Israeli company which "specializes in the research and development of advanced non-toxic, non-lethal scent-based repellents for law enforcement."[72]

According to the Material Safety Data Sheet (MSDS) provided by the company, the Skunk is composed of water, sodium bicarbonate and yeast and "at the pH level of sodium bicarbonate the yeasts synthesise some amino acids causing heavy odor".[73] Although the MSDS described the Skunk as a "repulsive order liquid", it stated that the "product has not been reported as [a] toxic material" and it had "no carcinogenic activity".[74] However, the MSDS did warn that the Skunk might cause skin irritation and can cause pain and redness in the eyes. If ingested, it can cause abdominal pain.[75]

The Israeli Ministry of the Environment and the Israeli Defence Force (IDF) Chief Medical Officer have approved use of the Skunk.[76] According to Ordortec Ltd, the Skunk has been "enthusiastically adopted by both the Israeli police and the...IDF".[77] According to the Israeli NGO, B'tselem, the malodorant, was first employed in August 2008 against demonstrators in the village of Ni'lin. On this occasion, the Skunk was dispersed by Border Police officers employing backpack spray devices.[78] Subsequently, the standard means of dispersal has been by truck mounted water cannon. According to *The Economist*, in July 2014 "the Internal Security Ministry placed a follow-on order for Skunk worth $45,000."[79] In addition to Odorotec Ltd, a second Israeli company, Tar Ideal Concepts, has promoted the Skunk.[80] There are no confirmed reports that the Skunk has been acquired by the military, security or police force of any other State.[81]

3.3. Overarching health and safety concerns relating to RCAs

3.3.1. Variations in effect of RCAs upon targets

Although RCAs are promoted as being safe when used according to manufacturers' instructions, serious health and safety problems have resulted from the use of certain agents in practice. Stoppford and Olajos have noted that: "...despite the low toxicity of modern RCAs, these compounds are not entirely without risk".[82] Potentially serious adverse effects of the most commonly employed RCAs include corneal damage, allergic contact dermatitis, blistering of the skin, burns, oedema, bronchoconstriction, bronchoneumonia and, in certain rare cases, death.[83] In addition to infants, young children and the elderly, research indicates that other population groups that are more likely to be detrimentally affected by RCAs include those with underlying heath issues such as asthma, bronchitis, cardiac disease or hypertension[84]; and those who take certain addictive drugs, notably cocaine.[85]

3.3.2. Variations in composition and concentration of RCA products

There are concerns around the composition, concentrations and strength of the active chemical agents employed in RCA products used by law enforcement officials and also those commercially available for the personal defence market. For example, an analysis by the US National Institute of Standards and Technology of 10 pepper sprays sold in the United States found that the strongest one had 40 times the amount of OC than the weakest.[86]

In addition to concerns about the toxicity of the RCA itself, consideration should also be given to the properties and potential health effects of the associated solvents, carrier solutions or propellants that may be employed. For example, concerns have been expressed about the potential carcinogenic effects of the solvent methyl isobutyl ketone (MIBK), used to deliver CS in certain sprays.[87] Likewise, investigations of OC and pepper sprays found that some contained toxic solvents. For example, use of a Russian-manufactured pepper spray containing unidentified solvents reportedly resulted in "severe chemical burns" and eye damage lasting more than six weeks,[88] whilst an individual who was exposed to a training spray that contained no active RCA (normally OC), but solely the solvent trichloroethylene, went on to develop corneal erosions, with alteration of vision that lasted two days. Health professionals have highlighted the potential carcinogenic and mutagenic properties of trichloroethylene, as well as its potentially adverse cardiovascular and respiratory effect.[89] Despite such concerns, it appears that as of 9th July 2015, certain companies have continued to promote RCA devices containing trichloroethylene.[90]

3.3.3. Information provision and independent testing

Concerns relating to the composition of RCA products are exacerbated by the partial, and in many cases unverified, information publicly available in this area. For example, with regard to OC, Sutherland has noted that "Most manufacturers do not disclose the exact composition of the product, and their material safety data sheets [which provide information about the substance's properties] also state that the composition is a trade secret."[91] Sutherland cited the case of DuPont Chemicals, "who describe[d] its Dymel 22 propellant in the following terms: 'the compound is untested for skin and eye irritancy and is untested for animal sensitization'".[92]

Furthermore, manufacturers' claims about the safety and health effects of their products are often not subject to independent analysis. A study by Hay and colleagues of an RCA employed by the Israeli Army against protesting civilians in the West Bank highlighted the secrecy surrounding its use and showed that resulting skin injuries could be "far more severe than the effects that the material safety data sheet for the product suggests".[93] Stoppford and Olajos have stated their belief that: "Clearly, there is a great need to conduct rigorous studies and to validate claims by manufacturers as to the efficacy and safety of their products. Rigorous studies similar to those utilized by the pharmaceutical industry for new drug development and approval need to be done."[94]

Other researchers have proposed a broader approach. In its report to the Scientific and Technology Options Assessment (STOA) Panel of the European Parliament, the Omega Research Foundation recommended the consideration of five criteria when examining the safety of chemical irritants used for law enforcement: the innate relative toxicity of the chemical used; the ability of security force personnel to use the dispersion mechanisms to deliver a measured dose which remains non-damaging and "non-lethal"; the relative toxicity and safe dose of any carrier, solvent or propellant used to deliver the chemical to target subject(s); the safety from blast damage or fire hazard of any pyrotechnically dispersed irritant munition; the professionalism and training of any operatives to ensure that such devices are used within the context of their training, codes of conduct and in accordance with manufacturers' instructions.[95]

Despite long-standing concerns, the effective and comprehensive application of international standards for the testing and regulation of the chemical safety of RCAs used for law enforcement does not presently occur.

3.3.4. Nature of use and misuse

According to Hill and colleagues, conditions affecting an individual's reaction to the effects of RCAs include excessive application of the agent, delivery in an enclosed space, prolonged exposure (as when the victim cannot flee), a high minute ventilation (as during a fight), and (for skin reactions) high temperature and relative humidity.[96]

3.3.4.1. Use in confined spaces

Exposure to RCAs in confined spaces with limited ventilation increases the detrimental effects of the active agent. Hu and colleagues reported that pulmonary function deterioration and respiratory complaints might be observed several months after the cessation of exposure to CS in such circumstances.[97] Karagama and colleagues reported that out of 34 persons exposed to CS in a closed area, 23 had respiratory complaints after one hour following the cessation of exposure. They tracked the cases and showed that respiratory symptoms persisted for ten months in five subjects.[98]

Sutherland has claimed that:

> High-level exposure can cause ocular, pulmonary and dermal injuries and the use of RCAs in enclosed spaces can produce toxic effects. There is a need for additional research to establish the biological and toxicological effects of RCAs, and this is especially true of the use of RCAs in law enforcement activities where they are often misused deliberately or through ignorance.[99]

In their 2012 report detailing employment of RCAs by law enforcement officials in Bahrain, Physicians for Human Rights stated:

> Toxic chemical agent vapors accumulate in low areas and do not disperse easily when canisters are detonated in confined spaces. As a result, health effects of exposure to these toxic chemical agents in confined spaces may be both prolonged and more severe. Consistent exposure to high doses of riot control agents in enclosed spaces has been shown to cause acute lung damage and death.[100]

3.3.4.2. Potential long-term effects of RCAs

While certain previous studies have found no long-term physical effects resulting from single or controlled exposures to CS or OC,[101] a 2015 research article by Arbak and colleagues detailing their examination of long-term respiratory effects of tear gas among subjects with a history of frequent exposure found an apparent increased incidence of respiratory complaints and increased risk of chronic bronchitis.[102] In addition, field research undertaken by Physicians for Human Rights and others in Bahrain, the Occupied Territories and South Korea appear to indicate that exposure to high concentrations of RCAs and/or multiple and repeated exposure to RCAs over long time periods may potentially be linked to increases in reported asthma, respiratory failure or lung function deterioration. Furthermore, such RCA exposure has also reportedly been associated with increased numbers of miscarriages, still-births, and other reproductive anomalies.[103] In addition, there has been some research indicating potential mutagenic and carcinogenic effects of certain RCAs, although the implications of such findings have been contested.[104] In the light of such concerns, health professionals have repeatedly called for long-term research in these areas.

In their 1989 study, Hu and colleagues concluded that:

> From a toxicological perspective, there is a great need for epidemiologic and more laboratory research that would illuminate the full health consequences of exposure to tear gas compounds such as CS. The possibility of long-term health consequences such as tumor formation, reproductive effects, and pulmonary disease is especially disturbing in view of the multiple exposures sustained by demonstrators and nondemonstrators alike in some areas of civilian unrest.[105]

Subsequently, in 2009, Carron and Yerson called for research into: "the alleged safety of existing riot control agents... the delayed toxic effects and potential pro-carcinogenic risk of repeated exposure... the safety criteria for riot control agents."[106] However, as Casey-Maslen, Corney and Dymond-Bass have observed: "data that could assist in such long-term studies is not routinely collected by law enforcement agencies, making such a study almost impossible."[107]

3.3.4.3. Kinetic impact injuries

The risk of further harmful effects, not directly related to the properties of the RCA and associated solvents and carriers, includes kinetic impact injuries causing penetrating trauma, which can be exacerbated by the presence of the chemicals in use.[108] There is substantial documentation of serious injuries – including blindness, head trauma, paralysis, and in some cases death – resulting from the inappropriate employment of weapons-fired RCA devices such as grenades, cartridges and other specially designed RCA projectiles – particularly when the devices are used at close range or fired directly at the sensitive parts of the target's body.[109]

3.3.4.4. The use of out-of-date or expired RCA products

Normally, RCA products have a shelf life of between three to five years, after which they should not be used. The US company Federal Laboratories, in its pamphlet entitled "Riot Control", has stressed that inappropriate storage of their RCA products and/or use after the recommended five-year shelf life can cause the equipment to malfunction. It stated that while out-of-date material should not be used for riot control, they can be used as training aids, as "in practice sessions it is of no great importance if a device should malfunction".[110] Although little research has been conducted into whether and how the properties of RCAs may change over time, there are indications that some RCAs at least may be affected.[111] This leads to concerns regarding potential deleterious health effects upon those exposed, as well as potential safety consequences of RCA delivery mechanism malfunction – for both the targeted populations and the law enforcement officials employing them.[112] In recent years, photographic and other evidence has been obtained indicating that expired RCA projectiles have been employed by law enforcement officials in Bahrain,[113] Egypt,[114] India,[115] Israel/Occupied Territories,[116] Turkey,[117] Venezuela[118] and Yemen.[119]

3.4. Employment of RCAs by State forces

3.4.1. Use and misuse of RCAs for law enforcement

RCAs, when used in accordance with manufacturers' instructions and in line with international human rights standards, can provide an important alternative to other applications of force more likely to result in injury or death, for example, firearms. They are legitimately employed by law enforcement officials for activities such as the dispersal of assemblies posing an imminent threat of serious injury, or the incapacitation of violent individuals. However, they are also open to misuse.

To provide a preliminary indication of the nature of the misuse of RCAs by law enforcement personnel,[120] an analysis of documentation produced by relevant UN and regional human rights monitoring bodies,[121] and leading non-governmental human rights organizations,[122] relating to reported human rights abuses over a five-year period, was undertaken.[123] The survey indicated that, from 1st January 2009 till 31st December 2013, law enforcement personnel had reportedly utilized RCAs to conduct or facilitate human rights abuses in at least 95 countries and territories. Table 3.2 provides a summary of the survey results; the key findings are subsequently explored, together with case studies illustrating the range of human rights abuses reportedly perpetrated by law enforcement officials utilizing RCAs during this period.

Table 3.2 Alleged inappropriate use of RCAs and their means of delivery by law enforcement officials from 1st January 2009 to 31st December 2013[124]

Country or Territory	Use of RCAs to facilitate or carry out torture, ill-treatment or punishment	Use of RCAs to facilitate suppression of freedom of expression, assembly or association	Employment of RCAs with arbitrary, indiscriminate, or excessive use of force or firearms	Aeorsolized RCAs employed in enclosed space	Deleterious health effects, serious injury or death associated with RCAs/means of delivery
Algeria[1]					•
Angola[2]		•	•		
Argentina[3]		•	•		
Azerbaijan[4]	•	•	•	•	
Bahrain[5]	•	•	•	•	•

[1] Algeria urged to allow peaceful protests, Amnesty International, News, 11th February 2011, available at: http://www.amnesty.ca/news/news-item/algeria-urged-to-allow-peaceful-protests (accessed 15th July 2015).

[2] Angola: Punishing dissent: Suppression of freedom of association and assembly in Angola, Amnesty International, AFR 12/004/2014, 13th November 2014, available at: https://www.amnesty.org/en/documents/AFR12/004/2014/en/ (accessed 15th July 2015), pp. 11–12; Angola: Police Disrupt New "Disappearances" Protest, No Credible Inquiry 1 Year After 2 Activists Abducted, Human Rights Watch, News, 31st May 2013, available at: https://www.hrw.org/news/2013/05/31/angola-police-disrupt-new-disappearances-protest (accessed 15th July 2015).

[3] Argentina entry, Annual report 2011, Amnesty International, POL 10/001/2011, available at: http://files.amnesty.org/air11/air_ 2011_full_en.pdf (accessed 15th July 2011), p. 65; Annual Report of the Special Rapporteur for freedom of expression 2013, Catalina Botero Marino, Special Rapporteur for Freedom of Expression, OAS General Secretariat, Organisation of American States, Washington D.C., OEA /Ser.L/V/II.149 Doc. 50, 31st December 2013, available at: http://www.oas.org/en/iachr/expression/docs/reports/2014_04_22_%20IA_2013_ENG%20_FINALweb.pdf (accessed 15th July 2015), paragraph 27.

[4] The Spring that never blossomed: freedoms suppressed in Azerbaijan, Amnesty International, EUR55/011/2011, http://www.amnestyusa.org/sites/default/files/azerbaijan_freedom_of_expression.pdf (accessed 15th July 2015), p. 19; Azerbaijan: INGOs call on Aliyev to investigate brutal attack on journalist, ARTICLE 19, 19th April 2012, https://www.article19.org/resources.php/resource/3050/en/azerbaijan:-ingos-call-on-aliyev-to-investigate-brutal-attack-on-journalist (accessed 23rd July 2015); Azerbaijan entry, Human Rights Watch Annual Report 2014, available at: https://www.hrw.org/world-report/2014/country-chapters/azerbaijan (accessed 15th July 2015).

[5] Bloodied but unbowed: Unwarranted state violence against Bahraini protesters, Amnesty international, MDE 11/009/2011, March 2011, available at http://www.amnesty.eu/content/assets/mde110092011en.pdf (accessed 15th July 2015), pp. 2–5; United Nations, Human Rights Council, Report of the Special Rapporteur on the promotion and protection of the right to freedom of opinion and expression, Frank La Rue, A/HRC/17/27/Add.1, 27th May 2011, available at: http://www2.ohchr.org/english/bodies/hrcouncil/docs/17session/A.HRC.17.27.Add.1_EFSonly.pdf (accessed 15th July 2015), paragraphs 80, 175, 177, 183, 184; Bahrain: Reform shelved, repression unleashed, Amnesty International, MDE 11/062/2012, November 2012, available at http://www.amnesty.org.uk/sites/default/files/ai-bahrain-report-mde110622012.pdf (accessed 15th July 2015), pp. 15, 17, 27; Bahrain: Violent repression of peaceful demonstrations. Public Statement, OMCT, 16th February 2011, http://www.omct.org/urgent-campaigns/statements/bahrain/2011/02/d21124/ (accessed 22nd July 2015); Bahrain: OMCT calls on the international community to urge Bahrain to put an end to the violent repression of peaceful demonstrations, OMCT, 21st March 2011, http://www.omct.org/urgent-campaigns/urgent-interventions/bahrain/2011/03/d21185/ (accessed 22nd July 2015).

Table 3.2 (Continued)

Country or Territory	Use of RCAs to facilitate or carry out torture, ill-treatment or punishment	Use of RCAs to facilitate suppression of freedom of expression, assembly or association	Employment of RCAs with arbitrary, indiscriminate, or excessive use of force or firearms	Aerosolized RCAs employed in enclosed space	Deleterious health effects, serious injury or death associated with RCAs/means of delivery
Bangladesh[6]		•	•		•
Belarus[7]		•	•		
Bolivia[8]		•	•	•	•
Brazil[9]	•	•	•	•	

[6] Bangladesh: Deep concern about the harassment of union leaders members of BCWS, Open Letter to Mr. Md. Zillur Rahman, President of Bangladesh, OMCT, 20th August 2010, available at http://www.omct.org/human-rights-defenders/urgent-interventions/bangladesh/2010/08/d20 831/ (accessed 22nd July 2015); Amnesty International, Bangladesh police crack down on peaceful protesters, 30th November 2010, available at: https://www.amnesty.org/en/latest/news/2010/11/bangladesh-police-crack-down-peaceful-protesters/ (accessed 15th July 2015); United Nations, Human Rights Council, Report of the Special Rapporteur on torture and other cruel, inhuman or degrading treatment or punishment, Juan E. Méndez, A/HRC/25/60/Add.2, 11th March 2013, available at: http://www.ohchr.org/EN/HRBodies/HRC/RegularSessions/Session25/Documents/A-HRC-25-60-Add2_EFS.doc (accessed 15th July 2015), pp. 16–17; Blood on the Streets, The Use of Excessive Force During Bangladesh Protests, Human Rights Watch, 1st August 2013, available at: https://www.hrw.org/report/2013/08/01/blood-streets/use-excessive-force-during-bangladesh-protests (accessed 15th July 2015).
[7] Tear gas fired and websites blocked as Belarus protesters are targeted, Amnesty International, 4th July 2011, available at: https://www.amnesty.org/press-releases/2011/07/tear-gas-fired-and-websites-blocked-belarus-protesters-are-targeted-2011-07/ (accessed 15th July 2015); Belarus: Cease Violence Against Peaceful Protesters, Over 300 Arrested in Minsk, Regions on Independence Day, Human Rights Watch, News, 5th July 2011, available at: https://www.hrw.org/news/2011/07/05/belarus-cease-violence-against-peaceful-protesters (accessed 15th July 2015).
[8] Bolivia: use of excessive force must be investigated, Amnesty International, public statement, AMR 18/002/2011, 27th September 2011, available at: https://www.amnesty.org/en/documents/amr18/002/2011/en/ (accessed 15th July 2015); Bolivia must investigate violence at disability protest, Amnesty International, 24th February 2012, available at https://www.amnesty.org/en/latest/news/2012/02/bolivia-must-investigate-violence-disability-protest/ (accessed 23rd July 2015); Word Report 2012, Bolivia chapter, Human Rights Watch, 22nd January 2012, available at: http://www.hrw.org/sites/default/files/reports/wr2012.pdf (accessed 15th July 2015), p. 213; Report of the office of the Special Rapporteur for freedom of expression, Dr. Catalina Botero, Special Rapporteur for Freedom of Expression, OEA/Ser.L/V/II.147, Doc. 1, 5th March 2013, General Secretariat, OAS, Washington, D.C., available at http://www.oas.org/en/iachr/expression/docs/reports/annual/Annual%20Report%202012.pdf (accessed 15th July 2015), paragraph 46.
[9] Amnesty International, over 500 families in São Paulo left homeless after forced evictions, 27th August 2009, at Available at http://webcache.googleusercontent.com/search?q=cache:j0shF7THmOOJ:https://www.amnesty.org/articles/news/2009/08/over-500-families-sao-paulo-homeless-forced-evictions-20090828/+ &cd=1&hl=en&ct=clnk&gl=uk (accessed 15th July 2015); Amnesty International, Brazil: Forced eviction, Urgent Action, AMR 19/011/2009, 23rd June 2009, UA 165/09, available at: https://www.amnesty.org/download/Documents/44000/amr190112009en.pdf (accessed 15th July 2015); Brazil: "They use a strategy of fear": Protecting the right to protest in Brazil, Amnesty International, AMR 19/005/2014, 5th June 2014, available at: http://www.amnesty.ch/de/laender/amerikas/brasilien/dok/2014/brasilien-spielregeln-gelten-auch-fuer-polizei/rapport-201cthey-use-a-strategy-of-fear201d-anglais-24p. (accessed 15th July

Table 3.2 (Continued)

Country or Territory	Use of RCAs to facilitate or carry out torture, ill-treatment or punishment	Use of RCAs to facilitate suppression of freedom of expression, assembly or association	Employment of RCAs with arbitrary, indiscriminate, or excessive use of force or firearms	Aerosolized RCAs employed in enclosed space	Deleterious health effects, serious injury or death associated with RCAs/means of delivery
Burundi[10]		•			
Cambodia[11]		•	•		
Canada[12]	•		•		
Chile[13]		•	•		

2015), pp. 3, 6, 10; Brazil: State response to protests indiscriminate and disproportionate, Article 19, 17th June 2013, available at https://www.article19.org/resources.php/resource/37110/en/brazil:-state-response-to-protests-indiscriminate-and-disproportionate (accessed 23rd June 2015); World Report 2014: Brazil Chapter, Human Rights Watch, 17th December 2013, available at: https://www.hrw.org/world-report/2014/country-chapters/brazil (accessed 15th July 2015); Brazil: Seize Opportunity to Curb Widespread Torture, New Official Body Will Need Adequate Resources, Political Support, 2nd August 2013, available at: https://www.hrw.org/news/2013/08/02/brazil-seize-opportunity-curb-widespread-torture (accessed 15th July 2015); OAS (31st December 2013), p. 594; Subcommittee on Prevention of Torture and Other Cruel, Inhuman or Degrading Treatment or Punishment, Report on the visit of the Subcommittee on Prevention of Torture and Other Cruel, Inhuman or Degrading Treatment or Punishment to Brazil, United Nations, CAT/OP/BRA/1, Optional Protocol to the Convention against Torture and Other Cruel, Inhuman or Degrading Treatment or Punishment, 5th July 2012, available at: http://www2.ohchr.org/english/bodies/cat/opcat/docs/CAT-OP-BRA-1_en.pdf (accessed 15th July 2015), p. 20.

[10] Burundi, 8 February 2013: Journalists demonstrate for the release of Hassan Ruvakuki, Newsletter: Freedom of Expression in Eastern Africa, Article 19, 14th March 2013, available at https://www.article19.org/resources.php/resource/3664/en/newsletter:-freedom-of-expression-in-eastern-africa (accessed 23rd July 2015); Burundi: Online Newspaper Forum Suspended, Government Should Allow Publication of Comments, Debate; Human Rights Watch, News, 3rd June 2013, available at: http://www.hrw.org/node/249912 (accessed 15th July 2015).

[11] Amnesty International, Hundreds left homeless in Cambodia after forced eviction, 27th January 2009, available at https://www.amnesty.org/press-releases/2009/01/cambodia-hundreds-left-homeless-after-forced-eviction-20090126/ (accessed 15th July 2015); Eviction of Dey Krahorm: Cambodia loses the battle to uphold the rule of law!, FIDH, 4th February 2009, available at https://www.fidh.org/International-Federation-for-Human-Rights/asia/cambodia/Eviction-of-Dey-Krahorm-Cambodia (accessed 25th July 2015); Cambodia: Activists robbed of their homes & voices, Article 19, 24th January 2012, available at https://www.article19.org/resources.php/resource/2936/en/cambodia:-activists-robbed-of-their-homes-&-voices (accessed 23rd July 2015); Questions and answers Crimes against humanity in Cambodia from July 2002 until present, FIDH/Global diligence, October 2014, https://www.fidh.org/IMG/pdf/qanda_cambodia_icc-2.pdf (accessed 22nd July 2105), p. 6.

[12] Those Who Take Us Away, Abusive Policing and Failures in Protection of Indigenous Women and Girls in Northern British Columbia, Canada, 13th February 2013, available at https://www.hrw.org/report/2013/02/13/those-who-take-us-away/abusive-policing-and-failures-protection-indigenous-women (accessed 15th July 2015).

[13] Amnesty International, Americas: Solutions to the historic violation of Indigenous rights will only be found through respectful dialogue, in good faith, with Indigenous peoples, AMR 01/004/2009, 7th August 2009, available at https://www.amnesty.org/download/Documents/44000/amr010042009en.pdf (accessed 16th July 2015), p. 5 ; Amnesty International Annual Report 2013, Chile, Amnesty International, AMR 22/002/2013, 23rd May 2013, available at:

Table 3.2 (Continued)

Country or Territory	Use of RCAs to facilitate or carry out torture, ill-treatment or punishment	Use of RCAs to facilitate suppression of freedom of expression, assembly or association	Employment of RCAs with arbitrary, indiscriminate, or excessive use of force or firearms	Aerosolized RCAs employed in enclosed space	Deleterious health effects, serious injury or death associated with RCAs/means of delivery
China[14]		•	•		
China/Hong Kong[15]		•	•		
China/Tibet[16]		•	•		
Colombia[17]		•	•		•
Cote d'Ivoire[18]		•	•		

https://www.google.co.uk/url?sa=t&rct=j&q=&esrc=s&source=web&cd=1&cad=rja&uact=8&ved=0CCIQFjAAahUKEwjKl5DOpN_GAhUwF9sKHZ2PA_4&url=https%3A%2F%2Fwww.amnesty.org%2Fen%2Fannual-report%2F2013&ei=bXKnVYqtEbCu7Aadn47wDw&usg=AFQjCNHOyWug6xkel163qaim3KyxRjO6uw&bvm=bv.97949915,d.ZGU (accessed 16th July 2015), p. 60; World Report, Human Rights Watch, Chile Chapter 31st January 2013, https://www.google.co.uk/url?sa=t&rct=j&q=&esrc=s&source=web&cd=1&cad=rja&uact=8&ved=0CCEQFjAAahUKEwjCqOKgp9_GAhWIOhQKHdUwCJg&url=https%3A%2F%2Fwww.hrw.org%2Fworld-report%2F2013&ei=M3WnVYLDHIj1UNXhoMAJ&usg=AFQjCNGDRYtLysTxs86AtatwYzV8gfGeJA (accessed 16th July 2015), pp. 209–210; Second report on the situation of human rights defenders in the Americas 2011, Inter-American Commission on Human Rights, OEA/Ser.L/V/II. Doc. 66 31st December 2011, www.oas.org/en/iachr/defenders/docs/pdf/defenders2011.pdf (accessed 16th July 2015), paragraph 296.

[14] Amnesty International, 5th February 2010, Stop human rights violations against Uighurs in China, available at: http://web.archive.org/web/20100208011522/http://www.amnesty.org/en/appeals-for-action/stop-human-rights-violations-against-uighurs-china (accessed 16th July 2015); China: "Justice, justice": The July 2009 protests in Xinjiang, China, Amnesty International, ASA 17/027/2010, 2nd July 2010, available at: https://www.amnesty.org/en/documents/ASA17/027/2010/en/ (accessed 16th July 2015), pp. 13–14; "Beat Him, Take Everything Away", Abuses by China's Chengguan Para-Police, Human Rights Watch, 23rd May 2012, available at: http://www.hrw.org/sites/default/files/reports/china0512ForUpload_1.pdf (accessed 16th July 2015), p. 86.

[15] Hong Kong: Investigate Police Actions at July 1 Rally, Protect Activists, Journalists from Unwarranted Police Obstruction, Human Rights Watch, News, 11th July 2011, available at: https://www.hrw.org/news/2011/07/11/hong-kong-investigate-police-actions-july-1-rally (accessed 16th July 2015).

[16] China's trade in tools of torture and repression, Amnesty International, ASA 17/042/2014, 23rd September 2014, available at http://www.amnesty.org.uk/sites/default/files/china-tools-of-torture-report.pdf (accessed 16th July 2015), p. 21.

[17] Americas: Indigenous people's long struggle to defend their rights in the Americas, Amnesty International, AMR 01/002/2014, 8th August 2014, available at http://www.refworld.org/pdfid/53e9c0364.pdf (accessed 16th July 2015), p. 7.; Catalina Botero Marino, Special Rapporteur for Freedom of Expression, OAS, OEA/Ser.L/V/II. Doc. 5, 4 March 2011, available at http://www.cidh.oas.org/annualrep/2010eng/RELATORIA_2010_ENG.pdf (accessed 17th July 2015), p. 57.

[18] Amnesty International, News, 21st December 2010, Defenceless people need urgent protection from escalating violence in Côte d'Ivoire, available at http://www.amnesty.ca/news/news-item/cote-divoire-defenceless-people-need-urgent-protection-from-escalating-violence (accessed 16th July 2015); Amnesty International, Côte d'Ivoire: Security forces

Table 3.2 (Continued)

Country or Territory	Use of RCAs to facilitate or carry out torture, ill-treatment or punishment	Use of RCAs to facilitate suppression of freedom of expression, assembly or association	Employment of RCAs with arbitrary, indiscriminate, or excessive use of force or firearms	Aeorsolized RCAs employed in enclosed space	Deleterious health effects, serious injury or death associated with RCAs/means of delivery
Cuba[19]		•	•		•
Cyprus[20]	•	•	•	•	
Denmark[21]		•			
Djibouti[22]		•	•		
Dominican Republic[23]	•	•			•

kill at least ten unarmed demonstrators, 16th December 2010, available at http://webcache.googleusercontent.com/search?q=cache:EbL4FYp6URMJ:https://www.amnesty.org/articles/news/2010/12/cc3b4te-de28099ivoire-security-forces-kill-least-nine-unarmed-demonstrators/+&cd=1&hl=en&ct=clnk&gl=uk (accessed 16th July 2015); Human Rights Watch, Cote d'Ivoire: Violence campaign by security forces, militias; Gbagbo should rein in supporters; rampant killings require robust UN response, 27th January 2011, available at https://www.hrw.org/news/2011/01/26/cote-divoire-violence-campaign-security-forces-militias (accessed 16th July 2015); "They Killed Them Like It Was Nothing", The Need for Justice for Côte d'Ivoire's Post-Election Crimes, Human Rights Watch, 30th September 2011, available at http://www.hrw.org/sites/default/files/reports/cdi1011WebUpload.pdf (accessed 16th July 2015), pp. 28, 30.

[19] Amnesty International, Urgent Action, UA 89/09 Cuba: Fear for safety, AMR 25/003/2009, 30th March 2009, available at http://www2.amnesty.se/uaonnet.nsf/7e65f5b0a8b73763c1256672003ecdef/9264ebf1c92cbffac125758a00310c4d?OpenDocument (accessed 16th July 2015); OAS, Inter-American Commission on Human Rights, Precautionary Measures 2011, PM 13/11 – Néstor Rodríguez Lobaina and Family, Cuba, available at: http://www.oas.org/en/iachr/decisions/precautionary.asp (accessed 16th July 2015); Annual Report of the Special Rapporteur for freedom of expression 2012, Catalina Botero Marino, Special Rapporteur for Freedom of Expression, OAS, OEA/Ser.L/V/II.147, Doc. 1, 5th March 2013, available at http://www.oas.org/en/iachr/expression/docs/reports/annual/Annual%20Report%202012.pdf (accessed 16th July 2015), p. 77.

[20] Report to the Government of Cyprus on the visit to Cyprus carried out by the European Committee for the Prevention of Torture and Inhuman or Degrading Treatment or Punishment (CPT) from 23 September to 1 October 2013, CPT/Inf (2014) 31, available at http://www.cpt.coe.int/documents/cyp/2014-31-inf-eng.pdf (accessed 16th July 2015), pp. 21, 28.

[21] Annual Report 2010, Amnesty International, Denmark entry, 28th May 2010, available at http://www.amnestyusa.org/research/reports/annual-report-denmark-2010 (accessed 16th July 2015).

[22] Djibouti: Article 19's Submission to the UN Universal Periodic Review, Article 19, 10th October 2012, available at https://www.article19.org/resources.php/resource/3472/en/djibouti:-article-19's-submission-to-the-un-universal-periodic-review (accessed 20th July 2015); Djibouti (2010–2011) Situation of Human Rights Defenders, updated May 2011, International Federation for Human Rights (FIDH), available at https://www.fidh.org/International-Federation-for-Human-Rights/Africa/djibouti/DJIBOUTI-2010-2011 (accessed 20th July 2015).

[23] Dominican Republic: 'Shut up if you don't want to be killed!': Human rights violations by police in the Dominican Republic, Amnesty International, AMR 27/002/2011, 25th October 2011 available at http://www.refworld.org/docid/4ea7d97d2.html (accessed 20th July 2015); 600 families face forced eviction from homes, Dominican Republic Urgent Action, Amnesty International, AMR 27/005/2013, 31st May 2013, available at http://www.amnesty.or.jp/en/get-involved/ua/ua/2013ua142.html (accessed 20th July 2015).

Table 3.2 (Continued)

Country or Territory	Use of RCAs to facilitate or carry out torture, ill-treatment or punishment	Use of RCAs to facilitate suppression of freedom of expression, assembly or association	Employment of RCAs with arbitrary, indiscriminate, or excessive use of force or firearms	Aerosolized RCAs employed in enclosed space	Deleterious health effects, serious injury or death associated with RCAs/means of delivery
DRC[24]	•	•	•	•	
Ecuador[25]		•	•		
Egypt[26]	•	•	•	•	•
Ethiopia[27]	•	•	•	•	
France[28]		•	•	•	

[24] DR Congo: 24 Killed since Election Results Announced, Security Forces Attack, Detain Protesters, Local Residents, Human Rights Watch, News, 22nd December 2011, available at https://www.hrw.org/news/2011/12/21/dr-congo-24-killed-election-results-announced (accessed 16th July 2015).
[25] So that no one can demand anything: criminalizing the right to protest in Ecuador? Amnesty International, AMR 28/002/2012, 17th July 2012, https://www.amnesty.org/download/Documents/20000/amr280022012en.pdf (accessed 16th July 2015), pp. 20, 21, 22, 28; Catalina Botero Marino, Special Rapporteur for Freedom of Expression, OAS, OEA/Ser.L/V/II. Doc. 5, 4 March 2011, available at http://www.cidh.oas.org/annualrep/2010eng/RELATORIA_2010_ENG.pdf (accessed 17th July 2015), p. 76.
[26] Egypt Unlawful killings in protests and political violence on 5 and 8 July 2013, Amnesty International, MDE 12/034/2013, July 2013, available at https://www.amnesty.org/en/documents/MDE12/034/2013/en/ (accessed 16th July 2015); Egypt's disastrous bloodshed requires urgent impartial investigation, Amnesty International, public statement, 16th August 2013, available at http://www.amnesty.ca/news/news-releases/egypt%E2%80%99s-disastrous-bloodshed-requires-urgent-impartial-investigation (accessed 16th July 2015); Egypt: Human rights in crisis: Systemic violations and impunity: Expanded Amnesty International submission to the UN Universal Periodic Review, October–November 2014, Amnesty International, MDE 12/034/2014, 1st July 2014, available at https://www.amnesty.org/download/Documents/8000/mde120342014en.pdf (accessed 16th July 2015); Egypt: Protester Killings Not Being Investigated, Impunity Encourages Excessive Force, Human Rights Watch, News, 2nd November 2013, available at https://www.hrw.org/news/2013/11/02/egypt-protester-killings-not-being-investigated (accessed 15th July 2015); Egypt: Security Forces Used Excessive Lethal Force, Worst Mass Unlawful Killings in Country's Modern History, Human Rights Watch, 19th August 2013, available at https://www.hrw.org/news/2013/08/19/egypt-security-forces-used-excessive-lethal-force (accessed 15th July 2015); Egypt: Deadly Clashes at Cairo University, Investigate Killings of Morsy Supporters and Opponents, Human Rights Watch, News, 5th July 2013, available at http://www.hrw.org/node/250374 (accessed 15th July 2015); Egypt: Violent repression of peaceful demonstrations, OMCT, 28th January 2011, available at http://www.omct.org/urgent-campaigns/urgent-interventions/egypt/2011/01/d21061/ (accessed 22nd July 2015); Egypt: police continues to use lethal weapons during demonstrations, 7th March 2013, FIDH, available at https://www.fidh.org/International-Federation-for-Human-Rights/north-africa-middle-east/egypt/Egypt-police-continues-to-use-12998 (accessed 23rd July 2015).
[27] World Report 2013, Ethiopia chapter, Human Rights Watch, 31st January 2013, available at https://www.hrw.org/world-report/2013/country-chapters/ethiopia (accessed 16th July 2015).
[28] Europe: "We ask for Justice": Europe's failure to protect Roma from racist violence, Amnesty International, EUR 01/007/2014, 8th April 2014, available at http://www.amnesty.eu/content/assets/Reports/08042014_Europes_failure_to_protect_Roma_from_racist_violence.pdf (accessed 16th July 2015), p. 16.

Table 3.2 (Continued)

Country or Territory	Use of RCAs to facilitate or carry out torture, ill-treatment or punishment	Use of RCAs to facilitate suppression of freedom of expression, assembly or association	Employment of RCAs with arbitrary, indiscriminate, or excessive use of force or firearms	Aerosolized RCAs employed in enclosed space	Deleterious health effects, serious injury or death associated with RCAs/means of delivery
Greece[29]	•	•		•	•
Georgia[30]	•	•			
Germany[31]		•			
Ghana[32]	•	•			
Guinea[33]	•	•		•	
Guinea-Bissau[34]	•				
Guatemala[35]		•		•	

[29] Tear gas fired as Greek police clash with Athens' protesters, Amnesty International, News, 29th June 2011, available at http://www.amnestyusa.org/news/news-item/tear-gas-fired-as-greek-police-clash-with-athens-protesters (accessed 16th July 2015); Don't beat protesters EU countries warned, Amnesty International, press release, 25th October 2012, available at https://www.amnesty.org/en/latest/news/2012/10/don-t-beat-protesters-eu-countries-warned/ (accessed 16th July 2015); Greece: A law unto themselves: A culture of abuse and impunity in the Greek police, Amnesty International, EUR 25/005/2014, 3rd April 2014, available at http://www.amnesty.ca/sites/default/files/greecereport3april14.pdf (accessed 16th July 2015), pp. 11, 12, 13, 14, 17, 18.

[30] Georgia: Article 19 Calls on the Georgian President to Condemn the Attack on Journalists, Article 19, 2nd June 2011, available at https://www.article19.org/resources.php/resource/1777/en/georgia:-article-19-calls-on-the-georgian-president-to-condemn-the-attack-on-journalists; World Report 2013, Human Rights Watch, Georgia chapter, 31st January 2013, available at https://www.hrw.org/sites/default/files/wr2013_web.pdf (accessed 15th July 2015), p. 445.

[31] Annual Report 2012, Amnesty International, Germany entry, available at http://www.amnestyusa.org/sites/default/files/air12-report-english.pdf (accessed 16th July 2015), p. 155.

[32] Amnesty International, Thousands facing forced eviction in Ghana, Urgent Action: UA: 251/10, AFR 28/001/2011, 26th January 2011, https://www.amnesty.org/en/documents/AFR28/001/2011/en/ (accessed 16th July 2015).

[33] United Nations, Report of the International Commission of Inquiry mandated to establish the facts and circumstances of the events of 28th September 2009 in Guinea, UN doc. S/2009/693, 18th December 2009 available at http://www.securitycouncilreport.org/atf/cf/%7B65BFCF9B-6D27-4E9C-8CD3-CF6E4FF96FF9%7D/Guinea%20S%202009%20693.pdf (accessed 15th July 2015); Amnesty International, Guinea: "You did not want the military, so now we are going to teach you a lesson", AFR 29/001/2010, February 2010, available at https://www.amnesty.org/download/Documents/36000/afr290012010en.pdf (accessed 15th July 2015); Human Rights Watch, "Guinea: September 28 Massacre Was Premeditated, In-Depth Investigation Also Documents Widespread Rape", 27th October 2009, available at https://www.hrw.org/news/2009/10/27/guinea-september-28-massacre-was-premeditated (accessed 15th July 2015); World Report, Human Rights Watch, Guinea chapter, 31st January 2013.

[34] Guinea-Bissau, Amnesty International's concerns following the coup in April 2012, Amnesty International, AFR 30/001/2012, available at http://reliefweb.int/report/guinea-bissau/amnesty-international%E2%80%99s-concerns-following-coup-april-2012 (accessed 16th July 2015), p. 7.

[35] Guatemala forcibly evicting indigenous farmers, Amnesty International, press release, PRE01/153/201123 March 2011, available at https://www.amnesty.org/press-releases/2011/03/guatemala-forcibly-evicting-indigenous-farmers/ (accessed 16th July 2015); Catalina Botero Marino, Special Rapporteur for Freedom of Expression, OAS, OEA/Ser.L/V/II. Doc. 5, 4 March 2011, available at http://www.cidh.oas.org/annualrep/2010eng/RELATORIA_2010_ENG.pdf (accessed 17th July 2015), p. 96.

Table 3.2 (Continued)

Country or Territory	Use of RCAs to facilitate or carry out torture, ill-treatment or punishment	Use of RCAs to facilitate suppression of freedom of expression, assembly or association	Employment of RCAs with arbitrary, indiscriminate, or excessive use of force or firearms	Aeorsolized RCAs employed in enclosed space	Deleterious health effects, serious injury or death associated with RCAs/means of delivery
Haiti[36]		•	•	•	•
Honduras[37]	•	•	•	•	•
India[38]		•	•		•
Indonesia[39]			•		

[36] Haiti: Submission to the UN Human Rights Committee: 112th Session of the UN Human Rights Committee, 7–31 October 2014, Amnesty International, AMR 36/012/20147, October 2014, available at https://www.amnesty.org/en/documents/AMR36/012/2014/en/ (accessed 16th July 2015), pp. 14–15.

[37] Recommendations to the new Honduran government following the coup of June 2009, Amnesty International, AMR 37/003/2010, January 2010, available at https://www.amnesty.org/en/documents/AMR37/003/2010/en/ (accessed 20th July 2015), pp. 9, 10, 13; Amnesty International, Honduras: Honduran lawyer at risk: Kenia Oliva Cardona, AMR 37/011/2010, 28th July 2010, available at http://ua.amnesty.ch/urgent-actions/2010/07/166-10?ua_language=en (accessed 20th July 2015); Human Rights Watch, After the Coup: Ongoing Violence, Intimidation and Impunity in Honduras, 20th December 2010, available at http://www.hrw.org/sites/default/files/reports/ honduras1210webwcover_0.pdf (accessed 20th July 2015), pp. 4, 13, 57; United Nations, Report on the visit of the Subcommittee on Prevention of Torture and Other Cruel, Inhuman or Degrading Treatment or Punishment to Honduras, CAT/OP/HND/1, 10th February 2011, available at http://www2.ohchr.org/english/bodies/cat/opcat/docs/ CAT.OP.HND.1.doc (accessed 20th July 2015), pp. 9, 11, 12, 16; Office of the special rapporteur expresses concern over new attacks against journalists and media in Honduras, Press release N° R96/10, OAS, Washington D.C., 20th September 2010; Second Report on the situation of human rights defenders in the Americas 2011, Inter-American Commission on Human Rights, OEA/Ser.L/V/II. Doc. 66 31 December 2011, available at www.oas.org/en/iachr/defenders/docs/pdf/defenders2011.pdf (accessed 20th July 2015), p. 122.

[38] India: Probe police violations against protestors in Andhra, Amnesty International, public statement, ASA 20/007/2011, 3rd March 2011, available at http://www.amnestyusa.org/news/press-releases/india-probe-police-violations-against-protestors-in-andhra (accessed 16th July 2015); India: Authorities must reopen investigation into 2010 killing of Tufail Mattoo, Amnesty International, public statement, ASA 20/028/2013, 17th June 2013, available at https://www.amnesty.org/en/documents/asa20/028/2013/en/ (accessed 16th July 2015); Human Rights Watch, India: Kashmir Arrest a Step for Accountability Government Should Also Investigate Higher Ranking Officers in Security Force Abuses, 11th February 2010, available at https://www.hrw.org/news/2010/02/11/india-kashmir-arrest-step-accountability (accessed 15th July 2015); United Nations, Human Rights Council, Report of the Special Rapporteur on the promotion and protection of the right to freedom of opinion and expression, Frank La Rue, A/HRC/17/27/Add.1, 27th May 2011, paragraph 1015.

[39] Indonesia: Arbitrary and excessive use of force and firearms in North Sumatra, Amnesty International, ASA 21/026/2011, 1st August 2011, available at https://www.amnesty.org/en/documents/document/?indexNumber= ASA21%2F026%2F2011&language= en (accessed 15th July 2015).

Table 3.2 (Continued)

Country or Territory	Use of RCAs to facilitate or carry out torture, ill-treatment or punishment	Use of RCAs to facilitate suppression of freedom of expression, assembly or association	Employment of RCAs with arbitrary, indiscriminate, or excessive use of force or firearms	Aerosolized RCAs employed in enclosed space	Deleterious health effects, serious injury or death associated with RCAs/means of delivery
Iran[40]		•	•	•	•
Iraq[41]		•	•		

[40] Iran: Release Students Detained for Peaceful Protests, Renewed Crackdown on Campus Activism, Human Rights Watch, 28th February 2009, available at http://www.hrw.org/news/2009/02/27/iran-release-students-detained-peaceful-protests (accessed 20th July 2015); Iran: Election contested, repression compounded, Amnesty International, MDE 13/123/2009, 10th December 2009, available at https://www.amnesty.org/en/documents/MDE13/123/2009/en/ (accessed 20th July 2015), pp. 2 ,3, 18, 26, 34, 36, 37; Iran: Iranian cleric released on bail: Further information, Amnesty International, MDE 13/014/2010, 2nd February 2010, available at http://www.amnesty.org.au/iar/comments/22493/ (accessed 20th July 2015); Iran: Iranian eco-activist detained without charge: Mahfarid Mansourian (f), Amnesty International, Urgent Action, MDE 13/022/2010, 15th February 2010, available at https://www.amnesty.org/en/documents/document/?indexNumber= MDE13%2F022%2F2010&language= en (accessed 20th July 2015); Additional student arrested, Amnesty International, Urgent Action, Further information on UA: 31/11 Index: MDE 13/021/2011, 22nd February 2011, available at http://ua.amnesty.ch/urgent-actions/2011/02/031-11/031-11-1/ua-031-11-1-english (accessed 20th July 2015); New wave of repression launched against ethnic communities, FIDH, 23rd May 2012, available at https://www.fidh.org/International-Federation-for-Human-Rights/asia/iran/New-wave-of-repression-launched (accessed 23rd July 2015); United Nations, Human Rights Council, Report of the Special Rapporteur on extrajudicial, summary or arbitrary executions, Philip Alston, A/HRC/14/24/Add.1, 18th June 2010, available at http://www.refworld.org/docid/4c29a7372.html (accessed 20th July 2015); Joint Letter to President Hassan Rouhani re: LGBT Rights, 20th December 2013, Amnesty International, Human Rights Watch, International Gay and Lesbian Human Rights Commission, Iranian Queer Organization-(IRQO), available at https://www.hrw.org/news/2013/12/20/joint-letter-president-hassan-rouhani-re-lgbt-rights (accessed 20th July 2015); Iran: Silenced, expelled, imprisoned: Repression of students and academics in Iran, Amnesty International, MDE 13/015/2014, 2nd June 2014, available at https://www.es.amnesty.org/uploads/media/FINAL_Silenced__expelled__ imprisoned_-_Repression_of_students_and_ academics_in_Iran.pdf (accessed 20th July 2015).
[41] Amnesty International, Eight reported killed as Iraqi forces attack Iranian residents of Camp Ashraf, 28th July 2009, available at https://www.google.co.uk/url?sa=t&rct=j&q=&esrc=s&source=web&cd=2&cad=rja&uact=8&ved=0CCgQFjABahUKEwiSw_Gp_t_GAhVHbxQKHXtgDoo&url=https%3A%2F%2Fwww.amnesty.org%2Farticles%2Fnews%2F2009%2F07%2Feight-reported-killed-iraqi-forces-attack-iranian-residents-camp-ashraf-20090729%2F&ei= gNCnVZKDHcfeUfvAudAI&usg=AFQjCNGR9HZ5x369tTwaqNPk5Ee5WB6F2A&bvm=bv.97949915,d.d24 (accessed 15th July 2016); United Nations, Human Rights Council, Report of the Special Rapporteur on extrajudicial, summary or arbitrary executions, Philip Alston, A/HRC/14/24/Add.1, 1st June 2010, available at http://www.refworld.org/docid/4c29a7372.html (accessed 15th July 2015), pp. 156–159; Iraq: Investigate Deadly Raid on Protest, Security Forces Open Fire at Sit-In, Human Rights Watch, News, 24th April 2013, available at https://www.hrw.org/news/2013/04/24/iraq-investigate-deadly-raid-protest (accessed 15th July 2015).

Table 3.2 (Continued)

Country or Territory	Use of RCAs to facilitate or carry out torture, ill-treatment or punishment	Use of RCAs to facilitate suppression of freedom of expression, assembly or association	Employment of RCAs with arbitrary, indiscriminate, or excessive use of force or firearms	Aeorsolized RCAs employed in enclosed space	Deleterious health effects, serious injury or death associated with RCAs/means of delivery
Israel/OT[42]	•	•	•	•	•
Jamaica[43]	•	•		•	•
Jordon[44]		•	•		
Kenya[45]	•	•	•	•	•

[42] Amnesty International, Bedouin village destroyed for ninth time, Urgent Action, Further information on UA: 236/10 Index: MDE 15/011/2011, 25th January 2011, available at http://ua.amnesty.ch/urgent-actions/2010/11/236-10/236-10-3?ua_language= en (accessed 20th July 2015); Amnesty International, Israel: Submission to the Human Rights Committee: 99th Session, July 2010, MDE 15/010/2010, 18th June 2010, available at http://web.stanford.edu/group/scai/images/amnestyiccpr.pdf (accessed 20th July 2015); United Nations Office of the High Commissioner for Human Rights, Illegal acts by Israeli authorities on the rise in the occupied West Bank – UN human rights expert, 14th January 2011; United Nations, Human Rights Council, Twelfth session, Agenda item 7, The grave violations of human rights in the Occupied Palestinian Territory, particularly due to the recent Israeli military attacks against the occupied Gaza Strip, Report of the United Nations High Commissioner for Human Rights on the implementation of Human Rights Council resolution S-9/1, A/HRC/12/37, 19th August 2009, available at http://www.refworld.org/docid/4ac1cdf32.html (accessed 20th July 2015), pp. 19, 23; Israeli military's killing of Nakba protesters must be investigated, Amnesty International, public statement, MDE 15/025/2011, 16th May 2011, available at https://www.amnesty.ie/news/israeli-military%E2%80%99s-killing-nakba-protesters-must-be-investigated (accessed 20th July 2015); Israel and Occupied Palestinian Territories: Trigger-happy: Israel's use of excessive force in the West Bank, Amnesty International, MDE 15/002/2014, 27 February 2014, available at http://www.amnesty.org.uk/sites/default/files/mde150022014en_0.pdf (accessed 20th July 2015), pp. 2, 17, 18, 26, 27, 32, 33, 35–40, 42–47, 49–51, 53, 55–59; Israel: Excessive Force against Protesters, Witnesses Say Horses Trampled, Police Beat Peaceful Demonstrators, Human Rights Watch, News, 18th July 2013, available at http://www.hrw.org/node/250510 (accessed 20th July 2015).
[43] United Nations, Human Rights Council, Sixteenth session, Report of the Special Rapporteur on torture and other cruel, inhuman or degrading treatment or punishment, Manfred Nowak, Addendum: Mission to Jamaica, A/HRC/16/52/Add.3, 11th October 2010, available at http://daccess-dds-ny.un.org/doc/UNDOC/GEN/G10/169/26/PDF/G1016926.pdf?OpenElement (accessed 17th July 2015), pp. 14, 26, 32, 41, 42; Not Safe at Home, Violence and Discrimination against LGBT people in Jamaica, Human Rights Watch, 22nd October 2014, available at https://www.hrw.org/report/2014/10/21/not-safe-home/violence-and-discrimination-against-lgbt-people-jamaica (accessed 17th July 2015), pp. 32, 45, 46.
[44] Jordan: Further information: Jordanian man and son released, Amnesty International MDE 16/001/2013, 12th April 2013, available at http://www.refworld.org/docid/5177d49bc.html (accessed 16th July 2015).
[45] Human Rights Watch, Kenya: Killing of Activists Needs Independent Inquiry, Lethal Force Against Students Protesting the Killing Underscores Need for Police Reform, 6th March 2009, available at http://www.hrw.org/news/2009/03/06/kenya-killing-activists-needs-independent-inquiry (accessed 17th July 2015); 400 families forcibly evicted, Amnesty International, urgent action, UA: 123/13 Index: AFR 32/004/2013, 15th May 2013, available at http://ua.amnesty.ch/urgent-actions/2013/05/123-13?ua_language=en (accessed 17th July 2015); "You Are All Terrorists",

Table 3.2 (Continued)

Country or Territory	Use of RCAs to facilitate or carry out torture, ill-treatment or punishment	Use of RCAs to facilitate suppression of freedom of expression, assembly or association	Employment of RCAs with arbitrary, indiscriminate, or excessive use of force or firearms	Aerosolized RCAs employed in enclosed space	Deleterious health effects, serious injury or death associated with RCAs/means of delivery
Kosovo (Serbia)[46]	•	•	•	•	
Kuwait[47]	•	•	•		•
Kyrgyzstan[48]	•	•			
Liberia[49]		•			
Libya[50]	•	•			
Madagascar[51]	•	•			
Malawi[52]	•	•			

Kenyan Police Abuse of Refugees in Nairobi, Human Rights Watch, 30th May 2013, available at https://www.hrw.org/report/2013/05/29/you-are-all-terrorists/kenyan-police-abuse-refugees-nairobi (accessed 17th July 2015), p. 38; High Stakes: Political Violence and the 2013 Elections in Kenya, Human Rights Watch, 8th February 2013, available at http://www.hrw.org/sites/default/files/reports/kenya0213webwcover.pdf (accessed 17th July 2015), pp. 5, 49, 53, 58.

[46] Amnesty International, Kosovo (Serbia): Vetevendosje! activists beaten during Kurti arrest, EUR 70/010/2010, 18th June 2010, available at http://www.amnesty.org.uk/press-releases/kosovo-serbia-vetevendosje-activists-beaten-during-kurti-arrest (accessed 15th July 2015); Amnesty International condemns excessive use of force by Kosovo police, Amnesty International, public statement, EUR 70/001/2012, 16th January 2012, available at http://www.amnesty.org.au/news/comments/30125/ (accessed 15th July 2015).

[47] Kuwait: Charges against Musallam al-Barrak must be dropped, Amnesty International, News, 1 November 2012, available at https://www.amnesty.org/en/latest/news/2012/11/kuwait/ (accessed 17th July 2015); Kuwait: The "Withouts" of Kuwait: Nationality for stateless Bidun now, Amnesty International, MDE 17/001/2013, 16th September 2013, available at https://www.amnesty.org/download/Documents/.../mde170012013en.pdf (accessed 17th July 2015), pp. 4–5.

[48] Kyrgyzstan: Violent unrest – Call to respect and protect human rights and fundamental freedoms; OMCT Secretariat, Urgent Intervention; 8th April 2010, available at http://www.omct.org/urgent-campaigns/urgent-interventions/kyrgyzstan/2010/04/d20631/ (accessed 17th July 2015).

[49] "No Money, No Justice" Police Corruption, Abuse and Injustice in Liberia, Human Rights Watch, August 2013, available at http://www.hrw.org/sites/default/files/reports/liberia0813_forUpload_0.pdf (accessed 17th July 2015), p. 31.

[50] Libya: Arrests, Assaults in Advance of Planned Protests, Halt Attacks on Peaceful Demonstrators and Free Those Arrested, Human Rights Watch, 17th February 2011, available at: https://www.hrw.org/news/2011/02/16/libya-arrests-assaults-advance-planned-protests (accessed 17th July 2015).

[51] Amnesty International, Madagascar: Urgent need for justice: Human rights violations during the political crisis, AFR 35/001/2010, 4th February 2010, available at https://www.amnesty.org/en/documents/AFR35/001/2010/en/ (accessed at 17th July 2015), p. 15.

[52] World Report 2012, Malawi Chapter, Human Rights Watch, 22nd January 2012, available at http://www.hrw.org/sites/default/files/reports/wr2012.pdf (accessed 17th July 2015), p. 140.

Table 3.2 (Continued)

Country or Territory	Use of RCAs to facilitate or carry out torture, ill-treatment or punishment	Use of RCAs to facilitate suppression of freedom of expression, assembly or association	Employment of RCAs with arbitrary, indiscriminate, or excessive use of force or firearms	Aeorsolized RCAs employed in enclosed space	Deleterious health effects, serious injury or death associated with RCAs/means of delivery
Malaysia[53]		•	•	•	•
Maldives[54]	•	•	•		•
Malta[55]	•				
Mauritania[56]		•	•		•
Mexico[57]	•	•	•		

[53] Human Rights Watch, Malaysia: More Rhetoric Than Reality on Human Rights False Promises and Persistent Abuses in 2009, 20th January 2010, available at https://www.hrw.org/news/2010/01/20/malaysia-more-rhetoric-reality-human-rights (accessed 17th July 2015); Malaysia: Police use brutal tactics against peaceful protesters, Amnesty International, News, 11th July 2011, available at https://www.amnesty.org/en/press-releases/2011/07/malaysia-police-use-brutal-tactics-against-peaceful-protesters/ (accessed 17th July 2015); Malaysia: FIDH condemns violent and massive crackdown on peaceful protesters, Paris-Bangkok, FIDH, 11th July 2011, available at https://www.fidh.org/International-Federation-for-Human-Rights/asia/malaysia/Malaysia-FIDH-condemns-violent-and (accessed 22nd July 2015); Malaysia: OMCT – SUARAM: Malaysia Stop the violent repression of demonstrations against the Internal Security Act (ISA)!, World Organisation Against Torture (OMCT)/Suara Rakyat Malaysia (SUARAM), 7th August 2009, available at http://www.omct.org/urgent-campaigns/urgent-interventions/malaysia/2009/08/d20173/ (accessed 22nd July 2015); Malaysia: Joint Submission to the UN Universal Periodic Review, Article 19, 11th March 2013, available at https://www.article19.org/resources.php/resource/3645/en/malaysia:-joint-submission-by-article-19-and-suaram--to-the-un-universal-periodic-review (accessed 23rd July 2015).

[54] Maldives: The other side of paradise: A human rights crisis in the Maldives, Amnesty International, ASA 29/005/2012, 5th September 2012, available at https://www.amnesty.org/en/documents/document/?indexNumber= ASA29%2F005%2F2012&language= en (accessed 17th July 2015), pp. 4, 6–11.

[55] Report to the Maltese Government on the visit to Malta carried out by the European Committee for the Prevention of Torture and Inhuman or Degrading Treatment or Punishment (CPT) from 26 to 30 September 2011, CPT/Inf (2013) 12, Strasbourg, 4th July 2013, available at http://www.cpt.coe.int/documents/mlt/2013-12-inf-eng.pdf (accessed 17th July 2015), pp. 22–23 and 29–30.

[56] Mauritania (2010–2011), situation of human rights defenders, FIDH, 27th January 2012, available at https://www.fidh.org/International-Federation-for-Human-Rights/Africa/mauritania/MAURITANIA-2010-2011 (accessed 23rd July 2015); The Observatory: Contribution to the 51st ordinary session of the African Commission on Human and Peoples' Rights, OMCT/FIDH, April 2012, available at http://www.omct.org/human-rights-defenders/statements/2012/04/d21763/ (accessed 22nd July 2015); Annual Report 2012, Amnesty International, Mauritania entry, available at: http://www.amnestyusa.org/sites/default/files/air12-report-english.pdf (accessed 17th July 2015), p. 233.

[57] Report on the visit of the Subcommittee on Prevention of Torture and Other Cruel, Inhuman or Degrading Treatment or Punishment to Mexico; Subcommittee on Prevention of Torture, United Nations, CAT/OP/MEX/1, 31st May 2010, available at http://tbinternet.ohchr.org/_layouts/treatybodyexternal/Download.aspx?symbolno=CAT%2fOP%2fMEX%2f1&Lang=en (accessed 17th July 2015), p. 23; Annual Report of the Special Rapporteur for freedom of expression 2010, Catalina Botero Marino, Special Rapporteur for Freedom of Expression, OAS, OEA/Ser.L/V/II. Doc. 5, 4 March 2011, available at http://www.cidh.oas.org/annualrep/2010eng/RELATORIA_2010_ENG.pdf (accessed 17th July 2015), p. 193.

Table 3.2 (Continued)

Country or Territory	Use of RCAs to facilitate or carry out torture, ill-treatment or punishment	Use of RCAs to facilitate suppression of freedom of expression, assembly or association	Employment of RCAs with arbitrary, indiscriminate, or excessive use of force or firearms	Aerosolized RCAs employed in enclosed space	Deleterious health effects, serious injury or death associated with RCAs/means of delivery
Morocco/Western Sahara[58]	•	•			
Mozambique[59]	•	•			
Myanmar[60]	•	•			
Nepal[61]	•	•			
Nicaragua[62]	•	•			
Niger[63]	•				
Nigeria[64]	•		•	•	•

[58] Morocco urged to investigate deaths in Western Sahara protest camp, Amnesty International, 11th November 2010, available at https://www.amnesty.org/en/latest/news/2010/11/morocco-urged-investigate-deaths-western-sahara-protest-camp/ (accessed 17th July 2015).

[59] Mozambique submission to the African Commission on Human and Peoples' Rights, 54th Ordinary Session, 22nd October–5th November 2013, Amnesty International, AFR 41/007/2013, available at https://www.amnesty.org/en/documents/AFR41/007/2013/en/ (accessed 17th July 2015), pp. 14–15.

[60] Burma: Investigate Violent Crackdown on Mine Protesters, Incident Tests Government's Claim to Respect Peaceful Assembly, Human Rights Watch, News, 1st December 2012, available at https://www.hrw.org/news/2012/12/01/burma-investigate-violent-crackdown-mine-protesters (accessed 17th July 2015).

[61] Nepal: Alleged torture and killing in police custody of Mr. Ram Sewak Dhobi, a 30-year-old permanent resident of Asuraina VDC-06, NPL 040913, Allegations of torture to death in police custody/ Risk of Impunity, OMCT International Secretariat, Geneva, 4th September 2013, available at http://www.omct.org/urgent-campaigns/urgent-interventions/nepal/2013/09/d22376/ (accessed 17th July 2015); Violent clashes between police, Maoists spark UN concern in Nepal, UN News Centre, 21st December 2009, available at http://www.un.org-apps-news-story.asp?NewsID=33308#.VcM_HrNVi1g (accessed 6th August 2015).

[62] Annual Report of the Special Rapporteur for freedom of expression 2013, Catalina Botero Marino, Special Rapporteur for Freedom of Expression, OAS General Secretariat, Organisation of American States, Washington D.C., OEA /Ser.L/V/II.149 Doc. 50, 31st December 2013, available at: http://www.oas.org/en/iachr/expression/docs/reports/2014_04_22_%20IA_2013_ENG%20_FINALweb.pdf (accessed 15th July 2015), paragraph 661.

[63] Amnesty International, Niger – Submission to the UN Universal Periodic Review, Tenth session of the UPR Working Group of the Human Rights Council, January 2011, AFR 43/001/2010, 5th July 2010, available at https://www.amnesty.org/en/documents/document/?indexNumber=AFR43%2F001%2F2010&language= en (accessed 17th July 2015), p. 6.

[64] Nigeria: Escalating fuel price protests – President must repeal Force Order 237 to prevent more casualties, Amnesty International, public statement, AFR 44/001/2012, 11th January 2012, available at: http://www.amnesty.nl/nieuwsportaal/pers/nigeria-escalating-fuel-price-protests-president-must-repeal-force-order-237-prev (accessed 17th July 2015); Nigeria: Authorities in Nigeria must not carry out any further executions of death row prisoners, Amnesty International, joint public statement, AFR 44/022/2013 28th August 2013, available at https://www.amnesty.org/download/Documents/12000/afr440222013en.pdf (accessed 17th July 2015); Nigeria: "Welcome to hell fire": Torture and other ill-treatment in Nigeria, Amnesty International, AFR 44/011/2014, 18th September 2014, available at https://www.amnesty.ie/sites/default/files/report/2014/09/Welcome-to-hell-fire-torture-and-other-ill-treatment-in-Nigeria-Amnesty-International-report.pdf (accessed 17th July 2015), p. 31.

Table 3.2 (Continued)

Country or Territory	Use of RCAs to facilitate or carry out torture, ill-treatment or punishment	Use of RCAs to facilitate suppression of freedom of expression, assembly or association	Employment of RCAs with arbitrary, indiscriminate, or excessive use of force or firearms	Aerosolized RCAs employed in enclosed space	Deleterious health effects, serious injury or death associated with RCAs/means of delivery
Oman[65]		•	•		
Pakistan[66]		•			
Palestinian Authority[67]		•			
Panama[68]		•	•		•
Peru[69]		•	•		•
Philippines[70]		•	•		
Puerto Rico/United States[71]		•	•		

[65] Word Report 2012, Oman Chapter, Human Rights Watch, 22nd January 2012, available at http://www.hrw.org/sites/default/files/reports/wr2012.pdf (accessed 17th July 2015), p. 609.

[66] Pakistan "Their Future Is at Stake" Attacks on Teachers and Schools in Pakistan's Balochistan Province, Human Rights Watch, December 2010, available at http://www.hrw.org/sites/default/files/reports/pakistan1210.pdf (accessed 17th July 2015), p. 19.

[67] Human Rights Watch, Gaza Crisis: Regimes React with Routine Repression Arab Governments, Iran, and Israel Ban Gaza Demonstrations; Protesters Beaten and Arrested, 21st January 2009, available at http://webcache.googleusercontent.com/search?q=cache:tulWQPGn0nMJ:https://www.hrw.org/news/2009/01/21/gaza-crisis-regimes-react-routine-repression+&cd=1&hl=en&ct=clnk&gl=uk (accessed 17th July 2015).

[68] Annual report 2011, Panama entry, Amnesty International, available at http://files.amnesty.org/air11/air_2011_full_en.pdf (accessed 16th July 2015), p. 256; Annual Report of the Special Rapporteur for freedom of expression 2010, Catalina Botero Marino, Special Rapporteur for Freedom of Expression, OAS, OEA/Ser.L/V/II. Doc. 5, 4th March 2011, available at www.oas.org/en/iachr/defenders/docs/pdf/defenders2011.pdf (accessed 16th July 2015), pp. 140–141.

[69] Amnesty International, Peru: Bagua, six months on: "Just because we think and speak differently, they are doing this injustice to us", AMR 46/017/2009, 2nd December 2009, available at https://www.amnesty.org/download/Documents/44000/amr460172009en.pdf (accessed 17th July 2015), pp. 7, 17, 27, 29–31, 36; Peru: Investigate Islay Province Killings 3 Dead, 31 Injured as Police Use Lethal Force on Protesters, Human Rights Watch, News, 8th April 2011, available at http://web.archive.org/web/20110415033105/http://www.hrw.org/fr/news/2011/04/08/peru-investigate-islay-province-killings (accessed 17th July 2015).

[70] Addressing the Economic, Social and Cultural Root Causes of Torture and Violence in the Philippines, A report on the implementation in the Philippines of the Concluding Observations and Recommendations of the United Nations Committee Against Torture and Committee on Economic, Social and Cultural Rights, OMCT, 1st October 2010, available at http://www.omct.org/files/2010/10/20939/philippines_follow_up_mission_report_final_01_10_10.pdf (accessed 22nd July 2015), p. 17.

[71] United Nations, Human Rights Council, Report of the Special Rapporteur on adequate housing as a component of the right to an adequate standard of living, and on the right to non-discrimination in this context, Raquel Rolnik, A/HRC/13/20/Add.1, 22nd February 2010, available at http://daccess-dds-ny.un.org/doc/UNDOC/GEN/G10/111/91/PDF/G1011191.pdf?OpenElement (accessed 17th July 2015), p. 46; Puerto Rico: Amnesty International calls for police restraint as student strike continues, Amnesty International, public statement, AMR 47/001/2010, 2nd June 2010, available at https://www.amnesty.org/en/documents/AMR47/001/2010/en/ (accessed 17th July 2015).

Table 3.2 (Continued)

Country or Territory	Use of RCAs to facilitate or carry out torture, ill-treatment or punishment	Use of RCAs to facilitate suppression of freedom of expression, assembly or association	Employment of RCAs with arbitrary, indiscriminate, or excessive use of force or firearms	Aeorsolized RCAs employed in enclosed space	Deleterious health effects, serious injury or death associated with RCAs/means of delivery
Romania[72]			•		
Senegal[73]		•	•	•	
Sierra Leone[74]			•	•	
South Africa[75]			•		
South Korea[76]	•	•	•		•
South Sudan[77]		•	•		
Spain[78]		•	•		

[72] Romania: Alleged excessive use of force during Bucharest demonstrations, Amnesty International, Public statement, EUR 39/001/2012, 26th January 2012, available at https://www.amnesty.org/en/documents/eur39/001/2012/en/ (accessed 17th July 2015); Policing demonstrations in the European Union, Amnesty International, EUR 01/022/2012, October 2012, available at http://www.amnesty.org.uk/sites/default/files/eu-police.pdf (accessed 17th July 2015), pp. 4–5.

[73] Senegal: President must rein in security forces as third person killed amid protests, Amnesty International, News, 1st February 2012, available at http://www.amnestyusa.org/news/news-item/senegal-president-must-rein-in-security-forces-as-third-person-killed-amid-protests (accessed 17th July 2015); Senegal: Intimidation and arrests of protestors one week before presidential elections, Amnesty International, public statement, AFR 49/002/2012, 17th February 2012, available at https://www.amnesty.org/download/Documents/.../afr490022012en.pdf (accessed 17th July 2015); Senegal: Campaign turns into repression, reflecting President Wade's "human rights" balance sheet, FIDH, 18th February 2012, available at https://www.fidh.org/International/Federation/for/Human/Rights/Africa/senegal/11338/senegal-campaign-turns-into-repression-reflecting-president-wade-s-human (accessed 22nd July 2015).

[74] Annual Report 2012, Amnesty International, Sierra Leone entry, available at http://www.amnestyusa.org/sites/default/files/air12-report-english.pdf (accessed 17th July 2015), pp. 296–297; Sierra Leone: Briefing on the events in Bumbuna (April 2012), Amnesty International, AFR 51/004/2012, August 2012, available at https://www.amnesty.org/download/Documents/.../afr510042012en.pdf (accessed 17th July 2015), pp. 6, 10, 16.

[75] South Africa: Police repeatedly turn on asylum-seekers amid xenophobia spike, Amnesty International, News, 29th May 2013, available at http://www.amnestyusa.org/news/news-item/south-africa-police-repeatedly-turn-on-asylum-seekers-amid-xenophobia-spike (accessed 17th July 2015).

[76] Amnesty International, South Korea: Call for unimpeded access to food, water and necessary medical treatment for Ssanyong striking workers, ASA 25/007/2009, 31st July 2009 available at http://www.amnesty.or.jp/en/news/2009/0731_1058.html (accessed 17th July 2015); South Korea: Stop arrests of trade union leaders and respect the rights of striking workers, Amnesty International, News, 24th December 2013, available at https://www.amnesty.org/en/latest/news/2013/12/south-korea-stop-arrests-trade-union-leaders-and-respect-workers-right-strike/ (accessed 17th July 2015).

[77] Annual Report 2012, Amnesty International, South Sudan entry, available at http://www.amnestyusa.org/sites/default/files/air12-report-english.pdf (accessed 17th July 2015), p. 310.

[78] Spain: The right to protest under threat, Amnesty International, EUR 41/001/2014, 24th April 2014, available at https://www.amnesty.org.uk/sites/default/files/spain_-the_right_to_protest_under_threat_0.pdf (accessed 17th July 2015), p. 56.

Table 3.2 (Continued)

Country or Territory	Use of RCAs to facilitate or carry out torture, ill-treatment or punishment	Use of RCAs to facilitate suppression of freedom of expression, assembly or association	Employment of RCAs with arbitrary, indiscriminate, or excessive use of force or firearms	Aerosolized RCAs employed in enclosed space	Deleterious health effects, serious injury or death associated with RCAs/means of delivery
Sri Lanka[79]		•	•		
Sudan[80]		•	•		
Swaziland[81]		•	•		
Sweden[82]			•		
Syria[83]		•	•		•

[79] United Nations, Human Rights Council, Report of the Special Rapporteur on torture and other cruel, inhuman or degrading treatment or punishment, Juan E. Méndez, Addendum: Observations on communications transmitted to Governments and replies received, A/HRC/25/60/Add.2, 11th March 2013, available at http://antitorture.org/wp-content/uploads/2014/03/Report_ Observations_Govt_Communications_Replies_2014.pdf (accessed 17th July 2015), p. 100.
[80] Human Rights Watch, Democracy on Hold Rights Violations in the April 2010 Sudan Elections, June 2010, http://www.hrw.org/en/reports/2010/06/29/democracy-hold (accessed 13th December 2011); United Nations, Human Rights Council, Fourteenth session, Agenda item 4, Report of the independent expert on the situation of human rights in the Sudan, Mohammed Chande Othman, A/HRC/14/41, 26th May 2010, available at http://www2.ohchr.org/english/bodies/ hrcouncil/docs/14session/A.HRC.14.41_en.pdf (accessed 17th July 2015), p. 6; Bloody repression of protests in Sudan FIDH and ACJPS call upon the African Union to Send an Urgent Commission of Inquiry, FIDH, 1st October 2013, available at https://www.fidh.org/International-Federation-for-Human-Rights/Africa/sudan/bloody-repression-of-protests-in-sudan-fidh-and-acjps-call-upon-the (accessed 22nd July 2015).
[81] World Report 2012, Human Rights Watch, Swaziland chapter, 22nd January 2012, available at http://www.hrw.org/sites/default/files/reports/wr2012.pdf (accessed 20th July 2015), p. 187; Annual Report 2013, Amnesty International, Swaziland entry, available at http://files.amnesty. org/air13/AmnestyInternational_AnnualReport2013_complete_en.pdf (accessed 20th July 2015), p. 255.
[82] Annual Report 2010, Amnesty International, Sweden entry, available at https://www.amnesty. org.nz/sites/default/files/Amnesty_International_Report_2010.pdf (accessed 20th July 2015), p. 311.
[83] Amnesty International, Boy killed and dozens of injured detained after Kurds clash with security forces in Syria, 26th March 2010, available at http://webcache.googleusercontent.com/search? q= cache:_Wx3RICuDbIJ:https://www.amnesty.org/articles/news/2010/03/boy-killed-and-dozens-injured-detained-after-kurds-clash-security-forces-syria-2010/+ &cd= 1&hl= en&ct= clnk&gl= uk (accessed 20th July 2015); Amnesty International, Hospital patients held incommunicado in Syria, Urgent Action 70/10, MDE 24/006/2010, 25th March 2010, available at https://www.amnesty. org/en/documents/MDE24/006/2010/en/ (accessed 20th July 2015); Independent investigation urged into Syria protest deaths, Amnesty International, News, 22nd March 2011, available at http://www.amnesty.org.au/news/comments/25149/ (accessed 20th July 2015); "We Live as in War", Crackdown on Protesters in the Governorate of Homs, Syria, Human Rights Watch, 11th November 2011, available at http://www.hrw.org/sites/default/files/reports/syria1111webwcover. pdf (accessed 20th July 2015), pp. 13, 18, 19, 24, 47.

Table 3.2 (Continued)

Country or Territory	Use of RCAs to facilitate or carry out torture, ill-treatment or punishment	Use of RCAs to facilitate suppression of freedom of expression, assembly or association	Employment of RCAs with arbitrary, indiscriminate, or excessive use of force or firearms	Aerosolized RCAs employed in enclosed space	Deleterious health effects, serious injury or death associated with RCAs/means of delivery
Tanzania[84]			•		•
Thailand[85]		•			•
Togo[86]	•	•		•	•
Tunisia[87]	•	•		•	•

[84] United Nations, Human Rights Council, Report by the Special Rapporteur on the situation of human rights and fundamental freedoms of indigenous people, James Anaya, A/HRC/15/37/Add.1, 15th September 2010, available at http://www2.ohchr.org/english/bodies/hrcouncil/docs/15session/A.HRC.15.37.Add.1.pdf (accessed 20th July 2015), p. 173; Newsletter: Freedom of Expression in Eastern Africa, Article 19, 5th October 2012, available at https://www.article19.org/resources.php/resource/3466/en/newsletter:-freedom-of-expression-in-eastern-africa (accessed 23rd July 2015).

[85] Human Rights Watch Annual Report 2014, Thailand entry, available at https://www.hrw.org/world-report/2014/country-chapters/thailand (accessed 20th July 2015); Human Rights Watch Annual Report 2010, Thailand entry, available at http://www.hrw.org/news/2010/01/20/thailand-serious-backsliding-human-rights (accessed 20th July 2015); Human Rights Watch, Descent into Chaos: Thailand's 2010 Red Shirt Protests and the Government Crackdown, May 2011, available at https://www.hrw.org/sites/default/files/reports/thailand0511webwcover_0.pdf (accessed 20th July 2015), p. 38, 52–55.

[86] Annual report 2011, Amnesty International, Togo entry, available at http://files.amnesty.org/air11/air_2011_full_en.pdf (accessed 20th July 2015), p. 323; Togo (2010–2011), Situation of human rights defenders, FIDH, 27th January 2012, available at https://www.fidh.org/International-Federation-for-Human-Rights/Africa/Togo,210/TOGO-2010-2011 (accessed 23rd July 2015); Annual Report 2013, Amnesty International, Togo entry, available at http://files.amnesty.org/air13/AmnestyInternational_AnnualReport2013_complete_en.pdf (accessed 20th July 2015), p. 269.

[87] Human Rights Watch, The Price of Independence Silencing Labor and Student Unions in Tunisia, 21st October 2010, available at http://www.hrw.org/sites/default/files/reports/tunisia1010w.pdf (accessed 20th July 2015), pp. 26–27, 48–49; Tunisia in revolt, State violence during anti-government protests, Amnesty International, MDE 30/011/2011, February 2011, available at https://www.amnesty.org/download/Documents/32000/mde300112011en.pdf (accessed 20th July 2015), pp. 2 ,9, 10, 15–26; Renewed outbreak of police violence, FIDH, 10th May 2011, available at https://www.fidh.org/International-Federation-for-Human-Rights/north-africa-middle-east/tunisia/Renewed-outbreak-of-police (accessed 23rd July 2015); Tunisia: Indefinite ban placed on protests in iconic avenue ends in violence, ARTICLE 19,10th April 2012, available at https://www.article19.org/resources.php/resource/3029/en/tunisia:-indefinite-ban-placed-on-protests-in-iconic-avenue-ends-in-violence (accessed 23rd July 2015); World Report, Tunisia chapter, Human Rights Watch, 21st January 2014, available at https://www.hrw.org/sites/default/files/wr2014_web_0.pdf (accessed 20th July 2015), p. 620; Tunisia: Protesters Describe Tear Gas Attacks, Beatings, Protect the Right to Peaceful Protest, Human Rights Watch, 29th July 2013, available at https://www.hrw.org/news/2013/07/29/tunisia-protesters-describe-teargas-attacks-beatings (accessed in 2015).

Table 3.2 (Continued)

Country or Territory	Use of RCAs to facilitate or carry out torture, ill-treatment or punishment	Use of RCAs to facilitate suppression of freedom of expression, assembly or association	Employment of RCAs with arbitrary, indiscriminate, or excessive use of force or firearms	Aerosolized RCAs employed in enclosed space	Deleterious health effects, serious injury or death associated with RCAs/means of delivery
Turkey[88]	•	•	•	•	•
Uganda[89]	•	•	•	•	•

[88] Human Rights Watch, Turkey: Combat Police Killings and Violence, New Wave of Shootings and Ill-Treatment, 20th April 2010, available at https://www.hrw.org/news/2010/04/20/turkey-combat-police-killings-and-violence (accessed 20th July 2015); Human Rights Watch, Letter to Turkish ministers in regard to treatment of human rights defenders from Pembe Hayat LGBTT, 18th October 2010, available at http://www.hrw.org/news/2010/10/18/letter-turkish-ministers-regards-treatment-human-rights-defenders-pembe-hayat-lgbtt (accessed 20th July 2015); Gezi Park Protests: Brutal denial of the right to peaceful assembly in Turkey, Amnesty International, EUR 44/022/2013, October 2013, available at https://www.amnesty.org/en/documents/EUR44/022/2013/en/ (accessed 20th July 2015); Turkey: End Incorrect, Unlawful Use of Teargas Dozens Injured as Police Fired Teargas Canisters Directly at Protesters, 17th July 2013, available at https://www.hrw.org/news/2013/07/16/turkey-end-incorrect-unlawful-use-teargas (accessed 20th July 2015); Turkey: Gezi, one year on, Witch hunt, impunity of law enforcement officials and a shrinking space for rights and freedoms, FIDH, the Human Rights Association (IHD) and the Human rights foundation of Turkey (HRFT), May 2014, available at https://www.fidh.org/IMG/pdf/turkey_avril_2014_uk_web.pdf (accessed 23rd July 2015), pp. 8–10, 16, 19–23.

[89] Uganda: Teargas and bullets used against peaceful protestors, Amnesty International, Public statement, AFR 59/008/2011, 15th April 2011, available at https://www.amnesty.org/download/Documents/28000/afr590082011en.pdf (accessed 20th July 2015); Uganda: Investigate Use of Force against Protestors, Amnesty International, public statement, AFR 59/012/2011, 16th May 2011, available at https://www.amnesty.org/download/Documents/28000/afr590122011en.pdf (accessed 20th July 2015); Uganda: Stifling dissent: Restrictions on the rights to freedom of expression and peaceful assembly in Uganda, Amnesty International, AFR 59/016/2011, 1st November 2011, available at https://www.google.co.uk/url?sa=t&rct=j&q=&esrc=s&source=web&cd=1&cad=rja&uact=8&ved=0CCIQFjAAahUKEwjdxJeQgurGAhXH1hQKHam4AfQ&url=http%3A%2F%2Fwww.amnestyusa.org%2Fnews%2Fnews-item%2Fuganda-government-backed-harassment-and-repression-of-critics-increasing&ei=qRKtVZ3MKsetU6nxhqAP&usg=AFQjCNEMthCzSRxe1U7Wss_MZpD8MFNEKg&bvm=bv.98197061,d.d24 (accessed 20th July 2015); Uganda: Launch Independent Inquiry Into Killings, No Lethal Force Was Needed in at Least 9 Fatal Shootings, Human Rights Watch News, 8th May 2011, available at https://www.hrw.org/news/2011/05/08/uganda-launch-independent-inquiry-killings (accessed 20th July 2015); 05 February 2013: Journalist's skin peels after being pepper sprayed by police, in: Article 19 (14th March 2013) op.cit.; Uganda: Violent attacks on protesters as letter crisis continues, Article 19, 28th May 2013, available at https://www.article19.org/resources.php/resource/3611/en/uganda:-violent-attacks-on-protesters-as-letter-crisis-continues (accessed 6th August 2015; "Where Do You Want Us to Go?", Abuses against Street Children in Uganda, Human Rights Watch, 17th July 2014, available at https://www.hrw.org/sites/default/files/reports/uganda0714_forinsert_ForUpload.pdf (accessed 20th July 2015), pp. 27, 28, 48.

Table 3.2 (Continued)

Country or Territory	Use of RCAs to facilitate or carry out torture, ill-treatment or punishment	Use of RCAs to facilitate suppression of freedom of expression, assembly or association	Employment of RCAs with arbitrary, indiscriminate, or excessive use of force or firearms	Aerosolized RCAs employed in enclosed space	Deleterious health effects, serious injury or death associated with RCAs/means of delivery
Ukraine[90]	•	•	•	•	•
United States[91]	•	•	•	•	•
Venezuela[92]			•		
Vietnam[93]		•	•		

[90] Ukraine: Euro 2012 Jeopardised by Criminal Police Force Amnesty International, media briefing, EUR 50/005/2012, 30th April 2012, available at https://www.amnesty.org.uk/sites/default/files/eur500052012en.pdf (accessed 20th July 2015); World Report 2013, Human Rights Watch, Ukraine chapter, 31st January 2013, available at https://www.hrw.org/sites/default/files/wr2013_web.pdf (accessed 20th July 2015); Buffeted in the Borderland, The Treatment of Asylum Seekers and Migrants in Ukraine, Human Rights Watch, 16th December 2010, available at https://www.hrw.org/report/2010/12/16/buffeted-borderland/treatment-asylum-seekers-and-migrants-ukraine (accessed 23rd July 2015); Letter to the General Prosecutor of Ukraine, Regarding Criminal Investigations into Events in Kyiv on November 30 and December 1, 2013, Human Rights Watch, 30th December 2013, available at https://www.hrw.org/news/2013/12/20/letter-general-prosecutor-ukraine (accessed 20th July 2015); Ukraine: A new country or business as usual?, Amnesty International, EUR 50/028/2014, 26th June 2014, available at https://www.amnesty.org/download/Documents/8000/eur500282014en.pdf (accessed 20th July 2015).

[91] Annual Report 2012, Amnesty International, USA entry, available at http://www.amnestyusa.org/sites/default/files/air12-report-english.pdf (accessed 20th July 2015); USA: Cruel isolation: AI's concerns about conditions in Arizona maximum security prisons, Amnesty International, AMR 51/023/2012, 3rd April 2012, available at http://www.amnestyusa.org/research/reports/cruel-isolation-amnesty-international-s-concerns-about-conditions-in-arizona-maximum-security-prison (accessed 20th July 2015), pp. 12–13; USA: In hostile terrain: Human rights violation in immigration enforcement in the US Southwest, Amnesty International, AMR 51/018/2012, 28th March 2012, available at http://www.amnestyusa.org/sites/default/files/ai_inhostileterrain_final031412.pdf (accessed 20th July 2015), p. 31; USA: Respect Rights of Protesters, Transparency Required in Investigations of Police Misconduct, Human Rights Watch, News, 18th October 2011, available at https://www.hrw.org/news/2011/10/28/us-respect-rights-protesters (accessed 20th July 2015); Callous and Cruel: Use of Force against Inmates with Mental Disabilities in US Jails and Prisons, 15th May 2015, Human Rights Watch, available at https://www.hrw.org/report/2015/05/12/callous-and-cruel/use-force-against-inmates-mental-disabilities-us-jails-and (accessed 23rd July 2015).

[92] OAS, Inter-American Commission on Human Rights, Precautionary Measures 2011, PM 219–11 – Relatives of Inmates at the Rodeo I and Rodeo II Prisons, Venezuela, available at: http://www.oas.org/en/iachr/decisions/precautionary.asp (accessed 2nd April 2015).

[93] Human Rights Watch, World Report 2011: Vietnam chapter, Events of 2010, January 2011, available at http://www.hrw.org/sites/default/files/related_material/wr2011_book_complete.pdf (accessed 20th July 2015), p. 387.

Table 3.2 (Continued)

Country or Territory	Use of RCAs to facilitate or carry out torture, ill-treatment or punishment	Use of RCAs to facilitate suppression of freedom of expression, assembly or association	Employment of RCAs with arbitrary, indiscriminate, or excessive use of force or firearms	Aeorsolized RCAs employed in enclosed space	Deleterious health effects, serious injury or death associated with RCAs/means of delivery
Yemen[94]	•	•		•	•
Zimbabwe[95]	•	•			

[94] Amnesty International, Yemen: Cracking down under pressure, MDE 31/010/2010, 24th August 2010, available at http://www2.ohchr.org/english/bodies/hrc/docs/ngos/Yemen-Cracking_down_under_pressureHRC101.pdf (accessed 20th July 2015), p. 64; Yemeni protesters killed in violent attacks, Amnesty International, 14th March 2011, available at http://www.refworld.org/docid/4d7f25071a.html (accessed 20th July 2015); Yemen: Impunity granted, transition at risk, The human rights violations committed during the repression of the protest movement February–December 2011, FIDH, 20th February 2012, available at https://www.fidh.org/IMG/pdf/yemen_report_en_vf.pdf (accessed 22nd July 2015), pp. 12, 13, 17, 18, 19; Unpunished Massacre, Yemen's Failed Response to the "Friday of Dignity" Killings, Human Rights Watch, 12th February 2013, available at http://www.hrw.org/sites/default/files/reports/yemen0213webwcover_0.pdf (accessed 20th July 2015), pp. 17, 21, 43; Letter to Minister of Interior on Abuses by Central Security Forces, Human Rights Watch, News, 15th February 2013, available at http://www.hrw.org/news/2013/02/15/letter-minister-interior-abuses-central-security-forces (accessed 20th July 2015); Yemen: Crackdown on Protest Leaves 13 Dead, Deadliest Clash Since Hadi Became President in 2012, Human Rights Watch, News, 13th June 2013, available at https://www.hrw.org/news/2013/06/12/yemen-crackdown-protest-leaves-13-dead (accessed 20th July 2015).

[95] Walk the talk: Zimbabwe must respect and protect fundamental freedoms during the 2013 harmonized elections, Amnesty International, AFR 46/009/2013, available at http://www.refworld.org/docid/51e500634.html (accessed 20th July 2015), p. 18.

3.4.1.1. Use of RCAs to facilitate or carry out torture, ill-treatment or punishment

Despite the absolute prohibition on torture and other cruel, inhuman or degrading treatment or punishment under international law, between 1st January 2009 and 31st December 2013, the misuse of RCAs by law enforcement officials to facilitate or carry out such activities was reported in at least 18 countries or territories. The majority of recorded cases involved the reported use of RCAs against individual prisoners or detainees, often by law enforcement or prison officials employing hand-held irritant sprays in a targeted fashion. However, there have also been reported incidents where RCAs were employed against groups or crowds in a manner which could be considered to be "collective punishment".

3.4.1.1.1. Brazil – reported use of pepper spray and tear gas during transportation and holding of prisoners. In its 2012 report analysing the treatment of persons deprived of their liberty in four Brazilian States,[125] the UN Subcommittee on Prevention of Torture and Other Cruel, Inhuman or

Degrading Treatment or Punishment (SPT) raised concerns regarding allegations of "severe ill-treatment and inhumane conditions of transportation" of detainees in vehicles of the Special Operations Services (*Serviços de Operações Especiais*, SOE). The alleged methods used by SOE personnel included locking-up a large number of detainees in uncomfortable positions, handcuffed, and with no ventilation, opening the vehicle and spraying pepper spray on them and then locking up the vehicle.[126] In addition the SPT also reported allegations of "ill-treatment and excessive use of force" by prison guards in the Viana II maximum security penitentiary, "especially the alleged use of tear gas in confined spaces, including cells". The SPT voiced "serious reservations about the use of irritant gases in confined spaces, as it may entail health risks and cause unnecessary suffering."[127]

3.4.1.1.2. Maldives – reported use of pepper spray in torture of an individual. Mariya Ahmed Didi, a member of parliament for the Maldivian Democratic Party (MDP), was among hundreds of MDP supporters arrested by police following the violent dispersal of a peaceful rally held on 8th February 2012. In testimony to Amnesty International, she detailed her ill-treatment whilst in arbitrary detention. "They...continued beating me with my handcuffs on," she said. "They were beating me with batons. Police and military officers then forcefully opened my eyelids. They went for the eye that had been injured [as a result of a police beating] the day before. They sprayed pepper spray directly into my eye. Then they did the same with my other eye. They then sprayed into my nose as they were also beating me. They then took me to a police station and continued to beat me there. I have bruises all over my body. At one point when they were beating me one of them shouted: 'Is she still not dead?'"[128]

3.4.1.1.3. Israel – reported use of RCAs as an element of collective punishment. National human rights NGOs such as B'Tselem[129] and the Association for Civil Rights in Israel (ACRI),[130] as well as international human rights NGOs, notably Amnesty International[131] have documented the frequent use of excessive force – including the misuse of "less lethal" weapons – by the Israeli Army and Border Police against Palestinians and others participating in the weekly protests held in those villages most directly affected by the erection of the fence/wall and the presence of Jewish-only settlements in the West Bank. In the village of Nabi Saleh, near Ramallah, Israeli forces have repeatedly used a range of "less lethal" weapons, including tear gas, pepper spray and malodorant water canon spray (the skunk) against largely peaceful demonstrators.[132]

In its February 2014 report, AI described how the Israeli army's response to the protests in Nabi Saleh had frequently impacted local people who were not involved in such activities. Villagers alleged that the army often fired large amounts of tear gas into residential areas, causing people breathing difficulties and putting them at risk of suffocation, and, sometimes, house fires. On several occasions Israeli forces sprayed malodorant "skunk water" at homes, causing damage and leaving families with a sickening smell that remained for days afterwards.[133]

In a January 2013 report, B'Tselem stated that:

> The actions of the security forces raise serious suspicions that the Skunk is used as a collective punitive measure against residents of villages where regular weekly demonstrations are held near the village's built-up areas, such as a-Nabi Saleh and Kafr Qadum. This was certainly the case on many occasions documented by B'Tselem at a-Nabi Saleh, in which security forces drove the Skunk truck down the village's main street and sprayed the foul-smelling liquid at homes far removed – sometimes even clear across the village – from the main location of the demonstrations and clashes between the military and Palestinian stonethrowers.[134]

AI has contended that "the combined impact of the army's repressive and restrictive policies and practices in Nabi Saleh and other villages near illegal settlements or along the fence/wall appears to amount to collective punishment, whereby the population as a whole is penalized, including those who play no active part in the weekly demonstrations and other protests against Israeli rule".[135]

Subsequently, ACRI have raised concerns about the continued misuse of RCAs in other parts of Israel. In a November 2014 letter to the Chief Commissioner of Israel Police, ACRI detailed how "in practice" the use of skunk spay in East Jerusalem constituted "collective punishment for all residents". According to ACRI:

> Since July 2014, many streets in almost every neighborhood of East Jerusalem have been covered in tremendous amounts of the skunk spray. As a result, the daily life of tens of thousands of East Jerusalem residents has been affected; they have been compelled to live for days at a time with a foul stench, which induces nausea. In addition to the physical side affects of skunk spray – difficulty breathing, eating and sleeping – the residents also report feelings of severe humiliation.[136]

3.4.1.2. Serious injuries or death from suffocation or toxicity of RCAs
A specific recurring concern raised by UN human rights monitors, international non-governmental human rights organizations and the medical community has been the employment of RCAs in excessive quantities, either in the "open air" (as illustrated in sections 3.4.1.2.1–2), or in confined spaces where the targeted persons cannot disperse (as highlighted in 3.4.1.2.3–5.).

In such situations serious injury or death, particularly of vulnerable individuals, can result from either the toxic properties of the chemical agents or due to asphyxiation.

3.4.1.2.1. Israel/Occupied Territories. On 14th January 2011, the UN Special Rapporteur on the Situation of Human Rights on Palestinian Territories Occupied Since 1967 expressed concern over what he described as a "series of illegal acts by Israeli authorities" in the Occupied Territories, including the killing of four Palestinians in the West Bank by Israel Defence Forces (IDF). One of the cases highlighted was that of Ms. Jawaher Abu Rahmeh, a 36 year-old Palestinian woman, who had been observing a demonstration against the separation Wall in the West Bank town of Bil'in on 31st December 2010.[137] Ms Abu Rahmeh was seriously injured following the employment by Israeli soldiers of tear gas against the demonstrators. She was later admitted to Ramallah Hospital "with very weak breathing as a result of inhaling a gas", its emergency department director, Dr Mohammed Eidi, told the *Reuters* news agency. "We put her on respiratory system. But she died this morning," he added.[138] The UN Special Rapporteur stated that her death was as a result of inhalation of tear gas fired by the IDF.[139]

3.4.1.2.2. Tunisia. On 10th January 2011, Khames Karmazi, his wife and their seven-month-old daughter, Yakin, were exposed to tear gas as they passed an area in Kasserine that was the site of protests and confrontations between security forces and youths. That night, Yakin had trouble sleeping and cried a lot, so the next morning her family rushed her to the emergency unit at Kasserine Hospital. She died at about 2pm that afternoon. A medical certificate signed by the hospital's chief paediatrician confirmed that Yakin died as a result of "exposure to very toxic tear gas". Her father told AI: "She was my only child. We have been trying to have children for five years... I want to know why this happened. Who is responsible?"[140]

3.4.1.2.3. Honduras. On 28th June 2009, a military *coup d'état* deposed President José Manuel Zelaya. UN and regional human rights bodies, and international human rights NGOs have documented the employment of RCAs to facilitate serious and widespread human rights violations by police and military officers against opponents of the coup. Reports by the UN Subcommittee for the Prevention of Torture,[141] and the Inter-American Commission on Human Rights (IACHR),[142] have documented the violent suppression by approximately 200 troops of the Preventive Police and the Engineers Battalion of a demonstration organized at Cuesta de la Virgen, in the department of Comayagua, on 30th July 2009, by several social associations belonging to the Frente de Resistencia Contra el Golpe (Coup Resistance Front), and villagers from the departments of Comayagua and La Paz. According to the IACHR, it received various statements describing how, in order to break up the demonstration, some 40 persons were loaded into a military truck, with very small windows; the truck's backdoor was closed and a soldier threw a tear gas grenade inside the vehicle, which caused those inside to cough and choke. "In their desperation, some people tried to jump out of the vehicle, while others stuck their heads through the windows to breathe. But the police beat them on the head to force them back inside. This tactic was allegedly repeated 5 times."[143]

3.4.1.2.4. Zimbabwe. Although occurring prior to the survey catchment period, a notorious incident highlighted by the UN Special Rapporteur on Extra Judicial, Summary or Arbitrary Executions[144] and non-governmental human rights organisations[145] concerned the employment of RCAs during the forced eviction by riot police, war veterans and youth "militia", of some 10,000 people from Porta Farm on 2nd September 2004. The legitimacy of this action as a law enforcement operation was undermined by the fact that the police were acting in defiance of a High Court order prohibiting the eviction.[146]

According to eye-witness testimony, the police fired tear gas directly into the homes of the Porta Farm residents. Eleven people died at Porta Farm following exposure to tear gas. Among the dead were five babies – the youngest just 1 day old. Many relatives and eye-witnesses, whose testimony was recorded by Zimbabwe Lawyers for Human Rights and AI, believe their deaths to be attributable to their exposure to the tear gas.[147] For example, Christine K., a Porta Farm resident, has described the deaths of her daughter and 5-month-old grandson:

> [T]he police fired...tear gas canisters that landed in our yard thereby clouding the inside and outside of our huts with a pungent of choking smoke. Myself, my...daughter and grandson inhaled the smoke;...[my daughter] immediately collapsed. She complained to me that her chest was full of the smoke and [she] had difficulties in breathing.... I dragged her in the hut and from that time she never recovered until death. [My grandson] was taken to Parirenyatwa hospital...[and] was spitting blood until the time of his death.[148]

Hundreds of other Porta Farm residents reported suffering ill-effects from the tear gas, including chest and stomach pains, nose bleeds and other health problems. Doctors who examined some of the Porta Farm residents told AI that they believed that those most seriously affected by the tear gas were particularly vulnerable due to pre-existing illnesses such as tuberculosis.[149]

3.4.1.2.5. Egypt. At around 6.30 a.m. on Sunday 18th August 2013, 45 prisoners, previously detained during the unrest following the military coup against President Morsi, were transferred from Heliopolis police station to

Abu Zaabal prison, in north-east Cairo. The prisoners, most handcuffed in pairs, were crammed into a police van which had very poor ventilation. Due to delays in unloading other prison transports, the prisoners were kept locked in the van in the prison courtyard, without adequate water as the temperature outside rose to over 31°C. One of the survivors, Mohamed Abdelmahboud, told the UK newspaper, *the Guardian*: "We started to get short of oxygen... and people started to shout for help. We started banging on the walls, we started screaming, but no one answered."[150] The commander of the Heliopolis police convoy, Lieutenant Colonel Amr Farouk, reportedly allowed the van door to be opened once for very a brief period but refused further requests until after 1pm. According to another survivor, Sayed Gabal, "People started passing out, one after the other. Of course, the elderly went down first. And the others started banging harder and harder. And outside they continued laughing and cursing Morsi."[151] The course of events following the opening of the van door is disputed: the majority of the police claim that the prisoners had rioted and captured a guard; the surviving prisoners and one of the police guards, Abdelaziz Rabia Abdelaziz, dispute this claim. According to *BBC News*, prosecutors found no evidence to support the police claim.[152] What is certain is that an unidentified police officer fired a hand-held tear gas canister – normally intended for self-defence – through one of the van's side windows. When the van was eventually unloaded, 37 of the 45 prisoners were found to have died.[153]

AI reported that autopsies carried out on the deceased concluded that the cause of death was suffocation.[154] Dr Hesham Farag, spokesperson for the mortuary where the autopsies conducted, stated in a written testimony to the *Guardian* that the men would still have been alive when the tear gas was fired into the van, as traces of the CS gas were found in each corpse's blood.[155] He doubted that the single tear gas canister contained enough CS to kill so many men on its own. However, its effects were exacerbated because of the conditions in the van. Dr Farag stated: "We decided that the police [are] responsible for all these casualties, because they loaded the vehicle with 45 prisoners, which is a very large number, because the vehicle should carry no more than 24 people... Therefore there was a lack of oxygen, which accelerated the death when tear gas was used."[156] The deaths sparked international condemnation, with UN Secretary General Ban Ki-moon saying he was "deeply disturbed" by the events.[157]

On 18th March 2014, a Cairo court found Lieutenant Colonel Farouk guilty of involuntary manslaughter and extreme negligence, and he was sentenced to ten years in prison; three subordinate officers were each sentenced to one-year suspended sentences for their role in the prisoners' deaths.[158] However, Reuters and Human Rights Watch (HRW) subsequently reported that an appeal court overturned the convictions on 7th June 2014 and referred the case to the Prosecutor General for further investigation. A retrial was scheduled to begin on 22nd January 2015.[159]

3.4.1.3. Use of RCAs for suppression of freedom of expression, association or assembly

UN human rights monitoring bodies and international human rights NGOs have reported the misuse of RCAs to intimidate or punish those involved in peaceful demonstrations in many countries. RCAs have reportedly been employed to suppress freedom of expression, association or assembly in at least 75 countries or territories from 1st January 2009 to 31st December 2013.

3.4.1.3.1. Bahrain. UN human rights monitors,[160] and international non-governmental human rights organizations,[161] have detailed the excessive use of force, including the employment of tear gas, to suppress peaceful demonstrations occurring throughout Bahrain during 2011 and subsequently. The UN Special Rapporteur on the Promotion and Protection of the Right to Freedom of Opinion and Expression highlighted allegations that security officials had reportedly used tear gas, rubber bullets, shotguns and live ammunition against peaceful demonstrators to contain the massive protests. He stated that: "Such excessive use of force had been carried out, at a short distance, against people who were not participating in the demonstrations, but were running away from the police in proximity to protests areas."[162]

The worst violence of this period occurred during an attack in the early hours of 17th February 2011 upon a demonstrators' camp at Manama's Pearl Roundabout. The attack began at 3 a.m. while most of those inside, including families with children, were reportedly sleeping.[163] During what AI called a "clearly planned and coordinated action", massed ranks of riot police stormed the area to evict the peaceful protesters, firing live ammunition and using tear gas, batons, rubber bullets and shotguns to disperse the crowd. Tanks and armoured vehicles then blocked access to the roundabout.[164]

Maryam al-Khawaja, who works with the Bahrain Center for Human Rights, was at a "media centre" tent across the street from the Pearl Roundabout when the attack occurred. She told HRW that there had been no warning or request to disperse. Al-Khawaja said the police surrounded the encampment from all sides, allowing no escape from the tear gas and conventional weapons fire. "If they simply wanted to disperse the gathering they would have left a way out", she said.[165]

Five people were fatally wounded and at least 250 were injured, some critically. Among the injured were people clearly identified as medical workers, who were targeted by police while trying to help injured protesters in or near the roundabout. Khadija Ahmed, an 18-year-old medical student who was volunteering at the medical tent in the roundabout, described to AI what happened early on 17th February: "After 3am we heard shots. Some injured arrived at the tent straight away with tear gas problems. Then police threw or fired two tear gas canisters inside the tent and pulled the flap down. People were crying 'Save me, save me from them'."[166]

3.4.1.3.2. Bangladesh. The UN Special Rapporteur on freedom of opinion and expression; together with the UN Special Rapporteurs on freedom of peaceful assembly; extrajudicial, summary or arbitrary executions; and torture have highlighted allegations of the excessive use of force against peaceful demonstrators in January 2013.[167] The demonstrations were organized by teachers, employees of non-governmental schools, colleges and technical education institutions, and concerned changes to the scope and nature of government control and funding. In correspondence to the Bangladesh Government, the four Special Rapporteurs highlighted the following incidents:

> On 10 January 2013, a group of teachers...conducting a peaceful hunger strike...was dispersed by police forces which used teargas and pepper spray, whose chemical composition was reportedly particularly toxic...According to sources, at least 20 teachers were injured in the course of the dispersal, of whom 10 had to be taken to the Dhaka Medical Hospital...Mr. Maulana Sekander Ali, a teacher who participated in the assembly and protested in a peaceful manner, died in Patuakhali five days after the police's intervention...On 12 January 2013, another group of hunger strikers reportedly gathered at the Central Shaheed Minar premises. Police forces allegedly dispersed them. Subsequently...they formed a human chain on a nearby road at Dhaka University, but they were dispersed again. Both dispersions were allegedly executed using batons and pepper spray...On 13 January 2013, some protesters were allegedly dispersed while peacefully assembling before the National Human Rights Commission's office premises. Police forces reportedly used a water cannon and pepper spray...On 15 January 2013...a group of teachers went to Sobhanbagh to stage a sit-in before the building Prince Plaza. It is alleged that police officers dispersed them by using tear [gas] shells and water cannons.[168]

The four UN Special Rapporteurs expressed "grave concern" about the dispersal of these peaceful demonstrations by law enforcement authorities, and the allegations of excessive use of force employed, especially in relation to the death of Mr. Ali.[169]

3.4.1.3.3. Turkey. On 30th May 2013, police in Istanbul broke up a demonstration by several hundred environmentalists, using tear gas, beating protestors and burning their tents. The heavy-handed tactics of the authorities sparked a country-wide reaction. Within days, tens of thousands of protestors had taken to the streets across the main cities of Turkey. By the middle of June, hundreds of thousands had taken part in "Gezi Park protests" that spanned almost every one of Turkey's 81 provinces.

UN human rights bodies,[170] and international non-governmental human rights organizations,[171] have highlighted the excessive use of force against peaceful demonstrators involved in the "Gezi Park protests", including the unnecessary and disproportionate employment of tear gas. On 18th June 2013, the UN High Commissioner for Human Rights stated that she was:

> [P]articularly concerned about allegations of excessive use of force by police against peaceful groups of protesters as this may have resulted in serious damage to health. Reports that tear gas canisters and pepper spray were fired at people from close range, or into closed spaces, and the alleged misuse of rubber bullets, need to be promptly, effectively, credibly and transparently investigated.[172]

AI reported the Turkish Government's announcement that 130,000 tear gas canisters had been used during the first 20 days of the protests alone, equivalent to a year's supply, and that orders would be made to replace the stocks. On 13th August 2013, it was reported that an order for 400,000 tear gas canisters had been placed. Previous annual procurement was reported to be 150,000 canisters.[173]

AI monitors witnessed tear gas being used repeatedly against peaceful protestors at demonstrations in a manner that was manifestly inappropriate, abusive and in violation of their rights. Widespread reports and photographic and video evidence also indicated the frequent employment of tear gas against protestors fleeing police, as well as the apparently random use against potential demonstrators and bystanders alike at the scene or close to demonstrations. Tear gas was also employed in confined spaces, including residential buildings and commercial premises where protestors had taken refuge, and health facilities where injured persons were receiving treatment.[174]

The 18-day occupation of Gezi Park ended on the evening of 15th June 2013 as a result of a violent police clearance operation. The police fired rounds of tear gas and plastic bullets in the park, including in an area clearly marked as a clinic. The police gave protesters a 20-minute warning. HRW noted that "the timing of the assault was particularly shocking given the large number of people in the park, swelled by supporters who included families and children".[175]

Soon after the police emptied the park, many people sought shelter in the adjacent Divan Hotel. Photographs show people taking refuge there or being cared for by doctors for injuries or overexposure to tear gas. The police threw tear gas canisters into or near the hotel entrance, engulfing the area in a thick fog. A *Radikal* newspaper journalist, İsmail Saymaz, who was in the hotel during the gas attack, described it to HRW:

After we were gassed out of the park we fled to the Divan Hotel. There were hundreds of us: women, children, older people. Being tear gassed in a confined space with no ventilation was a desperate experience. No one can help you and you can help no one. You feel you are drowning and around you are people fainting, vomiting, writhing around in pain.[176]

3.4.1.4. RCAs employed in conjunction with excessive use of force and/or firearms
UN human rights monitoring bodies and international non-governmental human rights organizations have regularly expressed concern regarding reports of the employment of RCAs as part of the indiscriminate, excessive or lethal use of force by law enforcement officials, particularly in crowd control situations. As well as the potential dangers to health due to the toxicity of the chemical agents employed, concerns have been raised that RCAs are used by law enforcement officials, in conjunction with other "less lethal" or lethal weapons to facilitate excessive force or even enhance the application of lethal force. Between 1st January 2009 and 31st December 2013, RCAs reportedly contributed to the excessive use of force in at least 83 countries or territories; in a number of reported incidents, RCAs were employed in conjunction with firearms.

3.4.1.4.1. Equatorial Guinea. The findings of an investigation conducted by an International Commission of Enquiry established by the UN Secretary General document how Guinean security forces used RCAs in combination with automatic weapons against opposition activists who were holding a rally in Conakry Stadium on 28th September 2009.[177] The incident was also investigated by AI,[178] and HRW.[179] According to HRW, the events at the stadium, which resulted in an estimated 150–200 opposition supporters being killed and dozens of women and girls being raped, were "organized and premeditated".[180]

At around 11:30 a.m., soon after the opposition leaders arrived at the stadium, a combined force of several hundred soldiers, police and civilian-clothed militias positioned themselves around the exits to the stadium. Anti-riot police then fired tear gas into the stadium from their vehicles, causing widespread panic. Minutes later, Presidential Guard soldiers (red berets), and a smaller number of other forces, stormed through the principal entrance, firing directly as they advanced forward into the packed and terrified crowd.[181]

A retired professor in her sixties, interviewed by HRW, stated:

All of a sudden, I heard these loud noises – boom, boom – it sounded like a war. That was the firing of the tear gas from outside the stadium. Then, within minutes, the red berets entered. They were everywhere. The youth were on the field. When the soldiers entered, they opened fire right away on that crowd. Everyone went into panic, people were running everywhere – I saw people jump from the top of the covered stands. There was screaming everywhere, screaming so loud, and the crowd started to stampede.[182]

Many reportedly died from the indiscriminate firing; others were beaten or knifed to death; and still others were trampled to death by the panicked crowd. According to the International Commission of Enquiry report, "dozens of people attempting to escape through the stadium gates either suffocated or were trampled to death in stampedes, which were compounded by the use of tear gas".[183]

Outside the main stadium, on the sports complex grounds, many more opposition supporters were killed as they tried to escape.[184] Evidence uncovered by AI strongly suggested that the RCAs and delivery mechanisms utilized by the Guinean security forces in this massacre were exported from France.[185]

3.4.2. Use of RCAs by State military forces in armed conflict

Scholars and research organizations have documented the previous (sometimes large-scale) employment of RCAs by a range of State military forces in armed conflicts from World War I onwards.[186] Since the coming into force of the Chemical Weapons Convention (CWC) in 1997, the stockpiling or use of RCAs for such purposes has become largely obsolete. However, as will be discussed in Chapter 5 of this publication, at least one CWC State Party, the United States, has policy permitting RCA employment in certain non-offensive actions during armed conflict. Furthermore, certain States, including Turkey and the United States, have reportedly sanctioned use of RCAs in counter-insurgency operations.

3.4.2.1. Reported employment of RCAs by US military in Afghanistan armed conflict

A number of researchers, most notably Fry,[187] have explored the reported use of RCAs by the US military to clear caves and tunnels of enemy combatants in Afghanistan as part of the armed conflict in that country, which began in October 2001. According to Fry:

> RCAs are useful in clearing caves and bunkers of insurgents by either incapacitating them long enough for military forces to gain control of the situation or driving the insurgents out of the cave into the open where the insurgents lose any advantages they would have enjoyed from a defensive posture. The fact that caves are an enclosed space, make RCAs even more effective because the walls keep the gas concentrated in the targeted areas.[188]

Fry has stated that use of RCAs to clear caves and tunnels of enemy combatants: "in the context of an international military operation is expressly prohibited under Article I.5 of the CWC, as well as implicitly prohibited by the Hague Regulations, the Hague Gas Declaration, and the Geneva Gas Protocol".[189] However, despite such prohibitions, Fry contended that: "Unfortunately, in reality, the United States has made the use of tear gas a fundamental part of its cave-clearing techniques in Afghanistan."[190]

Research by Bahmanyar, cited by Fry, has described the techniques employed by US military forces to deny enemy combatants access to recently captured tunnel complexes:

> If time or [explosive] materials are not available for immediate closure, CS-1 Riot Control Agent can be placed at intervals down the tunnel sharp turns and intersections. It must be emphasised, however, that the denial achieved by the use of CS-1 is only temporary in duration and used until demolitions are available to completely destroy the complex.[191]

Bahmanyar also described how RCAs could be used directly by US military against enemy combatants still present in tunnel complexes:

> Time constraints or the threat of enemy action can force a tunnel team to use the Mity Mite Portable Blower (RVN, MACV 1965) to flush the enemy from tunnels. The Mity Mite can be used in conjunction with burning type CS Riot Control Agent grenades (M7A2), which have the additional effect of producing smoke which in most cases helps identify hidden entrances and air vents ... After flushing an entrance with CS grenades, the Mity Mite can then blow powdered CS-1 into tunnel entrances, rendering the tunnel unusable to the enemy for a short period of time, at least as far as the first "firewall" within the tunnel system.[192]

It should be noted that previously, during the Vietnam War, the United States military employed similar techniques utilizing RCAs (or smoke pots) and Mity Mite blowers, to clear Vietcog (National Liberation Front for South Vietnam) fighters from caves and tunnels.[193]

The US Government has not explicitly confirmed or denied whether the US military ever employed RCAs to flush enemy combatants from caves in Afghanistan. However, it appears that the use of such tactics in Afghanistan was tacitly acknowledged in testimony given by US Secretary of Defense, Donald Rumsfeld, on 5th February 2003 to the House of Representatives Armed Services Committee. In his testimony Rumsfeld cited two instances when it was "perfectly appropriate" to use RCAs in a military context. The first was "when you are transporting dangerous people in a confined space [like an airplane]". The second scenario Rumsfeld cited was "when there are enemy troops, for example, in a cave in Afghanistan, and you know that there are women and children in there with them, and they are firing out at you, and you have the task of getting at them. And you would prefer to get at them without also getting at women and children, or non-combatants."[194]

Rumsfeld's statement was given following Committee questioning about the possibility of employing RCAs after an invasion of Iraq. Although the United States subsequently equipped some of its troops with RCAs and sanctioned use in certain circumstances, there has been no confirmed employment of such agents in armed conflict in Iraq.

3.4.2.2. Reported employment of RCAs by Turkish armed forces

On 27th October 1999, German TV broadcast allegations of the reported use of CS by Turkish armed forces against Kurdish armed fighters hiding in a cave near Balikaya, south-east of Sirnak, on 11th May 1999.[195] The military engagement resulted in the deaths of 20 Kurdish combatants. It is unclear whether they died from high concentrations of tear gas or whether they were shot when leaving the cave. Munition fragments reportedly collected from the cave were provided by a Kurdish member of the Red Crescent to a German television journalist.[196] An analysis of the munition fragments at the Institute for Forensic Medicine at the University of Munich identified the presence of CS on the samples.[197] The munitions used were reportedly identified as CS cartridges made in Germany and exported under licence to Turkey. [198] A Turkish Foreign Ministry spokesperson, Sermet Atacanli, subsequently countered the allegations made by German TV, stating that Turkey had assumed the obligation not to develop, produce, store or use chemical weapons, which it meticulously observed. He declared that "It is logical to infer that Turkey cannot use such weapons if they do not exist in Turkey."[199]

According to the Sunshine Project, subsequent video footage of training exercises by Turkish anti-terrorist forces aired on Turkish television in 2004 suggested that such forces continued to be trained to use tear gas in military combat, alongside lethal firearms fire and explosive grenades.[200] It is unknown whether this practice is still in place.

3.4.3. Use of RCAs by State military forces in peacekeeping operations

The deployment and use of RCAs by military forces during peacekeeping operations under the auspices of the UN or as part of regional or pluri-lateral peacekeeping forces have been widely reported. For example, according to a study by Fry of media and other open source information, UN forces were reported to have used RCAs on at least 40 separate occasions in 14 missions from 1951 to 2009.[201] Fry has acknowledged that "most instances of reported RCA use" by UN peacekeeping forces "do not seem that malignant on their face".[202] However, he has contended that "the exceptions...are the instances where young children, pregnant women and elderly people were involved and when RCAs were used in combination with lethal force in more offensive operations".²⁰³ The following three cases cited by Fry raise concerns about the potential scope for misuse of RCAs in such operations.

Former Yugoslavia

On 28th August 1999, US troops of the UN-authorized, NATO-led Stabilization Force (SFOR) in Bosnia reportedly dropped tear gas from helicopters as SFOR clashed with civilians in the Serb town of Brcko. The clashes arose after SFOR troops entered a police station looking for pro-Karadzic police, and this led to radio announcements calling for an attack on SFOR troops.²⁰⁴

Liberia

On 21st October 2004, troops from the United Nations Mission in Liberia (UNMIL) reportedly used tear gas against parents and students of a primary school who were protesting the closure of their school. Just over a week later, on 29th October 2004, UNMIL troops reportedly used tear gas against clashing Muslims and Christians to restore order in the Liberian capital of Monrovia: at least three people were killed by the UN forces' efforts to disperse the crowd.²⁰⁵

Haiti

On 6th July 2005, troops from the United Nations Stabilization Mission in Haiti (MINUSTAH) reportedly used tear gas, along with helicopters, tanks and machine guns, in a relatively aggressive operation that the UN claimed was "designed to rout gangs", but that human rights NGOs claimed was a massacre of unarmed civilians in one of Port-au-Prince's poorest neighbourhoods, Cité Soleil.²⁰⁶

3.5. Use of RCAs by non-State actors

It has proven difficult to obtain accurate information regarding the nature and scale of RCA use by non-State actors, as publicly available data in this area does not appear to be systematically compiled, is largely anecdotal and is problematical to verify. However, some limited analysis has been undertaken by the Centre for Non-Proliferation Studies (CNS) as part of their research of incidents involving sub-national actors and chemical, biological, radiological and nuclear material.²⁰⁷ In their initial study, CNS researchers recorded 29 cases for 1999 involving tear gas or pepper spray, which occurred in 14 countries and involved a diverse range of actors.²⁰⁸ In their subsequent study for 2000, this number rose slightly to 33 cases which occurred in 22 countries.²⁰⁹ On initial viewing, the CNS reports appear to indicate a lower but at least comparable geographic spread of instances of misuse by non-State actors to that of State law enforcement officials. However, on further analysis of this data it appears that the majority of the cases contained in both reports were either solely for possession of RCAs, hoaxes or were relatively small-scale use of RCAs by individuals for criminal, political or personal motivation. However, a minority of cases were serious, involving the use of RCAs against children,[210] the old or ill; in enclosed spaces resulting in injury or death;[211] or in conjunction with firearms.

3.5.1. Use of RCAs by armed opposition groups

There are relatively few publicly documented instances of armed opposition groups employing RCAs against State military or security forces. For example, in the CNS 1999 and 2000 chronologies only two cases are cited of the use of tear gas by armed opposition groups – an attack by Fuerzas Armadas Revolucionarias de Colombia (FARC) in Colombia,[212] and by Hezbollah in Lebanon.[213] In addition, two further cases are cited of possession of tear gas, both in 2000: by an armed opposition group in Fiji and by Hezbollah in Turkey.[214] Only one further credible case (detailed in Section 3.5.1.1) has been uncovered by the author following a preliminary investigation of media reports from 2000 to 2014.[215] This limited number of credible cases is somewhat surprising given the potential utility of such agents (e.g. for causing confusion amongst the enemy and acting as a force multiplier) in the types of small-scale military operations likely to occur during internal armed conflicts.

3.5.1.1. Reported use of tear gas by armed opposition group in Sri Lanka

In September 2008, the Sri Lankan Government claimed that the LTTE (the Tamil Tigers) used RCAs in defending their positions against attacks by Government forces. According to a statement from the Government's Media Centre for National Security: "A few soldiers in the Wanni [region of northern Sri Lanka] during their encounters with the enemy in the most recent past developed breathing difficulties following emission of a gas, believed to have been directed by the Tigers towards the advancing troops."[216] Army Commander Lt General Sarath Fonseka claimed that the LTTE had fired two canisters of CS in the Akkarayankulam and Vannivilankulam sector. He said a number of soldiers suffered burning effects and some other effects due to firing of CS canisters. They later recovered and operations were not affected. Lt General Fonseka also stated that: "If the LTTE fires CS canisters at us we have the capability [to] fire more powerful gases against them as we have the mandate as a sovereign state to make use of these gases to curb any terror activity of the LTTE."[217]

3.5.2. Use of RCAs by private military and security companies

Over the last two decades, a number of countries have relied on a disparate range of private contractors to support military operations in conflict situations. Such private military companies (PMCs) have provided services that traditionally were performed by national military, security or police forces, including interrogation of detainees, protection of military assets, training of local armed forces, collection of intelligence and the performance of defensive and even offensive military activities. In addition, Governments or non-State entities such as mining companies have employed private security companies (PSCs) to guard key installations.[218] In carrying out such activities, PMCs and PSCs have utilized and sometimes misused RCAs – as illustrated in the case studies in Sections 3.5.2.1 and 3.5.2.2.

3.5.2.1. Reported use of RCAs by private military company in Iraq

In January 2008, the *New York Times* reported that personnel working with the PMC Blackwater Worldwide had released CS from a helicopter and an armoured vehicle,[219] temporarily blinding drivers, passers-by and at least ten US soldiers operating a checkpoint in Baghdad in May 2005. Officers from the US Army's Third Infantry Division who were affected by the RCA stated that there had previously been no evidence of violence at the checkpoint that might have triggered such a CS release.[220] Instead, they claimed that the Blackwater convoy appeared to be stuck in traffic and may have been trying to use the riot-control agent as a way to clear a path. It is unclear whether permission was given for Blackwater to deploy or use CS under its contract with the US State Department. According to the *New York Times:* "Blackwater says it was permitted to carry CS under its contract at the time with the State Department. According to a State Department official, the contract did not specifically authorize Blackwater personnel to carry or use CS, but it did not prohibit it."[221]

The incident clearly appears to have been an inappropriate use of an RCA. In addition, it raises concerns as to the range and nature of actions that Blackwater personnel were permitted to conduct under the US State Department contract and with what oversight. For example, were Blackwater personnel permitted to engage in activities amounting to counter-insurgency and, if so, what regulations were in place to ensure that RCAs were not employed in such actions?

3.5.2.2. Reported use of RCAs by private security company in Papua New Guinea

The Porgera Joint Venture (PJV) has operated the Porgera gold mine in Papua New Guinea, and employed a PSC to protect the mine and its employees. PJV personnel patroled and guarded the mine site, including the waste dumps around the mine. They have also reportedly apprehended and detained illegal miners. HRW interviewed 21 people who said that they had been beaten, shot at or tear gassed by PJV personnel in 2009 and 2010. PJV personnel have often faced violent situations that justified responding with force. But in some cases they appear to have used force in circumstances that were not permitted by the international principles that form the basis for their own rules of engagement.[222]

One young man told HRW that he was looking for rocks on the Anawe waste dump with his father when both were surprised by a group of PJV security guards who fired tear gas at them without warning. "When I saw them I tried to run away and they shot me with tear gas...I fell down on the hard rocks. My father also fell down on the rocks." He said the guards caught him and that six stood around him and his father, kicking the pair in the ribs, stomach and back with steel-toed boots as they lay helpless on the ground. He was imprisoned and then released after paying a K1,500 ($570) fine.[223]

The importance of addressing utilisation of RCAs by PSCs and PMCs is heightened given the likelihood that the employment of PMCs and PSCs by both Governments and non-State entities will continue and may well increase in the coming years. Concern is further heightened because PSCs and PMCs are often inadequately integrated into State command structures, and their accountability for breaches of international humanitarian law,[224] and international human rights law,[225] remains disputed. Attempts have been made by certain States to explore and describe the application of international law to PSCs and PMCs, most notably under the so-called Swiss Initiative, which resulted in the agreement of the "Montreux Document" in 2008.[226] Subsequently, in 2010, the International Code of Conduct for Private Security Service Providers (ICoC) was created, which defines industry rules and principles based on human rights and international humanitarian law. As of April 2013, over 600 PSCs had signed the ICoC.[227]

3.6. Conclusions

As part of the first stage of an HAC analysis of RCAs, an examination was undertaken into their properties, effects and the nature of their use. They were found to be potent sensory irritants normally of low lethality that produce dose and time-dependent acute site-specific toxicity. Targeted mainly towards the respiratory and mucosal surfaces, their short-term effects are well defined, although the potential for deleterious long-term effects is contested. Although a variety of RCAs were previously employed in armed conflict by a number of States, such use by the majority of States is now obsolete. There is a small discrete range of agents widely stockpiled and regularly utilized by law enforcement officials in the majority of States for dispersal of crowds and/or incapacitating individuals. Credible reports from the UN and reputable international NGOs indicate that the misuse of RCAs by law enforcement officials for human rights abuse is geographically widespread. In contrast, publicly available data on the scope and nature of use and misuse by non-State actors is limited and problematical to verify. RCAs can be considered as a relatively mature and established technology, although one that is subject to continuing potential development. State-sanctioned research into these agents and associated delivery mechanisms continues to be reported in a number of countries, and new agents have been developed, such as the Skunk malodorant now employed in Israel.

4
Means of Delivering or Dispersing Riot Control Agents

4.1. Introduction

Whilst Chapter 3 concentrated upon the properties, effects and patterns of employment of RCAs, a second dimension of weaponization, namely the potential means of agent delivery and dispersal, needs to be examined in order to develop appropriate and effective regulatory mechanisms. As part of the HAC analytic process, this chapter will firstly explore the range of delivery mechanisms currently developed and employed in law enforcement, highlighting the current nature of use and potential misuse. The chapter will then explore the diversity of contemporary and emerging "wide area" means of delivery that have been developed or are under development for the dispersal of RCAs over wide areas or extended distances, highlighting the potential dangers in the proliferation and misuse of such devices.

4.2. RCA means of delivery with limited dispersal area deployed in law enforcement

4.2.1. Types of commonly employed RCA means of delivery

Although RCAs are commonly called "tear gas", this is a misnomer as none of these agents are gaseous at room temperature: OC is a crystalline to waxy compound, whilst CN, CS, CR and PAVA are crystalline solids at room temperature.[1] Consequently, for their effective dispersal and delivery they must either be converted into a particulate aerosol or delivered via a carrier liquid. A number of dispersal mechanisms have been developed for the delivery of relatively limited amounts of RCAs, in low concentrations, over discrete areas, as explored in Sections 4.2.1.1 and 4.2.1.2.

4.2.1.1. Hand-held liquid spray, cloud and fog dissemination
RCAs can be delivered as a liquid spray or foam, having been dissolved in a suitable solvent, the resulting irritant is then sprayed or expelled from a

range of devices. These include hand-held sprays (aerosols) which usually contain OC or PAVA, though some models utilise CS, CN, CR or a mixture of such agents. Typically, they range in capacity from a few millilitres up to 2 litres. Whilst the smaller sprays typically have maximum ranges between 3 and 5 m, certain larger hand-held devices have a maximum range of approximately 10 m. Hand-held spray devices are manufactured throughout the world and are widely employed by law enforcement officials to incapacitate individuals or small groups, and some forms are promoted to the general public for personal protection.[2]

RCAs can also be mixed with a solvent to produce a dense vapour fog or irritant cloud which is dispersed via hand-held "foggers". These devices, which have been predominately manufactured in the United States, and promoted for law enforcement, can disperse OC, CS or CN.[3]

4.2.1.2. Hand-thrown and weapons-launched single projectiles

RCA grenades and canisters which are activated by a fuse can be hand-thrown or weapon-fired via grenade launchers or conventional shotguns and rifles fitted with "launching cups". When hand-thrown their range is approximately 25–40 m, and between 50 and 300 m when weapons launched.[4] The RCA is dispersed via hot gases and smoke produced through pyrotechnic burning of the payload. Manufactured worldwide in countries such as Argentina, Brazil, China, India, South Korea and the United States, RCA grenades and canisters most commonly contain CS but some are available containing CN, CR and OC.[5]

RCA cartridges consist of a cartridge case housing either a projectile containing RCA which is expelled, or the RCA load itself. Depending on the nature of the cartridge, they can be fired from a variety of weapons including specially designed 37 mm or 40 mm launchers or 12 gauge shotguns. RCA projectiles generally have a range of up to 100 m and, once fired, a pyrotechnic charge initiates burning to expel the RCA, or a blast charge initiates to disperse powdered irritant agent. In contrast, "muzzle blast" cartridges expel powdered RCA directly from the muzzle of the weapon and are designed for firing at close range, up to approximately 10 metres. RCA projectiles and cartridges most commonly contain CS but are also available containing CN, CR and OC.[6]

Launched RCA grenades and canisters (and many additional RCA projectiles) are designed to be fired into the ground, where they release the RCA into the surrounding area. In contrast a range of "direct impact" RCA projectiles have been developed which are designed to be fired directly at an individual. On impact, these projectiles release a small localized burst of powdered RCA, thereby combining a kinetic impact with an irritant effect. Similarly, frangible direct impact RCA projectiles, typically containing OC or PAVA, have also been designed to be fired at an individual using a high-pressure air gun.[7] One range of such projectiles, widely employed by law enforcement and correctional personnel, particularly in the United States, has been developed by US company PepperBall Technologies, comprising a small gel ball containing powdered PAVA.[8] An alternative design developed by Belgian company FN Herstal for use with its FN 303 Less Lethal Launcher is composed of a fin-stabilized polystyrene projectile body with a non-toxic bismuth forward payload container. On impact the payload container splits open releasing the RCA or other potential payloads.[9]

A range of delivery devices and systems which have a narrow dispersal area and emit a limited quantity of RCA are widely employed by law enforcement officials. If such devices have been properly tested and trailed, their use should not raise undue concerns as long as it is in strict accordance with relevant human rights standards (as discussed in Chapter 8) and is effectively monitored.

However, the kinetic impact safety of certain weapons-launched pyrotechnic RCA grenades and canisters and other RCA projectiles not specifically designed for direct impact upon individuals may be of concern given limitations in their accuracy and their potential to cause trauma. Furthermore, the direct firing of such RCA devices at individuals has led to a number of serious injuries and deaths, particularly during public order situations (as illustrated in sections 4.2.1.2.1–3, and 8.7 of this publication). Between 1st January 2009 and 31st December 2013 individuals were reportedly struck – and in some cases wounded or killed – by RCA means of delivery fired by law enforcement officials in Algeria,[10] Bahrain,[11] Bangladesh,[12] Bolivia,[13] Egypt,[14] Greece,[15] Haiti,[16] Honduras,[17] India,[18] Iran,[19] Israel/Occupied Territories,[20] Malaysia,[21] Tanzania,[22] Thailand,[23] Togo,[24] Tunisia[25] and Turkey[26]. In addition, in isolated cases, the inappropriate employment of certain pyrotechnic RCA means of delivery against individuals in enclosed spaces has resulted in their death as a result of fires, for example in Jamaica[27] and Kenya.[28]

4.2.1.2.1. Turkey – serious injury caused by impact of RCA means of delivery. Turkish and international non-governmental human rights organizations have documented numerous allegations that police officers repeatedly employed tear-gas launchers as weapons – firing tear-gas canisters horizontally and directly at suspected demonstrators – during the 2013 Gezi Park protests. A significant proportion of people hurt at the scene of demonstrations received injuries through being struck by tear gas canisters or cartridges, many of them fired at close range. In its December 2013 "Medical Evaluation of Gezi Cases", the Human Rights Foundation of Turkey reported that, of the applications for treatment/rehabilitation made to their foundation, 40% were due to injuries caused by tear gas canisters or cartridges, a far higher proportion than they had recorded after previous protests.[29] HRW documented the cases of dozens of protesters seriously injured by such police tactics, including Burak Ünveren, a 31-year-old university lecturer at

Yıldız Technical University, who was shot directly in the face with a "tear-gas bomb": "I fell down, I never lost my consciousness. Some people I didn't know came and saved me, simply grabbed me, found a car and took me to a hospital." According to Mr Ünveren: "The second protester who grabbed me said 'Look at my face, boy, you have lost your eye. I also don't have one of my eyes. Life continues with one eye'."[30] Whilst he was in hospital, Mr Ünveren discovered three other patients who had similarly lost eyes as a result of injuries caused by police firing tear-gas canisters directly at the crowd.[31]

4.2.1.2.2. Iran – death due to kinetic impact of RCA means of delivery. The UN Special Rapporteur on Extrajudicial, Summary or Arbitrary Executions[32] and AI[33] have detailed the use of excessive force (including misuse of RCA means of delivery) by plain-clothed and armed personnel thought to be (pro-Government) Basij militia against those who took part in the 2009 pro-democracy demonstrations, despite the fact the protests were entirely peaceful. AI detailed testimony of a medical doctor present at a pro-democracy demonstration held in Tehran on 20th June 2009: "I saw from inside the ambulance how opposite the Navvab metro station, a couple of Basijis with Kalashnikov and G3 guns were shooting directly at people from the roof of Lolagar mosque." The doctor stated that he saw "only some meters from me, a young man was hit in the throat by a tear gas bullet that had been fired directly and purposely at him. Blood spurted from his throat and he dropped to the ground and died."[34]

4.2.1.2.3. Turkey – death due to kinetic impact of RCA means of delivery. A May 2014 report co-authored by the International Federation for Human Rights (FIDH), the Human Rights Association (IHD) and the Human Rights Foundation of Turkey (HRFT) documented wide-ranging human rights violations committed by Turkish law enforcement personnel, including serious injury and death from the misuse of tear gas and related means of delivery.[35] Among the cases highlighted was that of 14 year old Berkin Elvan. At around 7AM on the morning of June 16th 2013, Berkin left his house in Okmeydanı, a neighbourhood of Istanbul, to buy bread for the family breakfast, from a nearby shop. A large demonstration had taken place the previous night and police cars still remained in the area. While he was approaching the bakery, Berkin was hit in the back of the head by a tear-gas grenade, reportedly fired by a policeman who gave no warning before shooting. Berkin walked a few steps, vomited and collapsed unconscious. Although attempts were made to seek medical assistance through the emergency phone line, no medical support was sent to the scene, instead the response was "you did the demonstrations, you are responsible". The family rushed Berkin to hospital where medical personnel assessed that he had suffered from a skull fracture and brain haemorrhage. He remained in a coma for 269 days, until his death on 11th March 2014.[36]

4.2.1.3. Fixed-installation devices

A specific range of delivery mechanisms of potential concern are those fixed-position RCA "fogging" devices, spray dispensers or smoke generation units that have been developed and promoted for indoor installation in prisons, correctional centres or other places of detention. The placement of such devices in confined spaces or poorly ventilated rooms, or their use in situations where prisoners, detainees or other targets cannot leave the contaminated area rapidly due to limited exit routes, could result in the build-up of toxic chemicals, potentially leading to serious injury or death. The employment of such devices in larger enclosed areas such as prison halls also has the potential to lead to injuries resulting from panic and stampedes.[37]

In April 2015, AI and the Omega Research Foundation raised concerns that fixed-installation RCAs devices have "the potential for use in [an] arbitrary and indiscriminate manner that could amount to torture or other ill-treatment". Consequently, they have recommended "Ban[ning] fixed installation chemical irritant dispensing devices in places of detention or for use in other law enforcement contexts and from corresponding production and transfer."[38]

4.3. RCA means of delivery with a wide dispersal area for use in large-scale public order law enforcement

4.3.1. Overview

In contrast to the foregoing, a range of delivery mechanisms have been developed for crowd control and dispersal that deliver far larger amounts of RCAs over wider areas than could be delivered by hand-held sprays, individual hand-thrown grenades and canisters and other individually launched projectiles. Such devices include certain multiple RCA projectile launchers, large backpack or tank sprayers and fogging devices, water cannons and RCA dispensers attached to helicopters. Although certain types of such devices may have potential utility for use in law enforcement situations where there is large-scale disorder and risk of serious violence, for example, from rioting mobs, a number of these devices raise questions about the feasibility of their discriminate use, with the consequent danger of affecting bystanders.

In addition, given the potential quantities of agent dispersed by some of these mechanisms, questions arise as to their proportionality as area clearance devices, and also regarding the potential danger of serious injury due to agent toxicity. Such concerns are exacerbated by the fact that, unlike RCAs delivered through grenades and canisters, these devices generally deliver significant amounts of RCA in liquid or powder form, which will adhere to subjects and, without decontamination, will continue to deliver pain and irritation even when the subjects have moved away from the area. Certain forms of such "wide area" dispersal mechanisms may also be more open to intentional misuse by law enforcement officials on a large scale than are "narrow area" dispersal mechanisms. For example, the excessive or inappropriate use of water cannon-dispersing RCAs against crowds and demonstrations has been reported in a number of countries including Kenya,[39] Malaysia,[40] and Turkey,[41] while the use of helicopters to dispense RCAs to facilitate human rights abuses in Côte d'Ivoire,[42] South Korea,[43] Peru,[44] and Zimbabwe[45] has also been documented.

4.3.1.1. Concerns relating to the employment of water cannon to deliver RCAs

Water cannons are normally mounted on heavy land vehicles, though certain models can be attached to buildings, boats, or helicopters, and smaller portable backpack designs are also available. They are essentially high-pressure pumping systems designed to shoot jets of water at people. The pressure of the water can be varied from low pressure to soak and consequently deter or demoralize the target, to high pressure designed to impart blunt kinetic effects. Certain water cannon can be enhanced to fire small volumes ("bullets", "pulses", or "slugs") of water. The water can have additives, including marker dye (for later identification of individuals) or a range of RCAs.[46]

Many models of water cannon are manufactured with two or three separate tanks: the largest for the water and then one or two smaller additional tanks which can contain RCAs (predominantly CS or OC) and/or dye marking products, which are mixed with the water for dispersal. As most of the RCAs employed are unstable and insoluble in water, the chemical agent is injected into the water stream at the last moment. Water cannons with tanks for RCA dispersal are manufactured in a range of States including China, Israel, South Korea, the Russian Federation and the United States.[47]

In April 2015, AI and the Omega Research Foundation raised concerns about the inherently indiscriminate nature of these devices, which can affect bystanders as well as intended targets; in addition, "the use of a mixture of water and chemicals makes it impossible to deliver accurate targeted doses of the irritant".[48] Further concerns relate to the kinetic impact of the water, which can knock a person to the ground, push them into fixed objects or pick up loose objects and propel them as missiles. The eyes are particularly vulnerable to direct water impact, which can potentially result in severe injuries and permanent loss of sight.[49]

4.3.1.2. Turkey – Injuries from water-cannon dispersing unknown RCAs

AI has documented the inappropriate use of water cannons against peaceful demonstrators participating in the Gezi Park protests of 2013. This has included the apparently punitive employment of water cannon against those fleeing police, and the arbitrary use of such devices against demonstrators and bystanders alike at, or close to, the scene of demonstrations.[50] Furthermore, AI has claimed that there is "strong evidence" to suggest that RCAs were added to water used in water cannons during Gezi Park protests, and that this "add[ed] to the injuries resulting from the arbitrary use of water cannons".[51] Doctors reportedly informed AI that "water fired from water cannons had caused skin irritation and first degree burns". Furthermore, the Governor of Istanbul reportedly acknowledged that "medication" had been added to the water but denied that chemical materials had been added.[52] According to *Hürriyet Daily News*, the company that manufactured RCAs intended to be mixed with the water stated that the burns may have been caused by too high concentrations of the chemical irritants being mixed with the water.[53] Individuals hit by pressurized water at an incident at the Divan Hotel in Istanbul informed AI that, they suffered an immediate burning sensation which was accompanied by reddening of the skin that lasted for hours after the event.[54] A foreign journalist told AI that she suffered burns after being hit by pressurized water at the Divan Hotel and had not been previously exposed to tear gas. In a separate incident, an individual present at the Point Hotel in Taksim informed AI that, immediately after the pressurized water was sprayed at the doors of the hotel, the people in the lobby started coughing and gasping for air.[55]

4.3.1.3. Zimbabwe – Inappropriate use of helicopters to disperse RCAs
HRW has documented the use by military forces of RCAs in conjunction with live fire against unarmed civilians engaged in illegal mining activities in the Marange diamond fields. The military operation commenced at 7 a.m. on 27th October 2008, with five military helicopters carrying mounted automatic rifles flying over the fields around Chiadzwa, driving out the local miners. From the helicopters, soldiers indiscriminately fired live ammunition and tear gas onto the diamond fields and into surrounding villages. A ground operation with over 800 soldiers was also initiated.[56] One local miner caught up in the operation on the first day told HRW: "I first heard the sound and then saw three helicopters above us in the field...soldiers in the helicopters started firing live ammunition and tear gas at us. We all ...began to run towards the hills to hide...there were many uniformed soldiers on foot pursuing us. From my syndicate, 14 miners were shot and killed that morning."[57] According to HRW, the military operation continued every day for the next three weeks until 16th November 2008. Military helicopters would fire tear gas and live ammunition from the air to support soldiers shooting at miners on the ground.[58]

4.4. Survey of the development and marketing of "wide area" RCA means of delivery of potential concern

Research undertaken by the author (much of it in collaboration with colleagues from the Omega Research Foundation[59]) has detailed the development, testing, production, possession or promotion by State or commercial entities in at least 15 countries of over 40 "wide area" RCA means of delivery since the coming into force of the CWC in 1997. These "wide area" RCA means of delivery have included large RCA "smoke" generators and irritant sprayers; multiple munition launchers; rocket-propelled grenades (RPGs); automatic grenade launchers; mortar munitions; large-calibre projectiles; heliborne munition dispensers; cluster bombs; and RCA launchers fitted to unmanned aerial and ground vehicles – illustrative examples of which will be explored in sections 4.4.1–7. As will be discussed in Chapters 5 and 8, the international regulation of "wide area" RCA means of delivery currently appears to be inadequate. The potential consequences of the uncontrolled proliferation and misuse of certain "wide area" RCA means of delivery include:

(a) *Employment of inherently inappropriate munitions in law enforcement:* Whilst some types of "wide area" RCA means of delivery may have possible utility in large-scale public order law enforcement, other categories appear to be inherently inappropriate for such employment. Such means of delivery could never legitimately be used for law enforcement due to the dangers of serious injury or fatality to the targets and/or to uninvolved bystanders. Of particular concern are delivery mechanisms that disperse RCAs in quantities that entail a serious risk of asphyxiating or poisoning the targets; conflict with the principles of proportionality and discrimination with regard to the use of force in law enforcement; or risk causing serious injuries or death due to their design or physical characteristics not directly related to RCA toxicity, for example, consequences of high velocity impact with munition casing, sub-munitions, components or shrapnel.

(b) *Proliferation to and misuse by non-State actors:* Although, to date, there have been few confirmed cases of the use of RCA means of delivery by non-State actors in armed conflict, the current commercial availability of a wide range of such means of delivery raises the danger of their acquisition and employment by a range of non-State actors, including opposition armed forces, unregulated PMCs, PSCs and terrorist organizations.

(c) *Misuse to facilitate "large-scale" human rights abuses:* As discussed, the inappropriate employment of certain "wide area" RCA means of delivery potentially facilitates human rights abuses on a far greater scale. This could include the blanket application of RCAs against large peaceful gatherings or demonstrations, resulting in *en masse* infliction of cruel, inhuman or degrading treatment or punishment. Alternatively, such means of delivery could be employed as a "force multiplier" in conjunction with firearms, leading to potential violations of the principle of proportionality in the use of force in law enforcement.

(d) *Employment in armed conflict:* There is a long history, dating back to World War I, of the development and/or use of "wide area" RCA means of delivery by State military forces in armed conflict. Of particular note has been the extensive employment by the United States of "wide area"

RCA means of delivery in its conflict with Vietnam in the 1960s and early 1970s;[60] and the development in the 1980s of large-calibre RCA munitions by Iraq intended for use in its war with Iran.[61] In previous conflicts, "wide area" RCA means of delivery were employed to drive enemy combatants from entrenched, underground, enclosed or fortified positions; for subsequent area denial; to disable and incapacitate large numbers of combatants; or in conjunction with conventional arms as a "force multiplier". More recently, a range of contemporary "wide area" RCA means of delivery have been promoted for use in counter-insurgency operations or urban warfare.

(e) *Potential use in, or to conceal, chemical weapons programmes:* A range of "wide area" delivery mechanisms such as cluster munitions, mortar shells or large-calibre projectiles that are ostensibly designated as RCA munitions could instead be filled with other toxic chemicals and employed to disperse agents such as the incapacitating chemical agent BZ or "classic" chemical warfare agents. Given the limited declaration and transparency mechanisms applicable to RCA munitions under the CWC, there is a danger that certain States might seek to hide illicit chemical weapons production under the guise of law enforcement programmes.

4.4.1. High capacity RCA smoke generators, foggers and spraying devices

A range of high capacity RCA smoke generators, foggers and large back pack or tank spraying devices have been developed, some of which have the capacity to deliver significant amounts of RCAs over a wide area, potentially affecting a large number of people.

4.4.1.1. WDTC-Q38 trolley-style riot control sprayer (China)

Henan Weida Military and Police Equipment Co. Ltd has developed and promoted the WDTC-Q38 trolley-style riot control sprayer. According to the company's product manual, this sprayer employs "high pressure gas" to disperse "the stimulating agent [spread] quickly" over a large target area.[62] The "smoke-like small particle stimulating agent can be suspended in the air" two metres above the ground for a sustained period and with the aid of wind flow, can continue to spread.[63] The WDBF-Q38 can hold 18 kg of agent powder and can spray a continuous jet of agent for 76 seconds, it has a maximum range of 50 m and a coverage area of 2000 m² or more.[64] No information is currently available regarding the specific riot control agent and concentration employed in the WDTC-Q38. According to the company, the WDTC-Q38 "can be used to dispose of large scale emergencies" and in "counter terrorism operations".[65]

4.4.1.2. Afterburner 2000 smoke and RCA dispersal system (the United States)

Marketing material produced by US manufacturer MSI Delivery Systems Inc. has described the Multi-Mission Afterburner 2000 Aerosol Delivery System (AB2K-MMADS™) as a "robust multi-mission, multi-purpose smoke generator capable of rapidly blanketing large areas with dense smoke. The smoke solution can be mixed with specific chemicals to upgrade the mission requirements...".[66]

Company information detailing mission-specific formulations has stated that the Afterburner 2000 is capable of "dispensing many less-than-lethal formulations in a high density aerosol form". This included *"Standard non-toxic training smoke mixed with irritants such as OC, CS, or Pepper [that] upgrades the capabilities to include:* Crowd Control and Civil Unrest, SWAT Teams and Tactical Incursions, Corrections Dept. (Riots/Prisoner Extraction), Less-lethal Terrorist Suppression, *Urban Warfare (MOUT/COIN)* ..." (emphasis added).[67]

According to the company, the Afterburner 2000 can release over 1,500 cubic feet of smoke with a range greater than 100 feet (30 m) in one second.[68] The marketing material stated that "[the] standalone version" of the Afterburner 2000 "expels 50,000 cubic feet (1,416 cubic meters) of smoke on a single charge", whilst the "dependent version with high-capacity backpack expels 320,000 cubic feet (9,061 cubic meters) of smoke on a single charger".[69] According to the manufacturer's website, the Afterburner 2000 can "[be] mounted on walls, buildings and fixed on stationary structures on vehicles including small craft, military style boats, US military, law enforcement and Homeland Security vehicles, Humvees, riot control and other armor piercing vehicles".[70]

In addition, the manufacturer has stated that the Afterburner 2000 can also be "incorporated onto unmanned ground vehicles and aircraft, drones for deployment dependent upon size", and that it "has been tested for compatibility with military small craft boats".[71] A variation of the Afterburner 2000, called the AB2K-Robot Smoke Generator (AB2K-RSG), has also been "specifically designed by request for use with the Andros™ F6B Robot by Remotec a subsidiary of Northrop Grumman. The unit can be operated remotely from up to 4 miles distance. The unit is also tested and compatible with Black-I Robotics products."[72] Although there is no information publicly available concerning which (if any) law enforcement and military entities in the United States or elsewhere have purchased this product, the manufacturer has stated that "MSI Delivery Systems Inc... has commenced volume production and sales".[73]

4.4.2. Multiple "less lethal" munition launchers

A range of multiple launchers have been developed, some intended solely for firing RCA munitions, whilst others are capable of employing a variety of "less lethal" projectiles. Although these launchers can fire a limited number of projectiles in a fairly targeted fashion, they do also have the capability to rapidly discharge large salvoes of RCA projectiles and can be employed to blanket wide areas, cumulatively delivering significant amounts of RCAs and potentially affecting large numbers of people. They vary in the maximum number of projectiles launched, rapidity and mode of fire, range and area coverage, as well as in terms of the calibre, weight and agent fill of the munitions utilized.

4.4.2.1. IronFist 38 mm and Cobra 40 mm non-lethal weapon systems (the United States)

According to marketing material distributed in 2013 by US manufacturer, NonLethal Technologies,[74] the IronFist is "a new 38 mm weapon system with up to 36 barrels... [intended] to rapidly deploy a blanket of less lethal munitions into, or over, a hostile crowd".[75] The IronFist can employ "standard conventional 38 mm less lethal CS, flashbang, and colored smoke rounds... or... NonLethal Technologies' specially designed 10 inch 38 mm rounds with higher capacity CS...".[76]

When the IronFist 36 barrel system is loaded with high-capacity 10 inch CS rounds (each with 7 mini-grenades), *"it can rapidly deploy over 250 mini-grenades into the crowd within 2 minutes* from... up to 150 metres. *Two such configured systems mounted on one armoured vehicle can deploy over 500 CS mini-grenades*, or a mix of CS mini-grenades and flashbang-distraction projectiles downrange in that... time... now that is nonlethal firepower!" (emphasis added).[77]

The IronFist system is designed to be hard-mounted on a wide range of land vehicles and marine vessels, or to permanent structures "such as prisons, government buildings, *military base perimeters*, or embassy compounds" (emphasis added).[78] Nonlethal Technologies, Inc. has subsequently developed a 36 rifle-barrelled variant to this system, called Cobra40, which is capable of firing a range of low-velocity lethal 46×40 mm munitions for use against "enemy combatants".[79] When the "threat is one of civil unrest", Cobra40 can instead be "loaded with any of the [company's] less lethal 40 mm rounds, or Hi-Load™ 37/38–40 mm rounds", including a range of RCA munitions.[80] The company has subsequently promoted both IronFist and Cobra40 at the 2015 International Defence Exhibition and Conference (IDEX) held in the United Arab Emirates.[81]

4.4.2.2. VENOM 37 mm non-lethal tube launched munition system (NLTL/MS) and VENOM multi-caliber (37 mm, 40 mm and 66 mm) launching system (the United States)

The VENOM launcher has been developed by Combined Systems, Inc. (CSI) for use by a range of military, security or police forces in a variety of scenarios. According to CSI, VENOM is a high capacity non-lethal tube launched munition system (NLTLMS) which can be deployed remotely or on a wide range of land vehicles and marine vessels. The system is "ideal for warning signals at vehicle check points and for determining intent of approaching marine vessels." It is also an "effective force multiplier ... capable of precise delivery of munitions while enhancing the capabilities of area denial and force escalation in riot control situations."[82]

The VENOM is a modular launching system which accepts three cassettes, each loaded with ten 37 mm cartridges.[83] The 30 cartridges are contained in three levels at varying degrees elevation, and can be fired in immediate succession. Each cartridge is assigned an IP address allowing individual cartridge or desired sequence firing from a fire control panel, communicating via cable or wireless device. VENOM is capable of delivering a variety of "non-lethal" payloads including 37 mm "multi-7 smoke CS" munitions which have a maximum range of 150 m.[84]

CSI subsequently developed the VENOM Multi-Caliber (MC) Launching System and began promoting this product from January 2013. The VENOM MC is "modular and is available in single, double and triple bank configurations. Each bank can launch ten 37 mm or 40 mm grenades, and five 66 mm grenades".[85] CSI has promoted a range of associated payloads which "run the spectrum of non-lethal responses from flash and sound distraction, smoke obscuration, fast obscuration, smoke irritant, and blunt trauma individual or combination effects to OC vapor grenades."[86] According to the company: "These effects support escalation of force, early warning & determination of intent, crowd dispersal and area denial objectives."[87] In addition to the 37 mm Multi-7 CS smoke sub-munition, these payload options include a super long range 40 mm CS smoke sub-munition, with a maximum range of 450 m; a 66 mm high capacity CS smoke grenade, with a maximum range of 200 m; and a 66 mm high capacity OC vapour grenade, with a maximum range of 150 m.[88]

Although details of US or foreign military, security or police forces that have acquired VENOM are scarce, a contract to provide the US Marine Corp with 225 VENOM launchers and 75,000 (non-RCA) flash-bang stun munitions was awarded in June 2011[89]; and according to the company, the launchers were "deployed in Iraq by the US Marine Corp."[90] Furthermore, a 2013 report by B'Tselem has documented the employment of the VENOM system and associated RCA munitions by Israeli security forces in the West Bank since September 2011.[91]

4.4.3. Automatic Grenade launchers

Certain automatic grenade launchers can utilize a range of "less lethal" rounds including RCA projectiles. Given their high rate of fire, they are potentially capable of blanketing wide areas, cumulatively delivering significant amounts of RCAs and potentially affecting large numbers of people.

4.4.3.1. 30 mm grenade round and launcher (Russian Federation)

According to the 2009 English language version of the 2006 *"Ordnance and Munitions"* volume of *Russia's Arms and Technologies*,[92] a Russian company developed a 30 mm grenade round filled with irritant action pyrotechnic composition designed for the AGS-17 automatic grenade launcher. The 30 mm munition weighs 350 grams, and when employed in the AGS-17 has a maximum firing range of 1,700 m and a maximum rate of fire of between 350 and 400 rounds per minute. Reportedly, it can be used to "temporarily incapacitate armed lawbreakers on the open or rough terrain and those hiding in buildings, various structures and vehicles". Furthermore, "[i]t can also be used to harass armed offenders".[93]

4.4.3.2. 38 mm and 64 mm automatic grenade launchers (China)

In 2012, China Ordnance Equipment Research Institute (No. 208 Research Institute of China Ordnance Industries) began to promote a vehicle-mounted 38 mm automatic grenade launcher and a 64 mm automatic riot grenade launcher.[94] These two weapon systems were designed to "cope with (large-scale) mass events quickly and effectively"; they can be mounted on land vehicles or naval craft and can be remotely controlled by an operator inside the vehicle.[95] Both types of launcher are loaded by a belt-fed system and "the firing rate can be switched between single fire, interrupted fire, and continuous fire".[96] The 38 mm launcher has a maximum firing rate of 200 grenades per minute and a maximum effective range of at least 300 m. It is compatible with a range of 38 mm grenades, including 38 mm tear-gas grenades as well as smoke and stun grenades.[97] The 68 mm launcher has a maximum firing rate of 60 shots/minute and a maximum effective range of 600 m, which is, according to the company, "much further than similar equipment in the domestic and foreign market".[98] It is compatible with a range of 64 mm grenades, including 64 mm tear-gas grenades, smoke grenades, stun grenades and explosive dye projectiles.[99]

4.4.4. Rocket-propelled grenades

4.4.4.1. 105 mm munitions for RPG grenade launchers (Russian Federation)

According to the 2009 English language version of the 2006 *"Ordnance and munitions"* volume of *Russia's Arms and Technologies*,[100] a Russian company developed an RPG-7 grenade launcher round with a warhead filled with irritant-action pyrotechnic composition. The round is available in two models: one-piece and clustered. It was developed from the standard round fired by the RPG-7 grenade launcher. This 105 mm calibre munition weighs 4.3kg and has an effective range of between 400 and 600 m. A variant obstacle-penetrating 105 mm grenade filled with irritant-action pyrotechnic composition was also developed. It can be used to "suppress and temporarily incapacitate armed lawbreakers located in light field shelters, bunkers and city buildings... [and] to harass... armed offenders and as *an antitank weapon*" (emphasis added).[101]

4.4.5. Mortars, associated munitions and other large-calibre projectiles

A range of mortar munitions have been developed, either specifically designed to carry RCAs or else capable of carrying a variety of potential "less lethal" payloads. Such munitions can deliver significant quantities of RCA over wide areas and/or extended ranges, potentially affecting large numbers of people. They vary in terms of their calibre, weight, design, material construction, potential payloads, area coverage and range, as well as the purposes for which they have been promoted.

4.4.5.1. 82 mm PP87 tear-gas mortar projectile (China)

In 2002, marketing material from the Department of Scientific and Technical Development, Chinese People's Armed Police Force promoted the "82 mm PP87 tear gas mortar bomb":

> The 82 mm PP87 mortar-throwing tear bomb is made of non-metal shell body material, and is launched from the 82 mm mortar. It is powerful, reliable, user-friendly, and has a long firing range. The bomb can be detonated in the air, and spray lacrimatory agent in the shrapnel mode. It exerts effects over a large area, receives little influence from weather conditions, and will not inflict fatal injury on human body. It is a desirable device to remotely disperse large mass target.[102]

According to the company marketing materials, the "mortar bomb" weighs 1.35 kg has a range of between 200 and 350 m and has an effective area of coverage of more than 2,000 m². It is designed to explode at a height 30–50 m above the target. The "lacrimatory agent" is identified as CS, though no details of the total amount or concentration of the agent contained in each "bomb" are disclosed.[103] A second Chinese company, China Ordnance Industry Group, State-owned No. 672 Factory, also promoted what appears to be the essentially same projectile in marketing material distributed at the 2011 China (Beijing) International Exhibition and Symposium on Police and Anti-Terrorism Technology and Equipment (CIPATE 2011).[104]

4.4.5.2. 82 mm and 120 mm mortar projectiles (Russian Federation)

A Russian company reportedly developed an 82 mm mortar shell filled with irritant-action pyrotechnic composition for the Model 1937 and 2B14-1 mortars and the 2B9 automatic mortar. The mortar shell weighs 3.5kg and has a maximum firing range of 2,670 m. The round is available in two models: one-piece and clustered. It has been developed from the 82 mm standard mortar round using the S-8232S illumination shell.[105]

In addition, the company has also reportedly developed a 120 mm mortar shell filled with irritant-action pyrotechnic composition for Model 1938 and 2B11 mortars, and for 2S9, 2S23 and 2B16 artillery pieces. The mortar shell weighs 16 kg and has a maximum range of fire of 5.2 km (from 1938 model mortar), 6.8 km (from 2b11 mortar) and 6.6 km (from 2B16, 2S9 and 2S23 guns).[106] No further information concerning the manufacture, stockpiles, transfer or utilization of either the 82 mm or 120 mm mortar projectile has been made publicly available.

4.4.5.3. 120 mm mortar projectile (Turkey)

In November 2003, *Jane's Defence Weekly* reported that the Turkish (State-owned) arms manufacturer, Makina ve Kimya Endüstrisi Kurumu (MKEK) had developed a 120 mm mortar round – the CS MKE MOD 251 – filled with CS.[107] The CS MKE MOD 251 mortar round weighed 17.34 kg and had a maximum range of 8,132 m.[108] It was promoted by MKEK on their website,[109] and at international security exhibitions including the 7th International Defense Industry Fair (IDEF) held in Ankara, Turkey, in September 2005[110]; and at the Africa Aerospace and Defence (AAD) exhibition held in Cape Town, South Africa, in September 2010.[111]

In correspondence with the author and colleagues at the Omega Research Foundation and the Institute for Security Studies (ISS),[112] the Turkish Government stated that 1,000 CS MKE MOD 251 munitions had been produced in 1996, prior to Turkey's ratification of the CWC, with roughly 150 used for testing purposes during the initial R&D phase in 1997. The facility for their production was subsequently discontinued after 1997. At the time of ratification, there remained 850 pieces of CS MKE MOD 251 type munitions in the inventory of the Turkish Armed Forces. From that time and until 2011, Turkey stated that none of the remaining 850 munitions were used, but were stored at the Turkish Armed Forces ammunition destruction facility awaiting disposal. In July 2011, Turkey reported that such destruction had been completed.

4.4.5.4. XM1063 155 mm malodorant projectile (the United States)

According to information released by the US Government, General Dynamics Ordnance and Tactical Systems worked under the direction of the US Army's Armament Research, Development and Engineering Center (ARDEC) to develop a 155 mm artillery projectile called the XM1063.[113] According to General Dynamics, the XM1063 (also called the Non-Lethal Personnel Suppression Projectile) was designed to carry out three interrelated functions, to "separate combatants from non-combatants; suppress, disperse or engage personnel [and] deny personnel access to, use of, or movement through a particular area, point or facility".[114] The munition was intended to "Address...[the] need for Non-Lethal Options that is highlighted by current conflicts in Iraq and Afghanistan ... [the munition would] minimize...collateral damage, fatalities and permanent injury."[115]

The XM1063 was based upon the M864 artillery projectile,[116] and was intended to have a range of at least 20 km, and potentially up to 28 km.[117] The multiple sub-munitions would be released above the target area and then fall to the ground and disperse their payloads.[118] Estimates of the area covered vary between a minimum of 5,000 m^2,[119] to a reported maximum of 10,000 m^2.[120] Only limited details of the proposed payload have been made public, but the available documentation described it as a "liquid payload",[121] and a "non-lethal personnel suppression agent".[122] Payload agent effectiveness was apparently tested at Army Edgewood Chemical Biological Center,[123] indicating a chemical agent. Furthermore, a JNLWD reference book on "non-lethal" weapons, published in 2011, included a reference to a legal review conducted in 2007 of the "XM1063 Malodorant 155 mm Artillery Round"[124] that indicated that malodorant agents were considered for this munition.

According to a July 2008 article in the UK newspaper, the *Guardian*, testing of the XM1063 was completed successfully in 2007 and it was due for low-rate production from 2009.[125] According to the *Guardian*, ARDEC stated "that the production decision is on hold awaiting further direction from the program manager".[126] Information currently available from the General Dynamics website stated that "XM-1063 Non-Lethal Artillery has achieved TRL Level 6.1 through gun test firings as payload in 155 mm M483 rounds" and was "Prepared for Milestone B decision".[127] In his June 2012 *New Scientist* article, Hambling noted that although "the project is on hold, [it] has been developed by General Dynamics ... to the stage of test firings and could be reactivated".[128] No further information regarding the current status of the XM1063 R&D programme has been made public by the US Government.

4.4.6. Other aerial delivery munitions

4.4.6.1. Heliborne-dispensed RCA munitions and 500kg cluster munitions (Russian Federation)

According to the 2009 English language version of the 2006 *"Ordnance and munitions"* volume of *Russia's Arms and Technologies*,[129] a Russian company developed two aerial delivery munitions of particular concern. The first was a heliborne KMGV-type dispenser of packages of sub-munitions filled with irritant-action pyrotechnic composition. The publication stated that "These submunition packages are dispensed singly or all together from helicopters Mi-8MT and Mi-24 (four KMGV dispensers on external hardpoints) at an altitude of 50–300 m at a flying speed of 150–300 km/h. They can also be dropped in the helicopter hovering mode."[130]

In addition, the company also reportedly developed a 500 kg cluster bomb packed with sub-munitions charged with irritant-action pyrotechnic composition. The publication stated that:

> This cluster bomb has been developed from the standard 500 kg cluster bomb packed with smoke sub-munitions. It is dropped from a fixed-wing

or rotary-wing aircraft in an altitude span of 100–12,000 m at a speed of up to 1,200 km/h ... The bomb permits high concentrations of an irritant agent to be attained within a short time.[131]

No further information concerning the manufacture, stockpiles, transfer or utilization of either the heliborne-dispensed munitions or the cluster munition is currently publicly available.

4.4.7. Unmanned aerial vehicles and unmanned ground vehicles

4.4.7.1. *Skunk riot UAV (South Africa)*

The Skunk riot control copter is an unmanned aerial vehicle (UAV) which has been developed by South African company, Desert Wolf. According to the company website, as of August 2015, the UAV is designed to "control unruly crowds without endangering the lives of the protestors or the security staff".[132] It is equipped with four high-capacity paintball barrels which can fire solid plastic balls, dye marker projectiles or RCA pepperballs. Each barrel can fire up to 20 paintballs per second, consequently releasing "80 Pepper balls per second stopping any crowd in its tracks".[133] The company has stated that "the current hopper capacity of 4000 balls [combined] with [the] High Pressure Carbon Fiber Air system ... allows for real stopping power".[134] The UAV is also equipped with bright eye safe lasers, and on-board speakers enabling communication and warnings to the crowd. The company has also stated that the UAVs can be "operated in formation by a single operator [employing] the Desert Wolf Pangolin ground control station".[135]

The Skunk UAV was first promoted at the International Fire and Security Exhibition and Conference (IFSEC) in Johannesburg in May 2014 and subsequently at the AAD 2014 exhibition in September 2014. In an interview with the BBC, Desert Wolf's managing director Hennie Kieser stated that the company "received an order for 25 units just after [IFSEC]" from "an international mining house".[136] Mr Kieser claimed other potential customers included "Some mines in South Africa, some security companies in South Africa and outside South Africa, some police units outside South Africa and a number of other industrial customers."[137] According to media reports, in September 2014 and April 2015, Desert Wolf has been seeking to establish manufacturing facilities outside of South Africa to enable it to build at least a thousand Skunk UAVs a month, in response to what it claims is massive demand.[138]

4.4.7.2. *RiotBot (Spain)*

The Spanish company Technological & Robotics Systems (Technorobot) has developed the RiotBot – an "advanced security robot" especially designed for remote operation in areas considered to be too dangerous to deploy personnel.[139] Previous company promotional material stated that RiotBot employed a mounted PepperBall Tactical Automatic Carbine (TAC 700 launcher), which had been customized and adapted for use on the robot.[140] Subsequent information available on the Technorobot website, as of August 2015, has stated that "RiotBot employs a NLS 900 carbine modified and adapted for safe use in the robot with a shooting velocity of 900 balls per minute and a total capacity of 450 PAVA balls, a non-lethal ammunition. This carbine can only be shot using a remote control, which makes it impossible for unauthorized people to use it".[141]

Capable of speeds exceeding 20 km per hour, RiotBot can be deployed by a single operator either through direct viewing or through incorporated video equipment at distances of more than 1.5 km. The operator can remotely control the robot's movement, as well as the vertical and horizontal position of the gun turret before firing the carbine. RiotBot can be operated continuously for more than two hours.[142] According to Technorobot, RiotBot was developed for a "wide range of police, military and general security operations, mainly those in which the personal safety of the members of the intervention units is not fully guaranteed or could be in danger".[143] The company literature has stated that "some of the scenarios that have been studied for [RiotBot's] development include: 'Riot control...civil order...area denial...boundary defense and intervention...control point security...surrounding unit rescues ...*urban warfare*'" (emphasis added).[144]

4.4.7.3. Modular Advanced Armed Robotic System (MAARS) (the United States)
The US company QinetiQ North America, Inc. (QNA) has developed the Modular Advanced Armed Robotic System (MAARS), an unmanned ground vehicle developed through "partnership with various agencies in the Department of Defense".[145] It was "freshly created...to meet U.S. SOCOM [Special Operations Command] requirements".[146] MAARS has been "designed expressly for reconnaissance, surveillance, and target acquisition (RSTA) missions to increase the security of personnel manning forward locations".[147] According to the company, MAARS "provides remote options to commanders for reconnaissance, assaults, ambushes, hostage rescue, forced entry, booby-trapped areas, detainee riots, site security and improvised explosive device detection".[148] MAARS is "a-man-in-the-loop" system,[149] remotely controlled by an operator who can be "well over 1 kilometer"[150] and reportedly up to 3 km away.[151] According to QNA, MAARS can be "positioned in remote areas where personnel are currently unable to monitor their security, and can also carry either a direct or indirect fire weapon system".[152]

Furthermore, the company has stated that MAARS is the "first fully modular ground robot system capable of providing a measured response including non-lethal, less-lethal and even lethal stand-off capabilities".[153] As well as a M240B medium machine gun firing 7.62 mm ammunition, MAARS incorporates a four-barrelled 40 mm grenade launcher that has the capability to utilize either 40 mm high-explosive grenades or a range of "less lethal" ammunition, including 40 mm tear-gas grenades.[154] No details are available concerning the grenade launcher's rate of fire or range; nor of the area coverage, weight or fill of the 40 mm tear-gas grenades.

In June 2008, QNA announced that it had "shipped the first MAARS ground robot to the US military under a contract from the Explosive Ordnance Disposal/Low-Intensity Conflict (EOD/LIC) Program within the Combating Terrorism Technical Support Office (CTTSO)".[155] According to a November 2010 *New York Times* article, US Army Special Forces had bought six MAARS "for classified missions", and the National Guard had requested "dozens more to serve as sentries on bases in Iraq and Afghanistan".[156] Although there have been no subsequent reports of its field deployment by the US military, MAARS reportedly continues to be trailed and studied; for example in October 2013 at Fort Benning by the US Army Maneuver Center of Excellence's Battle Lab,[157] and in January 2015 by the US Marine Corps Warfighting Laboratory.[158]

4.5. Conclusions

There are certain means of delivery which have a narrow dispersal area, short range and emit a limited quantity of riot control agent – such as hand-thrown RCA canisters and grenades, hand-held spray disseminators and a range of individual weapons-launched projectiles. These "limited area" RCA means of delivery are widely employed in public order situations by law enforcement officials – potentially including police, security or military personnel. If such devices have been properly tested and trailed, their use should not raise concern, provided it is in strict accordance with the relevant human rights standards and is strictly monitored. However, if used by law enforcement officials in an inappropriate manner, such means of delivery can result in serious injury or death.

In addition, this chapter has also highlighted the development of a range of "wide area" means of delivery including certain high capacity RCA smoke generators, foggers or sprayers; water cannon; multiple "less lethal" munition launchers rapidly discharging large numbers of RCA projectiles; automatic grenade launchers; rocket-propelled grenades; large-calibre mortar munitions; large-calibre artillery munitions; cluster munitions; heliborne RCA munition dispensers; and certain RCA dispersal mechanisms attached to unmanned aerial or ground vehicles. Whilst some of these devices may have utility in large scale public order situations if employed in strict conformity with human rights standards; others appear to be intrinsically inappropriate for law enforcement and/or may have utility for use in either armed conflict or large-scale human rights violations.

5
Application of the Chemical Weapons Convention to ICA Weapons, Riot Control Agents and Related Means of Delivery

5.1. Introduction

The Chemical Weapons Convention (CWC) is a multilateral treaty that prohibits the development, production, stockpiling, transfer and use of chemical weapons and requires their destruction within a specified time period.[1] The CWC is of unlimited duration and is designed to be far more comprehensive in scope and application than any prior international agreement on chemical weapons. The CWC entered into force on 29th April 1997 and, as of August 2015, comprised 191 States Parties that had ratified, acceded or succeeded to it[2] – the highest number of any comparable arms control or disarmament treaty. The CWC States Parties represent about 98% of the global population and landmass, as well as 98% of the worldwide chemical industry.[3]

The CWC is implemented by its States Parties with the assistance of the OPCW.[4] The OPCW, which is headquartered in The Hague, Netherlands, is comprised of all CWC States Parties together with a Technical Secretariat with approximately 500 staff, headed by the Director General. The Technical Secretariat carries out the daily work of monitoring, verifying and facilitating implementation of the CWC.[5] It receives States Parties' declarations, detailing chemical weapons-related activities or materials and relevant industrial activities. After receiving declarations, the Technical Secretariat inspects and monitors States Parties' facilities and activities that are relevant to the CWC, aiming to ensure compliance.[6]

To date, most inter-governmental discourse regarding the regulation of ICA weapons, RCAs and related means of delivery has taken place within the context of the CWC and the attendant control regime. A study of the CWC and its implementation by States Parties is therefore an import focus of the second stage of the HAC analytical process.

5.2. Chemical Weapons Convention text

5.2.1. General obligations

Article I established the basic overarching obligations of States Parties under the Convention. Firstly, each State Party has undertaken "never under any circumstances" to "use a chemical weapon".[7] Article I also prohibited the development, production, stockpiling or transfer "directly or indirectly" of "chemical weapons to anyone".[8] The CWC does not explicitly prohibit research relating to chemical weapons (for example, for "protective purposes").[9] However, where research is an intrinsic part of a weapons development programme it clearly will fall within the scope of the Article I prohibition.[10]

States Parties are also prohibited from engaging in "any military preparations to use chemical weapons",[11] or to "assist, encourage or induce, in any way, anyone to engage in any activity prohibited to a State Party under this Convention".[12] In addition, Article I has also required that all existing stocks of chemical weapons,[13] and chemical weapons production facilities,[14] be destroyed.

In the 2014 legal Commentary to the Convention, Krutzsch has argued that the chapeau phrase to Article I.1 (a)–(d): "never under any circumstances" emphasized the comprehensive and totally binding nature of all the prohibitions set out under these paragraphs.[15] Krutzsch has contended that the wording of Article I has excluded any justification for the prohibited activities, and covered "all intents and purposes for such activities, independent of the character of the armed conflict, whether, an international or non-international one, whether the parties involved had recognized themselves or whether or not it is a civil strife".[16] The comprehensive nature of the obligations upon States under the CWC was further reinforced by Article XXII, which permitted no reservations by States Parties to the CWC.[17]

5.2.2. Defining toxic chemicals

The comprehensive nature of the CWC was further established under Article II.2, which defined a "toxic chemical" as:

> [A]ny chemical which through its chemical action on life processes, can cause death, temporary incapacitation or permanent harm to humans or animals. This includes all such chemicals, regardless of their origin or of their method of production, and regardless of whether they are produced in facilities, in munitions or elsewhere.[18]

Since this definition has included "any chemical, regardless of its origin or method of production", the CWC is not restricted to industrial pharmaceutical chemicals, but would also cover those produced naturally. In their 2014 Commentary on the Convention, Krutzsch and Trapp note that:

Toxins are toxic chemicals produced by living organisms... Although of biological origin, [and hence falling within the scope of the BTWC] toxins are chemicals and not living matter, and therefore are also covered under the definitions of Article II, paragraphs 1 and 2, of the CWC.[19]

Krutzsch and Trapp highlight the importance of the coverage of toxins under both the CWC and BTWC:

> This overlap recognizes that it is impossible to draw a line between toxins and other toxic chemicals: an increasing number of toxins can be synthesized in laboratories without resorting to organisms which produce them in nature, and a number of low-molecular-weight toxins are at the same time synthetic chemicals manufactured by industry.[20]

5.2.2.1. Scheduled chemicals

Article II.2 of the CWC also stated that: "For the purpose of implementing this Convention, toxic chemicals which have been identified for the application of verification measures are listed in Schedules contained in the Annex on Chemicals."[21] The CWC has consequently categorized certain individual toxic chemicals and families into three Schedules,[22] which should be reviewed and updated as required. Schedule 1 chemicals and precursors pose a "high risk" to the CWC and are rarely used for peaceful purposes.[23] Schedule 2 chemicals are toxic chemicals that pose a "significant risk" to the CWC.[24] In contrast, Schedule 3 chemicals may well be produced in large quantities for "purposes not prohibited" by the CWC but still pose a risk to the CWC; indeed, some of these Schedule 3 chemicals have previously been stockpiled as chemical weapons.[25] Although the CWC includes the three annexes of scheduled toxic chemicals detailed, these are specifically "identified for the application of verification measures",[26] and are not intended to delineate the full range of chemicals that may be considered to be "chemical weapons" under the CWC.

5.2.3. Defining chemical weapons

Article II.1 of the CWC defines chemical weapons. *inter alia* as: "(a) toxic chemicals and their precursors, *except where intended for purposes not prohibited under this Convention, as long as the types and quantities are consistent with such purposes*".[27] [emphasis added]. This highlighted text has become known as the General Purpose (GPC).

In the 2014 Commentary to the Convention, Krutszch and Trapp highlighted the implications of the open-ended nature of this provision:

> Under this concept, *all* toxic or precursor chemicals are regarded as chemical weapons *unless* they have been developed, produced, stockpiled, or

used for purposes *not* prohibited. The definition thus covers any toxic or precursor chemical if intended for chemical weapons purposes, irrespective of whether it has been listed in one of the Schedules.[28] [Emphasis in original]

Because the GPC establishes a prohibition based on intent rather than on a limited list of toxic chemical agents, it allows the CWC to accommodate and reflect developments in science and technology; consequently, as Meselson and Perry Robinson have highlighted, "even toxic chemicals whose existence is not yet known are covered" by its provisions.[29] Meselson and Perry Robinson also noted that the GPC has a second important function; it "protects legitimate uses of all toxic chemicals and chemicals from which they can be made".[30]

To determine whether the use of a toxic chemical would be in conformity with the CWC, the intention or purpose for its use needs to be established. The range of "purposes not prohibited" under the CWC has been defined under Article II.9 as:

(a) Industrial, agricultural, research, medical, pharmaceutical or other peaceful purposes;
(b) Protective purposes, namely those purposes directly related to protection against toxic chemicals and to protection against chemical weapons;
(c) Military purposes not connected with the use of chemical weapons and not dependent on the use of the toxic properties of chemicals as a method of warfare;
(d) Law enforcement including domestic riot control purposes.[31]

Consequently, any toxic chemicals held for purposes not provided for in Article II.9 are then chemical weapons and prohibited under the CWC. However, the use of toxic chemicals for "purposes not prohibited" would be acceptable only "as long as the types and quantities [of toxic chemicals] are consistent with such purposes".[32]

5.2.4. Implications of definitional lacunae

An analysis of the CWC shows that while many of the terms utilized in the Convention are defined in the text, several important concepts were left undefined by the negotiators.[33] For example, under Article II.9(d), toxic chemicals are permitted for "law enforcement including domestic riot control purposes"; however, neither the term "law enforcement" nor "domestic riot control purposes" has been further elaborated. Certain commentators have argued that such ambiguity was intentional, accommodating the interests of certain States. For example, the March 1994 editorial of the *Chemical Weapons Convention Bulletin* noted: "Some, by no means a majority, of the

5.2.5. Relevant operative provisions: Consultation, clarification and fact-finding mechanisms

Where a State Party is concerned about the possible non-compliance of another State Party it can initiate a range of consultation, clarification and fact-finding mechanisms under the Convention.[41] These range from informal bilateral consultations to full-fledged challenge inspections and investigations of alleged use of chemical weapons. They include on-site challenge inspections of any facility or location in the territory or in any other place under the jurisdiction or control of another State Party.[42] If such procedures fail to clarify the situation or do uncover evidence of non-compliance, the matter can be passed to the Executive Council (EC) or to a Special Session of the Conference of States Parties (CSP) for resolution.[43]

In theory, these mechanisms could be utilized, if States Parties were so minded, to raise concerns about, and potentially trigger investigations into, the development, stockpiling or employment of ICA weapons, RCAs or related means of delivery. If such agents were reportedly utilized by one State Party against a second State Party in the context of armed conflict, it is likely that the formal mechanisms would be employed by the targeted State Party. However, given the political sensitivities and contested application of the Convention in this area, it is currently uncertain whether the formal CWC investigatory mechanisms would be triggered by States Parties in response to reports of a State Party utilizing ICA weapons or RCAs internally against its own people to facilitate human rights violations. Such action would likely depend on the nature of the specific incident, the suspected agents utilized and the alleged perpetrating State.

5.2.6. Limitations of the control regime

A number of commentators have highlighted the long-standing limited ability of the OPCW to address certain important or pressing issues, even if delays in action could potentially seriously weaken the effectiveness of the Convention.[44] Factors considered to have contributed to this situation have included the OPCW's culture of decision-making by consensus and the consequent avoidance of difficult or controversial issues;[45] the wide disparity in resources and scientific and technical expertise available to State Party delegations and national authorities; the limitations on the autonomy of the Technical Secretariat, including its ability to receive and act on open source information;[46] limited transparency and accountability of the OPCW to civil society, and a reticence by the OPCW to receive information from and interact with relevant civil society organizations in a systematic manner.[47] In the last few years, the OPCW has examined some of these issues, particularly as it considers its future role, structure and activities following destruction of all existing declared chemical weapons stockpiles.[48] However, despite attempts to address some of the existing systemic failings, a number appear to have negotiating States wished to protect possible applications of disabling chemicals that would either go beyond, or might be criticized as going beyond, applications hitherto customary in the hands of domestic police forces."[34]

It can be argued that a degree of "constructive ambiguity" was useful and perhaps was indispensable in developing a CWC text that all negotiating States could sign up to, given the opposing positions of some States on certain issues.

Such textual inexactitude is by no means unusual in international agreements. Where such ambiguity exists, States have recourse to legal tools and guidelines to aid them in their interpretations of the text, most importantly the UN Vienna Convention on the Law of Treaties. The Vienna Convention has codified the customary international law on treaties between States.[35] Drafted by the International Law Commission of the United Nations, it entered into force on 27th January 1980, and, as of 2015, has 114 States Parties.[36]

Article 31 of the Vienna Convention established "a general rule of interpretation", which included the stipulation that "a treaty shall be interpreted in good faith in accordance with the ordinary meaning to be given to the terms of the treaty in their context and in the light of its object and purpose".[37] When undertaking such interpretation, States must take into account: any subsequent agreement between the parties regarding the interpretation of the treaty or the application of its provisions; any subsequent practice in the application of the treaty which establishes the agreement of the parties regarding its interpretation; and any relevant rules of international law applicable in the relations between the parties.[38] In addition, under Article 32 of the Vienna Convention: "recourse may be had to supplementary means of interpretation, including the preparatory work of the treaty and the circumstances of its conclusion...".[39]

However, even utilizing such interpretive tools, the absence of definitional clarity and the consequent textual ambiguity that can result if left unresolved over a lengthy period of time can become dangerous, leading to differing interpretations of the CWC by States Parties. This in turn could potentially lead to breaches of the CWC by some States Parties and also to an erosion of the stability and key prohibitions of the regime as a whole.

Despite these concerns, analysis of all publicly available OPCW documents shows that, as of August 2015, no policy-making body of the OPCW has issued guidance on the interpretation of the meaning of "law enforcement including domestic riot control purposes" and the nature and scope of activities covered under this term, nor established specific dedicated processes for doing so.[40] To date (August 2015), it has been left to States Parties to interpret the meaning of this term and the appropriate implementation of the CWC in this area. The implications of this lack of definitional clarity, for the regulation of ICA weapons, RCAs and related means of delivery are explored in the remaining sections of this Chapter.

influenced how the OPCW has reviewed and addressed the application of the Convention to ICA weapons, RCAs and related means of delivery.

5.3. Application of the CWC to riot control agents

5.3.1. Defining riot control agents

Article II.7 of the CWC defines RCAs as: "Any chemical not listed in a Schedule, which can produce rapidly in humans sensory irritation or disabling physical effects which disappear within a short time following termination of exposure."[49]

In June 2013, the OPCW Director General recognized that "the definition of RCAs in the Convention leaves some room for interpretation as to which chemicals can be considered as meeting the requirement specified in Article II(7)".[50] Consequently, in June 2014, the OPCW Technical Secretariat and Scientific Advisory Board (SAB) developed a non-exhaustive and indicative list of 17 toxic chemicals that corresponded to the Article II.7 definition and could therefore be considered as RCAs and subject to declaration under the CWC.[51]

In accordance with the definition under Article II.7, it is clear that RCAs are chemical agents that "through chemical action on life processes, can cause death, temporary incapacitation or permanent harm to humans", and thus fall within the ambit of "toxic chemicals" as defined by Article II, paragraph 2, of the CWC. Consequently, their use is permitted for a limited range of purposes defined under Article II, paragraph 9, and only "as long as the types and quantities [of toxic chemicals] are consistent with such purposes".

Although a range of toxic chemicals that fulfil the Article II.7 definition can be considered as "RCAs" under the CWC, this phrase belies the fact that RCAs have actually been used, in practice, for a variety of law enforcement applications beyond control or dispersal of rioting crowds, including for the incapacitation of violent individuals, personal defence and in counter-terrorist operations.

5.3.1.1. Malodorants

In addition to the "traditional" RCAs (i.e. tear gases and chemical irritant "pepper" sprays included in the Technical Secretariat June 2014 document), certain arms control experts believe that another group of chemicals – the malodorants – can also be classed as RCAs. Although no internationally agreed definition exists for such substances, malodorants have been described by one commentator as "chemicals designed to target human olfactory receptors in order to provoke a physiological response, ranging from simple aversion, to – in more extreme cases – symptoms such as nausea and vomiting".[52] According to Neill, the effects and duration of those malodorants that have been investigated and discussed in the open literature are similar to the effects and duration of some of the classical irritant and sternutating compounds.[53] Consequently, certain analysts therefore consider that malodorants should be grouped with RCAs, at least in terms of their regulation under the CWC.[54]

As of August 2015, this issue has not been addressed by any of the policy-making organs of the OPCW, and no State Party has formally expressed its view regarding the regulation of malodorants under the CWC.[55] However, there are indications that one State Party – the United States – may not consider at least some malodorants to be RCAs. In a June 2012 *New Scientist* article, Kelly Hughes, spokesperson for the Department of Defence Joint Non-Lethal Weapons Program, is reported as stating that "the CWC prohibits some temporarily disabling compounds on the basis of whether they activate the trigeminal nerve when people are exposed to it – those that do are classed as riot-control agents (RCA). The nerve conveys sensation from the face, cheeks and jaw, but does not control smell."[56] Consequently, according to Hughes, "If a particular malodorant is disseminated with a concentration that does not activate the trigeminal nerve, it may not require designation as an RCA under the CWC."[57]

This interpretation of the CWC appears to be reflected in a subsequent Broad Agency Announcement (BAA) issued by the US Office of Naval Research in June 2014, soliciting proposals for research into a range of "non-lethal weapon (NLW)" technologies for the fiscal year 2015, and including calls for research into malodorants. Under this solicitation, research "shall be limited to non-lethal malodorants that are not considered riot control agents, i.e., the malodorant must not cause 'sensory irritation' or any 'disabling physical effects'."[58] According to the BAA:

> The objective of these malodorants is to cause immediate repel effects to humans in open and confined spaces. This repel effect occurs with very small amounts (less than 8 ounces; 1/2 pint) of malodorant material applied within an open or confined space and often causes humans to: (1) leave a space (open or confined); (2) not enter a space (open or confined), and/or (3) discourage re-occupancy of humans in both open and confined spaces. This (repel) duration of effect can last 24 hours or more.[59]

No further details are publicly available as to whether research has been commissioned or initiated in this area, nor of the specific objectives and proposed use to which such research would be put. However, the indication that the United States does not consider certain malodorants to be RCAs is of potential concern, particularly if the corollary to this would be that such malodorants would not be regulated in the same manner as RCAs under the CWC.

5.3.2. Prohibition of use of RCAs in armed conflict

In addition to the overarching constraints on the use of RCAs that flow from the regulation of toxic chemicals, the CWC expressly prohibits the use of "riot control agents as a method of warfare".[60] A "method of warfare", although it is not formally defined in the CWC, is a well-understood term under international humanitarian law.[61] The term covers armed conflicts of a non-international character as well as international armed conflicts.

While the majority of CWC States Parties appear to hold a comprehensive interpretation of Article I.5 prohibiting all military use of RCAs in armed conflict, one State Party – the United States – maintains a long-held position, established in 1975 under Executive Order 11850, that RCAs can be legitimately used for a range of non-offensive actions, by military forces present in certain areas of armed conflict.[62] Under this Executive Order, use of RCAs is permitted: (a) in riot control situations in areas under direct and distinct US military control, to include controlling rioting prisoners of war; (b) in situations in which civilians are used to mask or screen attacks and civilian casualties can be reduced or avoided; (c) in rescue missions in remotely isolated areas, of downed aircrews and passengers, and escaping prisoners; and (d) in rear echelon areas outside the zone of immediate combat to protect convoys from civil disturbances, terrorists and paramilitary organizations. Use of RCAs in such circumstances would require approval of the US President or a senior officer designated by the President.[63] This position is subsequently enunciated in a number of US military handbooks or manuals, most recently in the US Department of Defense, Law of War Manual of June 2015[64]; and appears to be current US policy.[65]

Consequently, even under a permissive interpretation of the CWC, Article I.5 prohibits all use of RCAs for offensive military operations in a situation of armed conflict. However, certain previous reported uses of RCAs by Turkish and by US military forces, engaged in counter-insurgency operations (COIN), appear to have violated this prohibition.[66]

5.3.3. Use of RCAs for law enforcement

As discussed in Section 5.2.3., under Article II.9(d) toxic chemicals are permitted for "law enforcement including domestic riot control purposes".[67] Whilst this Article clearly allows the use of RCAs by law enforcement personnel, the scope of permissible activities – beyond that of domestic riot control – is not defined and is contested.

During the CWC negotiations, certain States outlined their interpretation of the meaning of "law enforcement". In May 1992, Ambassador Ledogar, head of the US CWC negotiating team, described the US position: "We understand the language 'law enforcement activities including domestic riot control' to mean that domestic riot control is a subset of law enforcement activities. We understand other law enforcement activities to include: controlling rioting prisoners of war; rescuing hostages; counterterrorist operations; drug enforcement operations; and non-combatant evacuation."[68]

However, at the conclusion of negotiations, no definition of the scope of "law enforcement" was agreed, and the ambiguity remained. As the *Chemical Weapons Convention Bulletin* noted: "[Article II.9 (d)] fully protects the use of chemicals such as tear gas for domestic riot control. But what is 'law enforcement?' Nowhere in the Convention is it defined. Whose law? What law? Enforcement where? By whom?"[69]

5.3.3.1. Use of RCAs by military forces in extra-territorial law enforcement

A number of legal scholars and analysts have argued that military forces can legitimately utilize RCAs when engaged in certain extra-territorial law enforcement activities.[70] Interpretations differ, but potentially permissible activities include:[71]

5.3.3.1.1. Controlling prisoners of war (POW). According to Fidler, international humanitarian law allows military forces to enforce laws against POWs and in extreme cases to use weapons against them. In his argumentation, he has cited an ICRC legal commentary, declaring that the detaining power may use force against POWs engaged in rebellious or mutinous behaviour, and that "Before resorting to weapons of war, sentries can use others which do not cause fatal injury...– tear gas, truncheons, etc."[72] In addition, Von Wagner has stated that during the CWC negotiations, "It was uncontradicted that such [riot control] agents could well be used in cases of a riot, for example, in a prisoner of war camp or in similar situations."[73] Although State practice is limited in this area, there have been contemporary reports of at least one State utilizing RCAs in this context.[74]

5.3.3.1.2. Extra-jurisdictional law enforcement. Fidler has contended that the CWC permits the use of RCAs for certain law enforcement purposes undertaken by military forces during military occupation or peacekeeping operations,[75] as long as they are directed against non-combatants. Fidler has argued that the following actions fall within this rubric: maintaining public order and safety in areas subject to their control; ensuring security of their members and property, the occupying administration and the lines of communication used by them; and enforcing the laws of the occupied territory and the laws promulgated by the occupying authority pursuant to its responsibilities under the international law of occupation.[76] Although Von Wagner has also stated his belief that the CWC permits use of RCAs in a police or peacekeeping action,[77] he has not elaborated upon the range of permissible actions.

The policy and practice of certain States indicates an apparent willingness to allow their military forces to deploy and use RCAs during certain peacekeeping operations under the auspices of the United Nations. For example, Fry has documented the relatively long history of RCA use by at least 14 UN peacekeeping forces drawn from many countries and operating in a diverse range of countries and situations.[78] However, from his analysis of the nature of RCA employment in practice, in such peacekeeping operations, Fry has concluded that "RCA use by UN forces is potentially illegal and certainly bad policy".[79] Consequently, he has argued that "it would be best for UN forces to foreswear all RCA use".[80] Krutzsch and Trapp have supported Fry's position and argued that UN troops "may lawfully use RCAs only when acting specifically as a domestic police force (and when authorized to act as such), and possibly within areas of their direct and complete control...".[81] However, as of August 2015, this issue has not been formally addressed by the OPCW policy-making organs, and, consequently, individual CWC States Parties are left to determine appropriate implementation of the CWC in this area.

5.3.3.2. Constraints upon the use of RCAs in law enforcement

Because neither "law enforcement" nor the types of activities covered by it have been defined under the CWC,[82] it is the obligation of each individual State Party to implement the relevant provisions of the CWC in good faith, taking into account "any relevant rules of international law applicable in the relations between the parties".[83] Of particular relevance are those rules and restrictions on the use of force by law enforcement officials – which can include police, security or military personnel[84] – that arise from international human rights law and related standards (addressed and discussed in Chapter 8 of this publication).

Consequently, if RCAs (or indeed other toxic chemicals) are intentionally employed by law enforcement officials to carry out serious human rights violations, including torture and other cruel, inhuman or degrading treatment or punishment; violent dispersal of lawful peaceful demonstrations; arbitrary, indiscriminate or excessive use of force; or in conjunction with firearms to make lethal force more deadly, then it can be argued that such actions should not be considered to be legitimate "law enforcement" activities as permitted under Article II, paragraph 9 (d), of the CWC.

Perry Robinson, for example, who has served as an advisor to the UK National Authority for the CWC, believes that such intentional misuse of RCAs would be inconsistent with the CWC:

> RCAs are designed and intended to work by causing people to move away from a particular area, to depart from a riotous assembly, to stop being a mob .... The intention behind the use is important. To use RCAs to intentionally disable and injure people is not consistent with the CWC. Such use would only be consistent if you had a weird interpretation of the Treaty.[85]

In a number of cases of reported human rights violations, RCAs have been employed in quantities or in a manner where the agent concentrations have become injurious or even fatal. In addition to contravening international human rights law and related standards, such use also appears to be in breach of the "types and quantities" provisions of Article II, paragraph 1, of the CWC.

According to Krutzsch and Trapp, RCAs can only legitimately be employed "under circumstances that allow people to leave the place of exposure early enough before those effects become irreversible and or more severe than sensorily irritating or physically disabling".[86] In certain situations, however, individuals are unable to leave the place of exposure in time, due to physical infirmity, age or because of the situation of the incident location (e.g. obstacles, RCA use in confined spaces or inside buildings). According to Krutzsch and Trapp: "If the time span of exposure was so long that the sensory irritation or disabling effects no longer disappear spontaneously (or other toxic effects manifest themselves), the chemicals used changed legally from 'RCA consistent with paragraph 9(d)' to 'chemical weapons prohibited under Article I'."[87] Consequently, Krutzsch and Trapp argue that States Parties have to "carefully apply rules to keep their use of RCAs in conformity with the 'types and quantities' rule of Article II, paragraph 1(a). Taken together with other bodies of law (in particular human rights law), this requirement affects the acceptable tactics and modes of use of RCAs in law enforcement."[88]

However, despite the apparent relatively widespread nature of the reported misuse of RCAs by law enforcement officials (as documented in Chapter 3), as of August 2015 no cases of such misuse have been raised publicly by States Parties as matters of concern in the context of the CWC. Nor has the nature and scope of "law enforcement" under the CWC been addressed by any CWC policy-making organ.[89]

5.3.4. Reporting obligations

The CWC requires that States Parties submit an initial declaration of all types of chemicals held for riot control purposes.[90] This obligation was intended as a confidence-building measure between States Parties, to increase transparency with regard to RCA holdings, and thereby to demonstrate that no State Party held or was seeking to develop toxic chemical weapons intended for warfare under the guise of law enforcement. Under this declaration procedure, all States Parties are required to supply the chemical name, structural formula, and Chemical Abstracts Service (CAS) registry number, for each chemical kept for riot control purposes.[91] States Parties are also required to provide an update of the initial declaration 30 days after any change has become effective.[92]

This declaration procedure, however, has significant limitations. For example, States Parties are not required to provide any information in their declarations about the quantities of RCAs that they hold, even though the bulk storage of RCAs by a State Party could become a matter of concern.[93] Similarly, the current RCA declaration procedure does not require States to detail the types and quantities of the RCA means of delivery they hold (e.g. whether they are in hand-held tear-gas grenades suitable for law enforcement purposes, or 155 mm artillery projectiles, cluster munitions, aerial bombs or other military munitions). In addition, States Parties are not required to provide any information on where RCA stocks are held, nor provide details of whether they are under military control or civilian law-enforcement agency control.[94] Without such information, the confidence-building utility of this system for alerting States to militarily significant levels of RCAs appears to be extremely limited.

In addition, once a State Party has submitted their initial RCA declaration there are no routine follow-up verification provisions to ensure that such declarations are full and accurate, that is, the Technical Secretariat has no authority to undertake routine inspections to verify RCA possession by States Parties.[95]

There are also limitations on the level and quality of public transparency in this area. Although States Parties have the right to obtain copies of the RCA declarations provided by other States Parties,[96] there is no mechanism for civil society to obtain this information. Instead, the only public document available is the summary of RCA holdings included in the OPCW Annual Report.[97] Whilst this provides information on the aggregate number of States Parties possessing various RCAs, it does not provide sufficient information to determine the RCA holdings of a specific individual State Party. Its utility to build public confidence of State RCA holdings is, therefore, negligible.

5.3.5. Application of CWC to RCA means of delivery

5.3.5.1. *Definition of means of delivery*

Under Article II.1 of the CWC, the definition of a chemical weapon has included:

> (b) *munitions and devices specifically designed* to cause death or other harm through the toxic properties of those toxic chemicals specified in subparagraph (a), which would be released as a result of the employment of such munitions and devices;
>
> (c) *any equipment specifically designed* for use directly in connection with the employment of the munitions and devices referred to in (b).[98]
>
> [Emphasis added]

Whilst the OPCW Technical Secretariat has provided States Parties, upon request, with technical guidance on this issue, it is the responsibility of States

Parties to interpret the meaning and appropriate application of the CWC in this area.

5.3.5.1.1. Prohibited and non-prohibited purposes. Since the use of "riot control agents as a method of warfare" is expressly prohibited under the CWC,[99] States would consequently be prohibited from developing, transferring, acquiring, stockpiling or employing RCA means of delivery intended for use in international and non-international armed conflict. This prohibition arguably extends to RCA means of delivery intended for use in certain military operations in urban terrain (MOUT) not of a law enforcement nature; as well as certain COIN or urban warfare. Consequently, any State or commercial entity manufacturing and promoting RCA means of delivery for such purposes would appear to be in contravention of the CWC. In contrast, CWC States Parties would be permitted to manufacture, promote, acquire, stockpile and employ delivery systems to disseminate appropriate types and quantities of RCAs for "law enforcement including domestic riot control" purposes.[100]

5.3.5.1.2. "Types and quantities" restriction. Although the CWC did not list the types of acceptable or non-acceptable delivery systems for use with toxic chemicals in law enforcement scenarios, it did place an important constraint upon such means of delivery through the Article II.1.(a) "types and quantities" restriction. In the 2014 Commentary on the Convention, Krutzsch has contended that: "The types and quantities restriction rule requires that, in order to be not prohibited, the technical characteristics and attributes of the means of delivery of RCAs must be consistent with the rules and conditions that apply to their use in law enforcement purposes under 9(d) of Article II."[101] Consequently, the means of RCA delivery must "be technically designed to enable law enforcement officers to use them safely for the permitted purposes. RCA delivery systems that are consistent with the permitted purpose of law enforcement including domestic riot control must allow for the limiting of exposure to a level acceptable for that purpose".[102]

5.3.5.2. Application of the CWC to "limited area" RCA means of delivery
As discussed in Chapter 4, there are a range of means of delivery which have a narrow dispersal area, short range and emit a limited quantity of RCA – such as certain hand-thrown RCA canisters and grenades, or hand-held spray disseminators – which do not appear to conflict with the "types and quantities" restriction of the CWC. These "limited area" RCA means of delivery are widely employed by law enforcement officials for example in public order situations. If such devices have been properly tested and trailed, their use should not raise concern, provided it is consistent with the "law enforcement" purpose under the CWC, and is in strict accordance with the relevant human rights standards.

5.3.5.3. Application of the CWC to "wide area" RCA means of delivery

5.3.5.3.1. "Wide area" RCA means of delivery of potential concern. In contrast to the foregoing, a range of delivery mechanisms have been developed that deliver far larger amounts of RCAs over wider areas and/or over greater distances than could be delivered by hand-held sprays and the like. As detailed in Chapter 4, such means of delivery include large sprayers and fogging devices, grenade launchers, multiple munition launchers and a variety of vehicle-mounted dispensers. Certain forms of such means of delivery, particularly those with extended ranges, may raise questions about the feasibility of their discriminate use with the consequent danger of affecting bystanders. In addition, given the potential quantities of agent dispersed by some of these mechanisms, questions of proportionality arise, as well as concerns regarding the potential danger of serious injury due to asphyxiation or agent toxicity. Certain forms of such "wide area" dispersal mechanisms may also be more open to intentional misuse by law enforcement officials on a large scale than are "limited area" dispersal mechanisms. However, depending on their specifications, certain "wide area" RCA means of delivery may have utility in large scale law enforcement situations, for example against violent crowds and riots, providing they meet the CWC "types and quantities" provision and are employed in strict conformity with human rights standards.

Whilst certain "wide area" RCA means of delivery may potentially be used for law enforcement, they could also potentially be employed in a variety of armed conflict scenarios. As detailed in Chapter 4, a range of such "wide area" RCA means of delivery have been promoted for use by security or military forces for peacekeeping, MOUT, COIN and urban warfare. The employment of such means of delivery for at least some of these proposed purposes would appear to be inconsistent with the CWC "law enforcement purpose" and/or breach the CWC prohibition on the use of RCAs as a "method of warfare".

5.3.5.3.2. Inherently inappropriate "wide area" RCA means of delivery. According to a number of international lawyers and arms control experts, a range of munitions containing RCAs which have military utility, such as cluster munitions, aerial bombs, mortar rounds and artillery shells, would be inherently unacceptable for use in law enforcement activities.[103] For example, Neill has argued that:

> [I]t is not appropriate to disseminate a non-lethal agent using a mechanism whose ancillary effects could easily be lethal (e.g. a large, high-velocity carrier shell or a bursting device producing shrapnel); or whose gross capacity and interoperability with conventional military equipment (e.g. in mortars, howitzers, rocket projectiles or by high-speed aircraft) would render it rapidly adaptable for use as a "method of warfare".[104]

Similarly, NATO's RTO has stated that:

> The employment of chemicals as NLT [non-lethal technologies] has to be compatible with use, thus demonstrating intent. For example, whereas CS in hand or baton round sized canisters would be considered legitimate law enforcement equipment, 155mm shells filled with CS would clearly be considered as preparation to use riot control agent in waging war, prohibited under the CWC.[105]

All such inherently inappropriate "wide area" RCA munitions and means of delivery would appear to breach the CWC "types and quantities" restriction and/or the prohibition on use of RCAs as a "method of warfare".[106] They should be considered to be chemical weapons and treated accordingly. All States Parties possessing such prohibited means of RCA delivery should declare these items to the Technical Secretariat as required under Article III.1, and verifiably destroy such means of delivery as required under Article I.2.

5.3.5.4. Developing a process to distinguish appropriate and inappropriate means of RCA delivery

In order to facilitate the appropriate and uniform application of the CWC in this area by all States Parties, the author has previously recommended that the OPCW should develop "criteria and a suitable process for determining which means of RCA delivery are inappropriate for law enforcement purposes and would breach Article II.1 and/or Article I.5 of the CWC".[107] Such mechanisms could then be utilized by the OPCW to develop a: "clarificatory document for States Parties detailing those means of RCA delivery that are considered inherently inappropriate for law enforcement purposes and breach Article II.1 and/or Article I.5 of the CWC".[108] Consequently, all States Parties "would be prohibited, under Article I.1, from developing, producing, stockpiling, marketing, transferring or using such means of delivery". Following its development, the clarificatory document should be reviewed regularly in an appropriate forum such as the Executive Council (EC) or Conference of States Parties (CSP) to determine whether additional items should be added in the light of developments in science and technology.[109]

Subsequently, Krutzsch, in the 2014 Commentary on the Convention, has also called for "the development of technical standards that are based on clearly defined technical parameters to ensure that the effects of the RCAs disseminated by these means of delivery will stay within the limits consistent with the purpose of law enforcement including domestic riot control".[110] Krutzsch has argued that "such standards should be developed on the basis of technical studies prepared by interested States Parties, the Technical Secretariat, or the Scientific Advisory Board".[111]

5.3.5.5. Current discourse on RCA munitions and means of delivery within the OPCW

Investigations by civil society organizations and the media have detailed a range of commercially available large-calibre RCA munitions and other "wide area" RCA means of delivery potentially in conflict with the CWC,[112] and concerted attempts have been made by the author and others (as detailed in Chapter 12) to highlight specific cases of concern and bring them to the attention of the OPCW.

Consequently, this issue is beginning to be examined by certain relevant OPCW bodies. For example, in its report to States Parties in preparation for the Third Review Conference, the Scientific Advisory Board (SAB) specifically highlighted RCA means of delivery, and stated that: "The SAB notes with concern isolated reports of the commercial availability of munitions apparently designed to deliver large amounts of riot control agents over long distances."[113]

To date, few States Parties have clarified their position regarding the regulation of RCA means of delivery under the CWC; one notable exception, however, has been Turkey. Following evidence uncovered by the author and colleagues of the international promotion of the 120 mm CS MKE MOD 251 munition by Turkish company MKEK, the Turkish Ambassador to the OPCW confirmed that such activities were prohibited under the CWC, and that such prohibition would extend to other "mortar ammunition containing tear gas or any other prohibited substance...".[114] This position was further elaborated by the Turkish Counsellor to the OPCW, who highlighted the activities of the Turkish Ministry of Defence to inform all licensed arms brokering companies in Turkey that "trading 120mm CS mortar ammunition is not permissible under Turkey's CWC obligations".[115] Turkey's corresponding actions in destroying all remaining 120mm CS MKE MOD 251 munitions, together with epoxy models and promotional materials, and its attempts to halt the trade, promotion and brokering of such munitions, have clearly underlined this position.[116] The information provided by the Turkish Government and its robust actions in this area have provided a powerful precedent for developing common understandings and approaches to this issue. However, as of August 2015, no OPCW policy-making organ had formally addressed the regulation of such RCA means of delivery under the CWC.

5.4. Application of the Chemical Weapons Convention to ICA weapons

5.4.1. Scope of the convention

Since those chemicals promoted for use as ICA weapons can "through chemical action on life processes...cause death, temporary incapacitation or permanent harm" to their targets, they are toxic chemicals within the meaning of the CWC as specified in Article II.2.[117] Such toxic chemicals would be deemed to be chemical weapons (and therefore prohibited) if they were used for purposes other than those exemptions stipulated under Article II.9 of the CWC,[118] or if their use was inconsistent with the types and quantities restriction of Article II.

5.4.2. Use of ICA weapons in armed conflict

The use in armed conflict of the toxic properties of chemical agents as weapons is absolutely prohibited, as is their development, production, acquisition, stockpiling, retention or transfer when intended for such purposes, under Articles I and II of the CWC. If States have undertaken programmes to develop ICAs and/or associated means of delivery for such purposes, they are required to halt such activities, declare any chemical weapons and chemical weapons production facilities they possess (under Article III[119]) and ensure they are verifiably destroyed (under Article I, and in accordance with Articles IV and V respectively[120]).

5.4.3. Use of ICA weapons in law enforcement

Among the "purposes not prohibited" listed in Article II.9 of the CWC are: "(d) Law enforcement including domestic riot control purposes".[121] However, toxic chemicals can only be employed for such purposes provided their use is consistent with the "types and quantities" restriction of Article II. Differing interpretations regarding the application of II.1 and Article II.9(d), have led to alternative views from arms control and legal scholars as to whether toxic chemicals promoted as ICAs can ever be used for law enforcement purposes.

One line of interpretation supported by Chayes and Meselson,[122] Krutzsch,[123] Krutzsch and Trapp[124] and Von Wagner[125] holds that only RCAs can be used for law enforcement activities under the CWC,[126] and, consequently, that the use of ICA weapons for law enforcement would be prohibited. Chayes and Meselson argue that:

> [A] toxic chemical used by virtue of its toxic properties is only of a type consistent with the purpose of law enforcement, in the sense of Article 2.1.a, if it meets the Convention definition of a riot control agent in Article 2.7. Thus such chemicals must be "not listed in a Schedule" and must "produce rapidly in humans sensory irritation or disabling physical effects which disappear within a short period of time following termination of exposure".[127]

Whilst Krutzsch and Trapp argue that:

> Article II, subparagraph 9(d) does not permit the use of toxic chemicals other than RCAs for law enforcement purposes. Accepting, in principle,

the use of certain toxic chemicals such as ICAs for law enforcement in exceptional situations, for instance during an anti-terrorist operation, would undermine the prohibition on Article I of the Convention and wrongly imply that there was a "choice" between applying binding rules of treaty interpretation and replacing them by explanations constructed as required by military interest.[128]

A second line of interpretation, as advanced by Fidler,[129] and supported by Neill,[130] contends that international law on treaty interpretation indicates that that the CWC does not limit the range of toxic chemicals that can be used for law enforcement purposes to RCAs. This line of interpretation may allow the use of ICA weapons for law enforcement in certain circumstances. However, Fidler notes that the "use of a toxic chemical for law enforcement purposes is still subject to the CWC requirements that the types and quantities of chemicals developed, produced, acquired, stockpiled, retained, transferred, or used must be consistent with such permitted purposes (Article II.1 [a.])".[131] According to Fidler, this restriction therefore:

> [R]equires scrutiny of the relationship between the chemical or biochemical agent and the law enforcement objective in question. The more difficult it is to control the effects of the use of a chemical or biochemical in a law enforcement operation, the more suspect such use becomes in terms of the agent being of a type or quantity consistent with a law enforcement purpose.[132]

Consequently, Fidler believes that: "For domestic law enforcement, use of incapacitating agents in contexts in which the government could control neither dosage nor the exposure environment would only be legitimate in extreme law enforcement situations."[133] Furthermore, Fidler contends that: "For extraterritorial law enforcement activities undertaken by military forces and sanctioned by international law, States can at present only legitimately use riot control agents, not incapacitating agents."[134]

A third interpretation, previously elaborated by the Office of the US Navy Judge Advocate General (JAG), suggested ways in which the use of ICA weapons might be consistent with the CWC and other international legal obligations.[135] First, it asserted that certain "convulsives and calmative agents may also be RCAs".[136] If there are ICA weapons that fall into this category then the JAG report considered that they, like RCAs, would only be "subject to Article I(5)'s limitation on the use of RCAs as a 'method of warfare', and are not be subject to Article II's proscriptions".[137] This interpretation with regard to RCAs has not been publicly supported by any other CWC State Party. Its potential application to "certain convulsives and calmatives", that is, ICA weapons, is highly problematic. In addition, the

JAG opinion highlighted an alternative legal route by which ICA weapons could in their view be legitimately used in operations other than war:

> Convulsives and calmatives may rely on their toxic properties to have a physiological effect on humans. If that is the case, and these two NLWs [non-lethal weapons] are not considered RCAs, in order to avoid being classified as a prohibited chemical weapon, they would have to be used for the Article I(9)(d) "purpose not prohibited", the law enforcement purpose. As discussed ... *the limits of this "purpose not prohibited" are not clear and will be determined by the practice of States* [Emphasis added].[138]

As will be discussed in section 5.4.3.3, recent statements by senior US officials, however, indicate a significant change of policy in this area, i.e. that the "development of [ICA weapons] for law enforcement purposes is incompatible with the Chemical Weapons Convention...".[139]

5.4.3.1. Implications of the Russian Federation use of an ICA weapon

Considerations of the potential implications of ICA weaponization, proliferation and (mis)use were heightened following the employment of such agents on 23rd October 2002 by the Russian Federation in their attempts to end the siege of the Dubrovka Theatre in Moscow by a group of heavily armed Chechen separatists, and to release over 900 hostages held.[140]

Whilst certain States may have had misgivings about the Russian Federation action, the Moscow incident was not met with any significant public expressions of disapproval from the international governmental community. Indeed, some leaders, such as US President George Bush,[141] and UK Prime Minister Tony Blair, supported Russia's actions without commenting explicitly upon the use of an ICA weapon.[142] Similarly, Denmark, which then had the presidency of the European Union (EU), praised Russia's actions. Denmark's Prime Minister, Anders Fogh Rasmussen, said the EU "commends the Russian Government for exercising all possible restraint in this extremely difficult situation".[143]

In its 2006 report, the NATO RTO also reviewed the Moscow incident favourably, specifically highlighting the use of an ICA weapon:

> Although it may seem excessive that 16% of the 800 hostages died from the "gas" exposure, still 84% survived. We do not know that a different tactic would have provided a better outcome. The use of a "sleeping gas" or "calmative" or "incapacitant" agent in this setting is a novel courageous attempt at saving the most lives. This counter-terrorist action showed on the other hand that chemical "non-lethal" weapons are not always non-lethal.[144]

The report further stated that: "It is significant that use of chemical incapacitants in hostage rescue situations appears to be acceptable, but only when there is a potential lethal threat to the hostages and the situation is very limited in time, location, and number of people involved."[145]

Non-governmental arms control analysts and legal experts were at the time divided (and remain divided) on whether the action of the Russian Federation was a breach of the CWC.[146] However, a leading commentator, Professor Mark Wheelis, has contended that "most analysts consider the Russian use of fentanyl derivative to have been legal under Article II.9(d) [of the CWC]".[147] In a subsequent interview with the author, Wheelis stated that:

> I think the Russian use was in conformity with the CWC, at least if you accept that agents other than RCA can be used for law enforcement ... I think incaps [ICA weapons] that deserve the name (none available currently) would have clear utility to police, and might be the best and most humane solution to certain situations (hostage taking, for instance, or capturing criminals or terrorists in crowded environments). In my reading of the CWC this would be legal. I do not believe it would be wise, though, and think nations should voluntarily eschew development of these weapons, given the abuses that will very likely occur.[148]

A Western Government expert interviewed by the author, in April 2010, stated his belief that the use of an ICA weapon by the Russian Federation was in conformity with the CWC. However, he raised concerns regarding the manner in which the ICA weapon was employed:

> There are a number of issues here: was the material used really appropriate for such an event given the high number of fatalities and long term adverse effects on some survivors? Why was medical assistance not given more promptly to survivors? What were the origins of the decision to use the material in question? Who controls such materials? Was this use a desperate ad hoc measure or did it draw on prior R&D and planning?[149]

A senior US State Department official (speaking in a personal capacity) interviewed by the author, in May 2010, was "not persuaded that" the actual use of an ICA weapon by the Russian Federation "violated the CWC *per se*".[150] However, the official did believe that: "The immediate availability of substantial quantities raises other questions... It raises questions about the quantities that have been produced. And whether those are consistent with requirements for ... quantities permitted under the CWC for law enforcement ... Because large quantities might provide a military capability".[151] Professor Wheelis voiced similar concerns: "Russia clearly has stockpiles; [it]

couldn't have assembled sufficient agent in a few days to saturate a theater, much less develop use strategies, prepare troops, etc."[152]

Despite the apparent public acceptance (or at least silence) by many in the international community of the Russian Federation's use of an ICA weapon, relevant international human rights bodies – specifically the UN Human Rights Committee[153] and the UN Special Rapporteur on Extrajudicial, Summary or Arbitrary Executions[154] – did raise concerns about the actions of the Russian Federation security forces. Concerns were also expressed by leading international non-governmental human rights organizations including AI,[155] and HRW.[156] Furthermore, the ICRC issued a statement concerning ICA weapons, on the margins of the First CWC Review Conference. Although it did not specifically mention the Russian Federation, the ICRC statement: "express[ed] its alarm at the increasing interest among police, security and armed forces in the use of incapacitating chemicals and the lack of expressions of concern about the implications of such development by States Parties to this Convention...."[157]

A public silence on this issue was maintained by the OPCW. An analysis of the OPCW website shows that there were no press releases or other public statements made by the Technical Secretariat or any other policy organ of the OPCW about the Russian Federation use of an ICA weapon.[158] Even in a presentation analysing Russian implementation of the CWC, delivered by the Director of External Relations on behalf of the Director General, and made just two weeks after the theatre siege, no mention was made of the Moscow incident.[159] Despite its public silence, however, there are indications that the OPCW Technical Secretariat was seeking to facilitate the greater provision of information about this incident by the Russian Federation. Indeed, according to the BBC, the public statement made by the Russian Health Minister on the 30th October 2002 came after a request for clarification about the ICA weapon from the OPCW Director General.[160] Furthermore, an OPCW spokesperson stated that, after the siege, the Russian Federation Government sent a letter to the OPCW, "explaining its actions", though no details of the letter's contents have been made public.[161]

There are also indications that individual CWC States Parties sought further information from the Russian Federation on its actions, through bilateral consultations. In his statement to the UK Parliament, Foreign Office Minister Mike O'Brien indicated that the Russian announcement followed "inquiries by the United Kingdom and others".[162] Similarly, a senior US State Department official stated that the United States sought additional information from the Russian Federation, but noted that "A lot of questions remain. We have gotten relatively little information".[163]

CWC States Parties, as a whole, do not appear to have formally been informed of the results of such bi-lateral consultations, nor were the details made available to the public. A review of all relevant OPCW documentary sources shows that no CWC State Party specifically raised this incident publicly in any of the CWC policy-making organs, nor initiated high-level multilateral consultation or investigatory mechanisms under the CWC.[164]

5.4.3.2. Has State policy and practice become established?
According to the UN Vienna Convention on the Law of Treaties, an important element that needs to be taken into account when interpreting meaning to specific obligations under treaties is "any subsequent practice in the application of the treaty which establishes the agreement of the parties regarding its interpretation".[165]

An important factor determining whether State policy and practice on ICA weapons is becoming, or has become, established will be the level and nature of contemporary research, development and use of such weapons by States. A review by the author and Professor Dando of publicly available information (as detailed in Chapter 2) has uncovered evidence of the possession and/or use of ICA weapons by China, Israel and the Russian Federation, whilst an additional small number of States have conducted dual-use chemical and life science research, since the coming into force of the CWC, potentially applicable to the study or development of ICA weapons.

Fidler has argued that the use of an ICA weapon by the Russian Federation and the subsequent "acquiescence of other CWC States Parties to such use ... provides some evidence of State practice that the CWC does not limit the range of chemicals that can be used under Article II.9(d) to RCAs".[166] Similarly, a UK official questioned by the author in November 2008 stated that: "The implication is that future use of toxic chemicals in similar law enforcement situations would also be considered as permitted under the Convention, but each case would of course need to be considered according to the individual circumstances."[167]

Other analysts, however, have stated their belief that State policy and practice has not yet become established as to the permissibility of ICA weapon use in law enforcement activities. In a presentation to the Open Ended Working Group of CWC States Parties preparing for the Second Review Conference, Dr Alan Pearson stated: "The nervousness and concern expressed by States and CWC experts about this incident [Russian Federation use of an ICA weapon] suggests that such agreement has not yet been definitively established among the States Parties."[168] He further added: "Indeed, although the agent may have seemed consistent with certain law enforcement purposes prior to its use [by the Russian Federation], States Parties might conclude upon examination that this agent (and other fentanyl type agents) may not be consistent with any valid law enforcement purpose under the CWC."[169] Similarly, a former OPCW official, interviewed by the author, stated his belief that State policy and practice had not yet become established on this issue. For although the issue was not openly addressed at the First Review Conference, "enough States Parties at the Second Review Conference stated that the issue needs to be discussed".[170] As described in Section 5.4.3.3., the subsequent debate within the OPCW (prior to, during and following the Third Review Conference) as to whether ICA weapons can be legitimately utilized for law enforcement purposes under the CWC, and, if so, in what manner and with what constraints – has shown that this issue has not been settled and that divergent positions are held by CWC States Parties.

5.4.3.3. Attempts of CWC States Parties to address ICA weapons

The potential threats arising from the development and use of ICA weapons led a growing number of CWC State Parties to call for this issue to be addressed by the OPCW at the Third Review Conference in 2013. Switzerland, which has repeatedly championed this issue,[171] was particularly vocal. In his opening statement, Ambassador Markus Börlin highlighted "Switzerland's fears that the silence and uncertainty surrounding the use of toxic chemicals for law enforcement purposes other than riot control agents risks eroding the Convention." He stated that: "a debate on this issue in the framework of the OPCW should no longer be delay[ed] until the next Review Conference, which is why my delegation has proposed language for this Conference's final document".[172]

Support for addressing ICA weapons appeared to have grown considerably since the previous Review Conference. Concerns regarding ICA weapons were raised by a number of States in their National Statements to the Review Conference, including Germany (which also submitted a working paper on this issue), Ireland, Norway, Romania, Slovakia, the United Kingdom and the United States, and from the ICRC.[173] For example, Ambassador Rolf Nikel of Germany noted that: "In the past years the issue of 'toxic chemicals for law enforcement' has been extensively discussed in various fora outside the OPCW... There is now a substantial body of scientific analysis on developments [regarding ICAs] that have taken place since the entry-into-force of the Convention."[174] Citing the CWC Review Conference's "specific mandate to 'take into account any relevant scientific and technological developments'", Germany consequently recommended that the Conference should "through its final declaration initiate discussions on the issue of toxic chemicals for law enforcement".[175]

Of particular importance were supportive statements from the United Kingdom and the United States, both of which had previously undertaken research into ICA weapons.[176] The US Acting Under Secretary for Arms Control and International Security, Ms Rose Gottemoeller, explicitly declared that: "the development, production, acquisition, stockpiling, or use of incapacitating chemical agents – or any other toxic chemicals – in types and quantities inconsistent with purposes not prohibited by the

Chemical Weapons Convention, is clearly prohibited by Article I of the Convention".[177] Ms Gottemoeller highlighted concerns that "illicit programmes could possibly be concealed under the guise of a legitimate treaty purpose, such as law enforcement", and further warned that States Parties "must all be vigilant to ensure that incapacitating chemical agents and other technologies do not jeopardise the twin goals of the Convention – the destruction of all chemical weapons and the prevention of the re-emergence of chemical weapons".[178]

In his statement on behalf of the United Kingdom, Mr Alistair Burt, Under Secretary of State for Foreign and Commonwealth Affairs, noted the UK's involvement in the "ongoing discussions on the place of incapacitating chemical agents in the Convention, particularly given scientific change and the absence of any definition or common understanding of law enforcement".[179] He highlighted the reports of the Royal Society and the SAB, noting that "[b]oth have set out the scientific position as well as advancing our understanding of the complex issues surrounding this topic", and declared that the OPCW should "address such relevant issues and show leadership".[180] He recommended that the OPCW "should work together to establish a norm to discourage the use of chemicals more toxic than Riot Control Agents for law enforcement and consider transparency measures or limitations".[181] Finally, he "unequivocally" stated that "the UK neither holds, nor is developing, any incapacitating chemical agents for law enforcement... and encourage[d] all other States Parties to state their positions on this question".[182]

However, despite widespread support, the Third Review Conference, which is a consensus-driven forum, failed to reach an agreement upon a process for the OPCW to initiate consultations on this issue. At the 19th Conference of States Parties, in December 2014, renewed calls for action were raised, most notably by Australia, which confirmed that it was not "developing, producing, stockpiling or intending to weaponise or use any CNS [central nervous system] acting chemicals such as anaesthetics, sedatives or analgesics for law enforcement purposes",[183] and called on other States Parties, which have not already done so,[184] "to make their positions known on the weaponisation of CNS acting chemicals... for law enforcement purposes".[185] As of August 2015, Australia,[186] Canada,[187] Germany,[188] Malaysia,[189] Poland,[190] Switzerland,[191] the United Kingdom[192] and the United States[193] have made formal statements in OPCW forums declaring either that they do not produce, stockpile, or use ICAs for law enforcement purposes; and/or that RCAs are the only toxic chemicals that can be employed for law enforcement in their countries. In addition, both the Czech Republic[194] and Iran,[195] in correspondence with the author, have stated that ICA weapons are prohibited for use in law enforcement in their countries.

In its Statement to the 19th Conference of States Parties, Australia also called for "consultations, in particular among members of the Executive

Council of the OPCW, with a view to commencing discussions as to whether weaponisation of CNS acting chemicals should be permitted for law enforcement purposes".[196] Subsequently a number of States, notably Australia,[197] Canada,[198] Germany,[199] Malaysia,[200] New Zealand,[201] Norway,[202] Switzerland[203] the UK,[204] and the US,[205] have raised the issue of ICA weapons within the Executive Council. In the July 2015 EC meeting, the Swiss Ambassador stated that "The [ICA] issue remains at the very heart of my delegation's concerns. I take note that on this topic the discussion has started in an irreversible manner and moves forward."[206] In addition, the US Ambassador "encourage[d] all delegations to consider and subscribe to the notion that the development of so-called incapacitation agents for law enforcement purposes is incompatible with the Chemical Weapons Convention and to put their views on the record in the Executive Council. My own delegation has clearly done so...".[207] In contrast, although not directly addressing ICA weapons use in law enforcement, the Russian Federation in its response to the Director General's discussion paper on the future of the OPCW[208] cited proposals for "a more stringent incapacitants regime" as one of "a number of very controversial points...which the OPCW has yet to reach a concensus on whether or not they truly fit into the object and purpose of the Convention."[209] Despite these important interventions, as of August 2015, no formal Organisation-wide consultation process on these issues has taken place. Consequently, no OPCW policy-making organ has made any interpretative statements regarding application of the CWC in this area, or issued guidance as to whether toxic chemicals intended for use as ICA weapons can be employed for law enforcement purposes and, if so, under what circumstances. It is therefore left to individual States Parties to interpret the scope and nature of their obligations in this area.

5.5. Conclusions

As part of the HAC stage two analysis, this chapter explored the regulation of ICA weapons, RCAs and related means of delivery under the CWC. Although RCAs are defined under the CWC, the scope and nature of their permissible use in situations of armed conflict and in law enforcement operations are ambiguously regulated, due in part to the CWC's failure to define "law enforcement".

There is evidence of widespread consensus amongst the vast majority of CWC States Parties that use of RCAs for both international and non-international armed conflict is prohibited under the CWC. However, one State Party – the US – has policy permitting RCA employment in certain non-offensive actions in areas of armed conflict. Furthermore, the reported practices of a small minority of States – including the United States and Turkey – raise concerns that such agents have been and may in future be utilized in certain non-international armed conflict situations, for example, COIN.

Although a range of international legal experts have explored the scope and nature of "law enforcement" activities utilizing toxic chemicals (including RCAs) that are permitted under the CWC both domestically and internationally, the issue has not been formally addressed by the relevant policy-making organs of the OPCW. Consequently, it is left to individual States Parties to interpret the application of the CWC in this area, taking into account the CWC's "types and quantities" limitations, as well as constraints arising from relevant international law, particularly that relating to human rights and the legitimate use of force.

ICA weapons – whether they are pharmaceutical chemicals or derive from living organisms (i.e. toxins and bioregulators) – are toxic chemicals and consequently come under the scope of the CWC. Whilst the employment of ICA weapons as a method of warfare is clearly prohibited under the CWC, there are divergent interpretations of the CWC amongst States Parties regarding the use of ICA weapons for law enforcement purposes. Concerns about the employment of ICA weapons for law enforcement were heightened following the Russian Federation use of such a weapon in 2002. However, despite the subsequent attempts by a growing number of States Parties, and actors such as the ICRC, to raise ICA weapons within the OPCW, as of August 2015, the organization has been unwilling or unable to adequately address these issues.

If this situation is not resolved by the OPCW policy-making organs, and the existing ambiguity continues, with State practice becoming a key determining factor influencing the scope of accepted activities, there is a danger that a permissive interpretation will evolve which will, in turn, allow further proliferation and the danger of misuse of ICA weapons. This could then lead, potentially, to the undermining of the CWC and prohibition on chemical weapons.

6
Arms Control and Disarmament Agreements Applicable to ICA Weapons and Riot Control Agents

6.1. Introduction

The Chemical Weapons Convention (CWC) appears to be the international arms control agreement most directly relevant to the regulation of ICA weapons, RCAs and their means of delivery. However, there are additional arms control and disarmament agreements and related regimes that are also potentially applicable, and these will be explored in this chapter, as part of the ongoing HAC analytical process. Examination of such agreements may reveal important additional regulatory mechanisms that impose obligations upon their State Parties (and more broadly if they are considered to constitute international customary law), which reinforce or supplement those stipulated under the CWC.

6.2. Geneva Protocol

6.2.1. Overview

The Geneva Protocol for the Prohibition of the Use in War of Asphyxiating, Poisonous or Other Gases, and of Bacteriological Methods of Warfare (Geneva Protocol) was signed on 17th June 1925 and came into force on 8th February 1928.[1] As of August 2015, the Protocol had 138 States Parties and 36 Signatory States.[2] Under the Geneva Protocol, the High Contracting Parties have acknowledged that "the use in war of asphyxiating, poisonous or other gases, and of all analogous liquids, materials or devices, has been justly condemned by the general opinion of the civilised world", and have further declared that "this prohibition shall be universally accepted as a part of International Law, binding alike the conscience and the practice of nations".[3]

The United Nations General Assembly (UNGA) subsequently adopted several Resolutions in which it called for strict observance by all States of the principles and objectives of the Geneva Protocol, condemned all actions contrary to those objectives and invited all States to accede to the Protocol.[4]

In December 1969, UNGA Resolution 2603A included an interpretation of the Geneva Protocol which reaffirmed the comprehensive nature of the agents covered by this prohibition, and specifically including "(a) Any chemical agents of warfare – chemical substances, whether gaseous, liquid or solid – which might be employed, because of their direct toxic effects on man, animals or plants;... ".[5] It should, however, be noted that this Resolution was not agreed by consensus. Instead, the Resolution was passed by an affirmative vote of 80 to 3 (Australia, Portugal and the United States voting against), with 36 abstentions.

6.2.2. Scope of agents covered

There has been disagreement concerning the scope of the Geneva Protocol, in particular whether RCAs were covered under this instrument.[6] Whilst the vast majority of States were of the opinion that the Geneva Protocol prohibited the use of all asphyxiating and poisonous gases and analogous materials, including RCAs, and applied it as such, certain States opposed this interpretation. In the late 1960s and early 1970s, Australia, Portugal and the United Kingdom – all Parties to the Geneva Protocol – changed their earlier comprehensive positions, subsequently stating that the Protocol did not apply to certain RCAs.[7]

Furthermore, the United States, which did not become a Party to the Geneva Protocol until 1975, maintained, prior to ratification, that the customary prohibition on chemical weapons did not apply to weapons with temporary effects (and continued to employ RCAs, most notably in Vietnam).[8] Even when the United States ratified the Geneva Protocol and renounced "as a matter of national policy... first use of riot control agents in war", it did so with reservations, enunciated under Executive Order 11850, allowing RCA use in a range of "defensive military modes to save lives".[9]

With the development and coming into force of the CWC, Australia, Portugal and the United Kingdom subsequently rejoined the majority of States in opposing all use of RCAs as a method of warfare in hostilities, without exception. However, as of August 2015, the United States maintained its position under EO 11850, which has guided its interpretation of both the Geneva Protocol and the CWC in this area.

Although the scope of the chemical and biological agents covered by the Geneva Protocol is considered by the international community to be very broad, and would appear to include RCAs (except for the United States in certain circumstances) and ICA weapons, the prohibition relates solely to their *use*; the Protocol does not address the development, production, transfer or stockpiling of such agents. Furthermore, the Protocol's prohibition on use is limited to situations of war (although it is now interpreted, through customary international law, to apply to all armed conflict).[10]

6.2.3. Effectiveness of the Geneva Protocol

Although the Geneva Protocol has widespread support, its prohibitions are somewhat weakened by reservations submitted by a number of countries when becoming Parties to the Protocol, declaring that they only regarded the non-use obligations as applying to other States Parties, and that these obligations would cease to apply if the prohibited weapons were used against them. Whilst at least 18 States Parties have subsequently removed their reservations to the Protocol, there are still at least 23 States Parties which maintain such reservations.[11] Consequently, the Protocol is considered by certain commentators to be reduced to a *de facto* no-first-use agreement amongst States Parties.[12] Requests for the removal of all outstanding reservations have repeatedly been made by the States Parties to the Biological and Toxin Weapons Convention (BTWC), notably during BTWC Review Conferences.[13] The Protocol's potential for impacting directly upon the regulation of weapons employing chemical and biological agents is further constrained due to its lack of verification measures, and the absence of an organization to facilitate and monitor implementation.

Despite the limitations of the Geneva Protocol, legal scholars consider that the prohibition on the use of chemical and biological agents as a method of warfare enshrined in this instrument has now become part of international customary law, applicable to both international and non-international armed conflicts.[14]

6.3. Biological and Toxin Weapons Convention

6.3.1. Overview

The Convention on the Prohibition of the Development, Production and Stockpiling of Bacteriological (Biological) and Toxin Weapons and on their Destruction – commonly known as the Biological and Toxin Weapons Convention (BTWC) – entered into force on 26th March 1975.[15] As of August 2015, the BTWC has 173 States Parties and 9 signatories;[16] there are 14 states which have neither signed nor ratified the Convention.[17]

The object and purpose of the BTWC is set out in the treaty's preamble: "Determined for the sake of all mankind, to exclude completely the possibility of bacteriological (biological) agents and toxins being used as weapons, Convinced that such use would be repugnant to the conscience of mankind and that no effort should be spared to minimize this risk, ..."[18]

Article I of the BTWC established the principal obligations and prohibitions of the Convention. It declared that:

Each State Party to the Convention undertakes never in any circumstances to develop, produce, stockpile or otherwise acquire or retain:

- Microbial or other biological agents, or toxins, whatever their origin or method of production, of types and in quantities that have no justification for prophylactic, protective or other peaceful purposes.
- Weapons, equipment or means of delivery designed to use such agents or toxins for hostile purposes or in armed conflict.[19]

Because of the prohibition on use in the Geneva Protocol, the prohibitions set out in Article I of the BTWC did not include a direct ban on use. However, although the BTWC did not *explicitly* prohibit the *use* of biological and toxin weapons, such prohibition was *implicit* through bans on the development, production, stockpiling or acquisition or retention by other means of these agents. It would only be after one or other of these prohibitions had been breached that use could occur. The BTWC prohibition on use has subsequently been affirmed by the BTWC State Parties at the Fourth,[20] Sixth,[21] and Seventh[22] Review Conferences. A further, and arguably more important, omission in the range of activities prohibited under the BTWC was that of research, particularly that involving biological weapons agents, with the consequent weakness of the BTWC control regime to adequately address dual-use research activities that could be misused.[23]

6.3.2. Scope of agents covered
6.3.2.1. Incapacitating chemical agents
Since a range of ICAs are either components or products of biological organisms or are synthetic biologically active analogues of such substances, they appear to fall within the scope of Article I of the BTWC, particularly if one considers the term "other biological agents" broadly. As Chevrier and Leonard have noted: "The ordinary meaning of 'other biological agents', unmodified by any restricting adjectives or clauses, implies a very broad scope, provided that broad scope is consistent with the object and purpose of the Convention."[24] As well as being captured under the term "other biological agents", some ICAs may also be covered by the Article I language that related to toxins. Although toxins were not defined under the Convention nor has a definition been collectively agreed subsequently by BTWC States Parties, they are commonly taken to denote chemical compounds produced by living organisms that are toxic or harmful to another living organism.[25]

A Western Government expert interviewed by the author has stated that: "The BTWC applies to incapacitating biological and toxin agents. SEB ... would be [an] incapacitating agent ... weaponised in [a] past offensive programme...."[26] However, the full extent of the range of incapacitating agents covered by the Convention is contentious. The expert has noted that although certain States believe the term "other biological agents" extends to bioregulators, this "has never been tested fully in a BTWC context such as a Review Conference".[27]

When considering the range of incapacitating agents that may be covered under the BTWC, it is important to take account of the "additional understandings and agreements" reached by the BTWC Review Conferences. For example, the Final Declarations of Seventh Review Conference declared that:

> [T]he Convention is comprehensive in its scope and that all naturally or artificially created or altered microbial and other biological agents and toxins, as well as their components, regardless of their origin and method of production and whether they affect humans, animals or plants, of types and in quantities that have no justification for prophylactic, protective or other peaceful purposes, are unequivocally covered by Article I.[28]

Consideration should also be given to the background papers prepared by Member States for Review Conferences, many of which have highlighted the relevance of peptides, bioregulators and their analogues to the BTWC, and have highlighted the potential dangers that misuse of such agents could pose. For example, in its submission to the Implementation Support Unit (ISU) review of "new scientific and technological developments relevant to the Convention",[29] conducted in advance of the Seventh Review Conference, the United Kingdom included a discussion of neuroscience, in which it stated that:

> Developments in this area could also result in the identification of compounds with potential for misuse as biological or toxin weapons agents since drugs acting on the brain to produce toxic or incapacitating effects could also have utility in a BW programme. Methods to facilitate delivery of such agents could also be exploited for harmful purposes, for example, to facilitate the entry of peptide neurotoxins across the BBB [blood–brain barrier]. However, the prohibitions of Article I of the BTWC fully cover all such biological agents and toxins, whether naturally occurring or artificially created, regardless of their origin or method of production.[30]

Similar concerns were highlighted by the ISU in its synthesis report of scientific and technological advances prepared for the Seventh Review Conference. Considering neurobiology amongst the range of "advances with potential for weapon applications", the ISU stated that: "Since the last review conference, there have been advances in: understanding the role of neuroregulators; how to influence psychological states and alter physical performance...."[31]

Furthermore, certain States, at least at the national level, have explicitly recognized the applicability of both the BTWC and the CWC to ICA weapons. For example, in August 2009, as part of its response to the Fourth Report from the UK House of Commons Foreign Affairs Committee, the UK Government stated that: "Development, production, retention, acquisition or use of 'Incapacitating biochemical weapons' are prohibited by both [BTWC and CWC] Conventions. Use of the word 'weapons' here is crucial'."[32]

6.3.2.2. Riot control agents

The majority of commonly used RCAs are chemicals not of biological origin. They could not, therefore, be classified as "microbial, or other biological agents, or toxins" and consequently do not appear to fall within the scope of the BTWC. However, OC, which is derived from the pepper plant and its relatives, is clearly of biological origin and could be considered to be a toxin. It therefore does appear to be covered by the Convention. Similarly, since Article I of the Convention included "toxins, whatever their origin or method of production", a case can be made that PAVA, a synthetic capsaicinoid, may also be covered under the BTWC. Both OC and PAVA are employed in a variety of spray devices and projectiles used widely by police and other security agencies for law enforcement activities in many parts of the world.

Only one BTWC State Party – the United States – has released information indicating its position with regard to such agents. In 1998, a legal review of OC undertaken by the Office of the US Navy Judge Advocate General (JAG) concluded that: "neither the 1925 Geneva Gas Protocol nor the 1972 Biological Weapons Convention prohibit the acquisition or employment of oleoresin capsicum".[33] Furthermore, the JAG legal review also stated that "OC... falls outside the BWC definition. It is, in fact, used as an additive in foodstuffs and pharmaceutical products."[34]

The JAG review is troubling on both counts. Whilst it is clear that neither the Geneva Protocol nor the BTWC would *per se* prohibit the acquisition or employment of OC, both agreements do appear to constrain the use of this agent, the former prohibiting use as a "method of warfare" and the latter prohibiting development, production, stockpiling, acquisition or retention of OC for use in "armed conflict" or for "hostile purposes". Furthermore, the stark JAG statement that the BTWC does not cover OC is also of concern, particularly as it is not supported by any detailed argumentation in the review. It should be noted that this legal review was undertaken in 1998 and it is unclear whether the position outlined is maintained by the current US Administration. Although there have been two subsequent legal reviews of "less lethal" weapons incorporating OC/PAVA,[35] in 2003 and 2008, the full contents of neither have been made public. Whilst the US, in the June 2015 DoD Law of War Manual,[36] detailed the prohibitions and restrictions relevant to toxin weapons, it has not specifically clarified its current position on their application to weapons employing OC and PAVA.

As of August 2015, the applicability of the Convention to OC, capsaicin or their synthetic analogues does not appear to have been addressed collectively by the BTWC States Parties during the negotiation or subsequently through additional agreements or interpretation. It is therefore for individual BTWC States Parties to determine whether and how the Convention applies to RCAs of biological origin and their synthetic analogues.

6.3.3. Scope of activities regulated

Although the BTWC does appear to cover certain incapacitating agents (including peptides, bioregulators, toxins and their synthetic analogues) and, potentially, RCAs of biological origin, as well as "weapons, equipment or means of delivery",[37] there are ambiguities regarding the nature and scope of such coverage.

Article I prohibits all activities with biological and toxin agents except those that can be justified for "prophylactic, protective or other peaceful purposes". During the *travaux préparatoires* States Parties clarified the term "prophylactic" as including medical activities, such as diagnosis, therapy and immunization, and also agreed that "protective" activities would cover development of protective masks and clothing, air and water filtration systems, detection and warning devices and decontamination equipment, and that these applications must not be interpreted as to permit possession of biological agents and toxins for defence, retaliation or deterrence.[38] There has, however, been no such definition agreed for "other peaceful purposes",[39] though Goldblat believed that "one can assume that it includes scientific experimentation".[40] Although these phrases have been used in BTWC Final Declarations, no interpretative or common understandings were further elaborated or agreed at the Review Conferences. These ambiguities have led to concerns by some commentators that bio-defence programmes may potentially be employed to mask prohibited activities.[41] An additional area where greater clarity from the BTWC States Parties would be beneficial is whether "law enforcement" purposes are considered to fall within the ambit of "other peaceful purposes", and if so, what is the nature and scope of permissible activities?

Further ambiguity surrounds the Article I prohibition on "weapons, equipment or means of delivery designed to use such agents or toxins for hostile purposes or in armed conflict", as the term "hostile purposes or in armed conflict" has been left undefined in the Convention. It is unclear whether the term "armed conflict", as used in the BTWC, was originally intended to refer solely to international armed conflicts or also covers non-international conflicts. However, in their authoritative review of international humanitarian law, Henckaerts and Doswald-Beck stated that they considered that "State practice establishes this rule [the use of biological weapons meant to affect humans is prohibited] as a norm of customary international law applicable in both international and non-international armed conflicts".[42]

Since Article I used the phrase "hostile purposes *or* in armed conflict", this construction implies that there are a range of activities other than "armed conflict" where the use of biological and toxin weapons would be prohibited under the BTWC. The range of activities covered by "hostile purposes" is not elaborated in the Convention nor has it subsequently been determined by the BTWC States Parties during BTWC Review Conferences. It is unknown whether the employment of weapons, equipment or means of delivery utilizing biological agents or toxins (including ICA weapons and RCAs of biological origin), for breaches of international humanitarian law or human rights violations, could be considered as "hostile purposes" and prohibited under the BTWC. This is an area where further study would be fruitful.

Since the terms "hostile purposes or in armed conflict" and "other peaceful purposes" have not been fully delineated, it is unclear how the use of ICAs or RCAs of biological origin for military operations other than war (MOOTW) would be regulated by the BTWC. Publicly available discourse on these issues, even amongst civil society experts, is sparse and a common position lacking. A Western Government expert, interviewed by the author, has contended that the use of ICA weapons (or RCAs of biological origin) for counter-insurgency operations or MOOTW would be prohibited under the Convention. However, the expert stated that the use of such agents for counter-terrorist activities or law enforcement more generally is a grey area, the acceptability of which would be dependent on how such activities were defined.[43] As of August 2015, however, there have been no determinations of these issues by the BTWC States Parties.

The situation with regard to the use of RCAs of biological origin for law enforcement activities seems to be clearer, given the substantive State Practice in this area. The employment of pepper spray, OC and PAVA is widespread amongst law enforcement agencies around the world and such use *per se* does not appear to have been raised publicly as an issue of concern in the context of the BTWC by any State Party. However, the acceptability of agent use in a specific instance would depend on the type and quantity of agent employed, the means of delivery or dispersal, as well as the nature and context of use.

6.3.4. Effectiveness of the BTWC

Important limitations on the value of the BTWC (and its control regime) as a tool to regulate ICA weapons and RCAs of a biological origin and their means of delivery arise from its current lack of effective verification and compliance mechanisms and also the absence of an international organization comparable to the OPCW to coordinate such activities and facilitate implementation by States Parties.[44]

Although lacking formal verification measures, the BTWC does contain two Articles that were intended to address non-compliance concerns by other means. Article V encouraged States Parties to "consult one another and to cooperate in solving any problems which may arise in relation to the objective of, or in the application of the provisions of, the Convention".[45] Although these provisions have been employed by States Parties, such bilateral consultations are necessarily confidential in nature, and consequently it is hard to ascertain how frequently such consultations have been utilized and how effective they have been. As well as such bilateral consultations, Article V also allows for consultations to be conducted on a multilateral basis "through appropriate international procedures within the framework of the United Nations and in accordance with its Charter".[46] As of August 2015, this provision has been utilized once, resulting in a consultative meeting in 1997.[47]

In addition, Article VI of the BTWC allowed any State Party to lodge a complaint with the United Nations Security Council (UNSC) about non-compliance by another State Party. The UNSC may then launch an investigation, with which all parties are enjoined to cooperate.[48] Hampson has, however, highlighted the danger that "using the Security Council as a means of enforcement exposes such a complaint to the risk of the exercise of the veto".[49] Indeed, ever since the BTWC came into force, no State Party has made such a request under Article VI.

Despite these major limitations, the BTWC States Parties have introduced an important mechanism which could be utilized to monitor advances in science and technology that could be applicable to the development of new forms of ICA weapons or RCAs of biological origin. The Seventh BTWC Review Conference decided to include, as part of its 2012–2015 intersessional programme, a standing agenda item to review developments in the field of science and technology related to the Convention.[50] Consequently, in 2014: "advances in the understanding of pathogenicity, virulence, toxicology, immunology and related issues" were reviewed, whilst in 2015 "advances in production, dispersal and delivery technologies of biological agents and toxins" will be considered.[51]

This structured review by the BTWC States Parties presents an important opportunity for individual States Parties and civil society to highlight R&D trajectories of potential concern, including those relating to the development of incapacitating chemical and toxin weapons, RCAs of biological origin and related means of delivery. Furthermore, the recognition by BTWC States Parties of the need to develop and strengthen mechanisms to monitor scientific and technological developments of potential threat to the BTWC mirrors, and will hopefully complement, similar initiatives taking place in the OPCW. It is hoped that such processes signal a growing willingness by both BTWC and CWC States Parties to address the convergence of the life sciences in areas of concern to both control regimes.

6.4. UN Secretary General's investigation mechanism

6.4.1. Overview

On 12th December 1980, the UN General Assembly granted the UN Secretary General a mandate under UNGA Resolution 35/144C to investigate allegations of chemical weapons use.[52] On 13th December 1982, a second UNGA, Resolution 37/98D, broadened this mandate to include "activities that might constitute a violation of the Geneva Protocol", and thus extended the scope of this authority to cover allegations of the use of biological and toxin weapons as well as chemical weapons.[53]

The last revision of the mandate was that of UNGA Resolution 42/37c, 1987, which expanded the mandate further still to encompass "the possible use of chemical and bacteriological (biological) or toxic weapons that may constitute a violation of the 1925 Geneva Protocol or other relevant rules of customary international law".[54]

This Resolution was also important in that it empowered the UN Secretary General to launch, on his or her own authority, an immediate field investigation of any credible complaint of alleged chemical or biological weapons use. The UN Security Council adopted a similar resolution in 1988. According to Tucker, these changes made "the secretary general the sole arbiter of which allegations to investigate and the level of effort devoted to each investigation".[55] Prior to this, an UNGA or UNSC Resolution formally requesting the Secretary General to launch an investigation had been required.

The mandate given to the Secretary General, as enunciated in the 1987–1988 Resolutions, was subsequently reaffirmed in UNGA Resolutions each year up to 1991.[56] It did not include an expiration clause and therefore continues to be in force to this day.

From 1981 to 1992 there were investigations or attempted investigations of alleged chemical/toxin weapon use by Vietnam in Cambodia/Kampuchia, Laos and Thailand (1981–1982), the Soviet Union in Afghanistan (1982), Iraq and Iran, as well as the use by Iraq against the Kurdish population within its borders (1984–1988), Armenia in Azerbaijan (1992), and a non-state actor (the Mozambican National Resistance – RENAMO) against the Mozambican Government forces (1992).[57] From 1993 until 2013, the mechanism had lain dormant.

Support for revivifying the Mechanism began to grow in the mid-2000s with three UN reports,[58] recommending it be strengthened so as to enhance UN capability to investigate potential use of biological weapons attacks. Furthermore, in 2006, the UNGA Global Counter-Terrorism Strategy and Action Plan encouraged the Secretary General to "update the roster of experts and laboratories, as well as the technical guidelines and procedures, available to him for the timely and efficient investigation of alleged use [of chemical, biological and toxin weapons]".[59]

Such an update was undertaken in 2007 by a group of experts from interested UN Member States and representatives of international organizations, including the World Organisation for Animal Health (OIE), the World Health Organization (WHO) and INTERPOL. As of June 2010, the updated roster of experts contained nominations of some 130 biological experts from

40 Member States and also from international organizations. Furthermore, there were some 30 diagnostic and analytical laboratories nominated by Member States, including one mobile laboratory for field deployment.[60] These laboratories can be called upon by the Secretary General to participate in interlaboratory calibration studies in order to establish validity and accuracy. In January 2011, the UN and the WHO signed a memorandum of understanding to facilitate further cooperation and enhance technical capabilities for investigating alleged use of chemical, biological and toxin weapons.[61] Training courses for experts on the UN Secretary General's roster were organized by Sweden in May 2009 and by France in November 2012.[62] Both training courses were undertaken in close cooperation with specialized international organizations, such as the OPCW, WHO, OIE and INTERPOL. Subsequently (as described in Section 6.4.4), in March 2013, based on the authority given to him by UNGA resolution A/RES/42/37C, the Secretary General established a UN Mission to investigate allegations of the use of chemical weapons in Syria.

6.4.2. Potential limitations of the UN Secretary General's mechanism

Under the initial 1980 mandate, reaffirmed in 1981, the Secretary General had authority to investigate allegations brought to him from any source of chemical weapons use. In 1982, however, the Secretary General's authority was limited to investigating allegations raised by UN Member States. This remains the position as of August 2015. Consequently, non-state actors, such as armed opposition groups, or indeed civilian communities that have been the targets of an attack utilizing chemical or biological weapons, cannot directly call for an investigation. They can, however, attempt to alert a receptive UN Member State and seek its assistance in bringing their allegations before the attention of the UN Secretary General. A second potential route by which non-State actors could attempt to trigger the Mechanism is by "request[ing] assistance [directly] from the United Nations, where a plea for aid could be presented to the President of the UN Security Council or to the UN General Assembly (UNGA)".[63]

Although the UN Secretary General's mechanism can be initiated by the UN Secretary General at the request of a single Member State and does not require any further body – such as the P5,[64] or the Security Council – to sanction such action, in reality, whether this mechanism would be triggered would probably be influenced by a variety of "geo-political" factors. Such factors could include the identity of the State or States requesting use of the Mechanism, the State allegedly employing the chemical or biological weapon, the target of such action, the circumstances surrounding the alleged use, as well as the perceived strength, independence and freedom of action of the UN Secretary General at the time. A similar mix of "political" considerations would no doubt influence the effectiveness of the resulting investigations and also whether and how the UNSC or the UNGA acted upon the findings of such investigations.

6.4.3. Scope of coverage

Since the UN Resolutions establishing and enhancing the UN Secretary General's Mechanism did not make any distinction between lethal or "less lethal" chemical or biological weapons, the Mechanism would be applicable in situations where ICA weapons, and also RCAs (at least in certain circumstances) were used in breach of the Geneva Protocol and "other relevant rules of customary international law". Although the Mechanism could be employed to investigate the *use* of ICA weapons (and potentially also RCAs), it could not be utilized to study the development, stockpiling, deployment or transfer of such weapons. There has been some State practice, albeit limited, which supports the future potential employment of the Mechanism to examine allegations of the use of an ICA weapon.[65]

The case for employing the Mechanism for the alleged use of RCAs in armed conflict is more problematical given the contested interpretation of the Geneva Protocol's coverage in this area. However, it should be noted that, in at least one instance, a UN Group of Experts acting under the auspices of the UN Secretary General Mechanism investigated the use of RCAs in Afghanistan in the early 1980s.[66]

Although initially designed to investigate breaches of the Geneva Protocol, subsequent UNGA Resolutions extended the Mechanism's scope to include "other relevant rules of customary international law".[67] However, whilst the Mechanism can potentially be employed in such circumstances, the range of "relevant" rules of customary international law has not been determined by the UNGA or the UNSC.

Although the UN Secretary General's Mechanism has previously been employed to investigate the use of chemical weapons, with the coming into force of the CWC, the OPCW would presumably now take the lead in cases of alleged use of chemical and toxin weapons against a CWC State Party. In such circumstances, and following a request from the CWC State under attack or a second CWC State Party, the OPCW could initiate the CWC investigation and assistance machinery.[68] However, the Secretary General's Mechanism provides a possible fall-back option which could potentially be used in certain situations where the CWC investigation and assistance mechanisms were not initiated.

One situation where the UN Secretary General's Mechanism could potentially be triggered is where there have been allegations regarding the use of RCAs or ICA weapons employing toxins or other biological agents derived from living organisms, and where the CWC States Parties could not agree that such substances fall within the ambit of the Convention. In such situations, the UN Secretary General's Mechanism could potentially be employed, if requested by a UN Member State, as the Mechanism's coverage explicitly includes both chemical and biological weapons.

An alternative scenario could arise, for example, where ICA weapons or RCAs have reportedly been used by and/or against CWC Non-State Parties. The OPCW's limited ability to act "[i]n the case of alleged use of chemical weapons involving a State not Party to this Convention or in territory not controlled by a State Party" is recognized in the Verification Annex to the CWC.[69] In such situations, the UN Secretary General would take the lead and "the Organization [OPCW] shall closely cooperate with the Secretary-General of the United Nations. If so requested, the Organization shall put its resources at the disposal of the Secretary-General of the United Nations."[70] In 2001, the OPCW and the UN further formalized and strengthened cooperation in such cases through agreement of a memorandum of understanding.[71] Although such scenarios may have seemed unlikely given the large majority of States now party to the Convention, there have been some particular "hold-out" States of concern. For example, there have been reports that Israel has previously undertaken research into and in fact used an ICA weapon (as detailed in Chapter 2). Furthermore, Egypt[72] and Israel[73] both possess and have reportedly misused RCAs for serious human rights violations. The utility of the Secretary General's Investigatory Mechanism was brought starkly to light following widespread reports, from early 2012 onwards, of the use of chemical weapons (allegedly including ICA weapons) against the civilian population and against combatants fighting a civil war in Syria (which was not then a party to the CWC).

6.4.4. Case study: Investigation of allegations of chemical weapons in Syria

Since the 1970s, Syria had reportedly acquired and/or developed and stockpiled quantities of a range of chemical weapons, including blister agents and nerve agent precursors, as well as associated means of delivery.[74] On 23rd July 2012, the Syrian Ministry of Foreign Affairs stated that Syria possessed chemical weapons and that "All of these types of weapons are in storage and under security and the direct supervision of the Syrian armed forces and will never be used unless Syria is exposed to external aggression."[75] From early 2012 there were repeated but, as of August 2015, unconfirmed allegations that the Syrian Government armed forces employed a range of toxic chemicals including ICAs, during the ongoing conflict with armed opposition forces.[76]

On 21st February 2012, the Istanbul *Hürriyet Daily News* reported allegations by Lt Abdulselam Abdulrezzak, who, the paper claimed, "used to work in the chemical weapons department in the Syrian army and defected to Turkey last week". It was claimed that: "chemical weapons were used against civilians during the military offensive of the Syrian security forces in Bab Amr [a neighbourhood in Homs]".[77] Abdulrezzak reportedly stated that: "BZ-CS, Chlorine Benzilate, which damages people's nerves and makes them fade away, is being used in Bab Amr."[78] Perry Robinson noted that: "this allegation seems to be the first occasion that incapacitating agent BZ has been mentioned [in the open literature] as an element of Syrian capability".[79] Perry Robinson further noted that "it is not obvious why 'BZ-CS' should have been glossed as 'Chlorine Benzilate'. Nor is it obvious that either agent would have brought about the signs and symptoms described."[80]

On 15th January 2013, the *Cable* (the online blog of the US magazine *Foreign Policy*) carried an article by Josh Rogin that stated that the US Consul-General in Istanbul had sent a secret cable to the State Department detailing his "investigation into reports from inside Syria that chemical weapons had been used in the city of Homs on Dec 23".[81] According to Rogin, an unnamed "Obama administration official who had reviewed the cable" stated that "We can't definitely say 100 percent, but Syrian contacts made a compelling case that Agent 15 was used in Homs on Dec 23."[82] Concerns about the veracity of the allegations contained in the *Cable* article soon emerged. On 15th January 2013, the White House National Security Council spokesman stated that: "reporting we have seen from media sources regarding alleged chemical weapons incidents in Syria has not been consistent with what we believe to be true about the Syrian chemical weapons program".[83]

On 27th April 2013, the Dubai-based television news channel *Al Arabiya* broadcast excerpts from an interview with a Syrian army defector, Brigadier Zaher Al-Saket, described as "former head of chemical warfare in the 5th division", in which he stated:

> When the demonstrations started, the regime used harassing agents, like any country in the world using tear gas to disperse demonstrations. As for [other types of chemical weapon]...the regime used incapacitating agents at first, but when the world remained silent about this, and the regime thought that the international community did not care, it used lethal weapons in more than 13 locations.[84]

Although it was apparent that some form of chemical weapon had been employed in Syria, the nature of the agent and the perpetrators were unclear and fiercely contested by the parties to the conflict.

On 19th March 2013, the Syrian Government reported the alleged use of chemical weapons against its forces in the Khan al-Asal area of the Aleppo Governorate. The following day Syria formally requested the UN Secretary General to launch an urgent investigation under the auspices of his Mechanism. The UN Secretary General agreed to do so and contacted the OPCW and the WHO, requesting their cooperation in mounting an investigation.

Other governments came forward alleging the same and other incidents of chemical weapons use, and the UN Secretary General announced that all credible allegations would be investigated. A special team for this purpose was put together, comprising qualified experts from the OPCW and WHO. Professor Åke Sellström of Sweden was appointed by the UN Secretary General to head the UN Mission.[85]

The team was despatched to investigate three of the reported incidents, including Khan al-Asal. They arrived in Damascus on 18th August 2013. Shortly after their arrival, reports emerged that there had been a major attack on 21st August 2013, again allegedly involving chemical weapons, in the Ghouta area of Damascus. A number of UN Member States wrote to the Secretary General requesting an urgent investigation making use of the Mission already in Syria. On 22nd August, the Secretary General determined that the investigation team should, as a priority, investigate this incident. On 25th August, permission was obtained from the Syrian Government, and the investigation mission commenced its onsite work the following day.

Following its investigations, the UN Mission produced two reports; the first (interim) report was restricted to "ascertain[ing] the facts" related to "the alleged use of chemical weapons in the Ghouta area of Damascus on 21st August 2013".[86] It concluded that "on 21 August 2013, chemical weapons have been used in the ongoing conflict between parties in the Syrian Arab Republic, also against civilians, including children, on a relatively large scale".[87] The report further concluded that "the environmental, chemical and medical samples we have collected provide clear and convincing evidence that surface-to-surface rockets containing the nerve agent Sarin were used in Ein Tarma, Moadamiyah and Zamalka in the Ghouta area of Damascus".[88] The second (final) report,[89] in addition to documenting the Ghouta attacks, also stated that the UN Mission collected "credible" information corroborating allegations of the use of a chemical weapon – apparently an organophosphorus compound – against soldiers and civilians in Khan al-Asal on 19th March 2013. The report also documented alleged chemical weapons attacks in Jobar, Saraqueb, Ashrafiah Sahnaya, Bahhariyeh and Sheik Maqsood. Neither

UN Mission report made any reference to the use of ICAs during these attacks.

On 14th September 2013, Syria deposited the instrument of accession to the UN Secretary General, requesting to join the Chemical Weapons Convention, and formally acceded to the CWC on 14th October 2013.[90] As required under the Convention, Syria declared its existing stockpile of chemical weapons and agreed to facilitate their verification and subsequent destruction under the supervision of the OPCW.

Although continued allegations have been made of the use of ICA weapons in Syria,[91] no evidence of the development or use, by either the Syrian Government or the armed opposition, of weapons employing ICAs has been made public. Instead, a fact-finding mission (FFM) established by the OPCW Director General found information constituting "compelling confirmation" that a toxic chemical – chlorine – was used "systematically and repeatedly" as a weapon in villages in northern Syria in 2014.[92] Although the FFM reports have not been formally released to the public, there appears to be no reference made to the use of any ICA weapons during these attacks.[93]

Despite its limitations, the UN Secretary General's Mechanism does provide a potential avenue of last resort for the investigation of alleged chemical and biological weapons (including ICA weapons and, potentially, RCAs in certain circumstances) when other mechanisms have not been triggered. As Littlewood has stated: [O]ne of the beauties of the Secretary General's Mechanism, in my view, is that it allows you to have an option on the table that will cover most eventualities. It might not be the ideal solution, but it is nevertheless one tool that could be used if other tools are deemed inappropriate.[94]

6.5. United Nations Security Council Resolution 1540

6.5.1. Overview

In April 2004, the UNSC adopted United Nations Security Council Resolution (UNSCR) 1540,[95] which required that UN Member States:

[A]dopt and enforce appropriate effective laws which prohibit any non-State actor to manufacture, acquire, possess, develop, transport, transfer or use nuclear, chemical or biological weapons and their means of delivery, in particular for terrorist purposes, as well as attempts to engage in any of the foregoing activities, participate in them as an accomplice, assist or finance them.[96]

6.5.2. Scope of coverage

Whilst the Resolution was specifically intended to combat nuclear, chemical and biological weapons proliferation to non-State actors, it also required the introduction and enforcement of broader non-proliferation measures in the areas of material accountancy and security; physical protection; border controls; and export and trans-shipment controls applicable to State as well as non-State actors.[97]

Under UNSCR 1540, States were called upon "to present a first report no later than six months from the adoption of this resolution...on steps they have taken or intend to take to implement this resolution".[98] UNSCR 1540 also established a Committee of the Security Council which was tasked to review Member State reports and to inform the Security Council of the implementation of this Resolution.[99]

Kelle, Nixdorff and Dando noted that, for CWC and BTWC States Parties, UNSCR 1540 was "largely a reiteration and specification of commitments already undertaken" under those Conventions. The importance of UNSCR 1540 lay in the fact that the Resolution "makes these stipulations binding on all states, including those outside the prohibition regimes and in that sense presents a useful step forward in ensuring universality of the norms against CBW".[100]

Since UNSCR 1540 made no reference to weapons of mass destruction, and did not differentiate between lethal and "less lethal" chemical and biological weapons, the Resolution appears to cover ICA weapons, RCAs and their means of delivery.

6.5.3. Effectiveness of UNSCR 1540

Although UNSCR 1540 was adopted by the UNSC under Chapter VII of the UN Charter and is therefore binding on all UN Member States, a range of arms control experts have raised concerns about its effectiveness in practice. Kelle, Nixdorff and Dando have noted that "the wording of the resolution... allows for loopholes in implementation".[101] In 2008, Arnold detailed how: "After four years, confusion and scepticism still reign among member states regarding compliance, prioritization, and applicability of Resolution 1540."[102]

Analysts also raised concerns over the limited resources and temporary nature of the mechanisms that were intended to monitor and facilitate implementation of the Resolution – most notably the 1540 Committee, which originally only had a lifespan of "no longer than two years".[103] Although the Committee's existence was subsequently extended, this was initially done in a spasmodic and limited fashion by two subsequent UN Resolutions for consecutive periods of two years.[104] Subsequently, however, there was a substantial reinforcement of the mechanisms in place to facilitate and build capacity to implement UNSCR 1540, with the UNSC adoption of UNSCR 1977 on 20th April 2011.[105] Under this Resolution, the Security

Council agreed to extend the Committee's mandate for a full ten years until 25th April 2021;[106] established a "Group of Experts" to assist the Committee;[107] and encouraged the Committee to form partnerships with regional and intergovernmental organizations to promote universal implementation of UNSCR 1540.[108] The 1540 Committee has been mandated to conduct two "comprehensive" reviews of the status of implementation of UNSCR 1540 (2004), submitting its reports to the Security Council; "the first review should be held before December 2016".[109]

A further important development has been the adoption, following the use of chemical weapons in Syria and the subsequent UN investigative report, by the UNSC of Resolution 2118 (2013).[110] In addition to instituting mechanisms to facilitate and verify the "expeditious destruction of Syria's chemical weapons programme", the Security Council also decided that: "Member States shall inform immediately the Security Council of any violation of resolution 1540 (2004), including acquisition by non-State actors of chemical weapons, their means of delivery and related materials in order to take necessary measures therefore."[111]

The potential importance of UNSCR 1540 as a regulatory mechanism to combat the proliferation and misuse of ICA weapons, RCAs and related means of delivery by both State and non-State actors is reinforced by its application to all UN Member States regardless of whether they are States Parties to the CWC or BTWC. Furthermore, its potential utility has been further strengthened by the international community's recent efforts to improve the implementation and monitoring of the Resolution. As of August 2015, however, the application of UNSCR 1540 to ICA weapons, RCAs and related means of delivery has not been formally raised by any UN Member State nor considered by the UN Committee 1540.

6.6. Conclusions

It is clear from the results of the HAC stage-two survey summarized in this chapter that there are a number of international arms control and disarmament instruments beyond the CWC that are applicable to the regulation of (certain) RCAs and ICA weapons, most notably the Geneva Protocol and the BTWC. Furthermore, the employment of investigation mechanisms established under the auspices of the UN should also be considered in cases of alleged misuse of ICA weapons or RCAs, particularly in those cases where the OPCW is unable to act.

7
International Humanitarian Law Applicable to ICA Weapons and Riot Control Agents

7.1. Introduction

As part of the second stage of the HAC analytical process, this chapter will explore the constraints imposed upon States regarding the use of ICA weapons, RCAs and related means of delivery, as a result of their obligations under international humanitarian law (IHL), the body of law that applies during situations of armed conflict with the aim of protecting civilians and others who are no longer participating in hostilities, and regulating the conduct of such hostilities. Amongst its provisions are those regulating the means of conflict (including the weapons employed) and also the methods of warfare (how such weapons are employed). IHL is comprised of two elements, the first being IHL treaty law. This is binding *only* on those States that are party to the specific agreements. Furthermore, a treaty is only binding on a State if it has ratified it. Mere signature, indicating a future intent to be bound, is not sufficient, but a State that has signed a treaty is not free to undermine its objects and purposes.[1] In addition to the weapons-specific agreements prohibiting development, possession or use of chemical and biological weapons (i.e. BTWC, CWC and the Geneva Protocol), there are a number of generally applicable IHL treaties – dealing more broadly with the conduct of armed conflict – that are of potential relevance to ICA weapons and RCAs and related means of delivery, particularly the four Geneva Conventions of 1949,[2] and two of their three Additional Protocols.[3]

A second body of IHL derived from customary international law is also applicable. The Statute of the International Court of Justice described customary international law as "a general practice accepted as law".[4] According to Henckaerts and Doswald-Beck:

> [I]t is generally agreed that the existence of a rule of customary international law requires the presence of two elements, namely State

practice (*usus*) and a belief that such practice is required, prohibited or allowed, depending on the nature of the rule, as a matter of law (*opinio juris sive necessitatis*).[5]

Where customary international law has been deemed to be established, it is binding on *all* States whether or not they are parties to relevant treaties. The most recent definitive account of this area can be found in the 2005 ICRC Study of Customary International Humanitarian Law.[6]

7.2. Scope of IHL

IHL is applicable to international armed conflicts which can be described as conflicts that arise between two or more States, even if a state of war is not recognized by one of them. It also covers all cases of partial or total occupation of the territory of a State, even if the said occupation meets with no armed resistance.[7] In addition, IHL is applicable to armed conflicts of a non-international character, which can be considered as conflicts between the armed forces of a State and organized armed groups (or between two or more such groups) which are under responsible command, control territory and carry out sustained military operations. An armed conflict is not considered to include a situation of internal disturbances and tensions, such as riots, isolated and sporadic acts of violence and other acts of a similar nature.[8] The determination of whether a specific situation should be considered as an internal armed conflict or as internal disturbance requiring law enforcement can be highly contentious.[9] As well as regulating the activities of States in armed conflicts, IHL, as certain commentators have highlighted, is applicable to armed non-State actors that meet the requisite criteria.[10]

Hampson has noted that, while international armed conflicts are subject to both "more rules and more precise rules" than are non-international armed conflicts, a significant body of customary international law has developed to regulate non-international conflicts.[11] Particularly important is such developments – though in the sphere of international criminal law rather than IHL – have been the Statute of the International Criminal Court and the judgements of the *ad hoc* criminal tribunals for the former Yugoslavia and Rwanda.[12] Of particular note is the *Tadic* decision of the Appeals Chamber of the International Criminal Tribunal for the former Yugoslavia, which concluded that:

> [E]lementary considerations of humanity and common sense make it preposterous that the use by States of weapons prohibited in armed conflicts between themselves be allowed when States try to put down rebellion by their own nationals on their own territory. What is inhumane, and consequently proscribed, in international wars, cannot but be inhumane and inadmissible in civil strife.[13]

7.3. Obligations of relevance to the regulation of ICA weapons, RCAs and related means of delivery

There are a number of obligations which derive from IHL – either from specific treaties or under customary international law – which constrain or prohibit the development and utilization of ICA weapons, RCAs or related means of delivery by States.[14] These include:

(i) The recognition that the right of the parties to an armed conflict to choose their methods or means of warfare was not unlimited: This is a basic tenet of IHL, stipulated, for example, in the Hague Regulations,[15] and subsequently in Additional Protocol I,[16] and in the Convention on Cluster Munitions.[17] As Casey-Maslen has noted, this was a general restatement "of the international legal reality that certain weapons can never be lawfully used, while other weapons can be used subject to the restrictions imposed by applicable international law".[18]

(ii) The protection of persons considered hors de combat: Common Article 3 of the Geneva Conventions has stated that "Persons taking no active part in the hostilities, including members of armed forces who have laid down their arms and those placed *hors de combat* by sickness, wounds, detention, or any other cause, shall in all circumstances be treated humanely."[19] In addition, certain actions against the above-mentioned persons are prohibited, including "(a) violence to life and person, in particular murder of all kinds, mutilation, cruel treatment and torture; (b) taking of hostages; (c) outrages upon personal dignity, in particular humiliating and degrading treatment...".[20] There is also a positive obligation to ensure that "the wounded and sick shall be collected and cared for".[21] Similar obligations are also deemed to be part of international customary humanitarian law.[22]

The ICRC has highlighted its concerns that the use of ICA weapons in armed conflict would "make it difficult or impossible to determine when a combatant is 'out of action' and thereby afforded protection and assistance. An incapacitated combatant would probably not appear to be injured and may be unable to show a sign of surrender."[23] Consequently, the ICRC has argued that: "It would be difficult to train soldiers to distinguish whether an enemy were incapacitated or remained a threat. The resulting combination of incapacitants and lethal force could significantly increase the lethality of armed conflicts."[24]

Further disquiet has been voiced regarding the potential use of certain ICAs in the coercive treatment of enemy combatants rendered *hors de combat*. The Royal Society has highlighted the importance of IHL in constraining such activities, noting in particular that the prohibition against "outrages upon personal dignity, in particular humiliating and degrading treatment" (as enunciated under Additional Protocol I) "could extend...to the use of neuropharmacological agents to control or alter behaviour".[25] In addition, the Society argued that "the use of neuropharmacological agents or stimulation technologies to coerce or interrogate prisoners of war" would also be prohibited under the Third Geneva Convention.[26]

(iii) Constraints upon the activities of medical personnel: Under Additional Protocol I, it is prohibited to subject prisoners or detainees to "any medical procedure which is not indicated by the state of health of the person concerned and which is not consistent with generally accepted medical standards".[27] This specifically includes "medical or scientific experiments", even with the person's consent.[28] In addition, "any wilful act or omission which seriously endangers the physical or mental health or integrity of any person who is in the power of a Party" is considered to be "a grave breach" of Additional Protocol I.[29] Furthermore, under Additional Protocol II, medical personnel shall not be compelled to "perform acts or to carry out work contrary to...the rules of medical ethics or other rules designed for the benefit of the wounded and sick, or this Protocol".[30] Consequently, the Royal Society has concluded that: "The involvement of medical professionals in the administration of neuropharmacological agents for purposes other than those consistent with generally accepted medical practice is therefore prohibited by international law."[31]

(iv) The prohibition upon the employment of means and methods of warfare of a nature to cause superfluous injury or unnecessary suffering (SIRUS): The SIRUS prohibition has been enunciated in a number of IHL treaties,[32] and is contained in certain weapons-specific treaties,[33] and in the Rome Statute.[34] It is considered to be part of customary IHL.[35] The 2005 ICRC Study noted that, although the existence of the SIRUS prohibition is not contested, "views differ as to whether the rule itself renders a weapon illegal or whether a weapon is illegal only if a specific treaty or customary rule prohibits its use".[36] Furthermore, there is no international consensus regarding an objective means of determining what constitutes "superfluous injury or unnecessary suffering", nor on the criteria which can be used to judge whether specific weapons potentially breach the SIRUS prohibition.

In 1996, the ICRC established the SIRUS Project, which attempted to develop objective criteria in this area. Although a number of States were highly critical of this approach and the project was suspended in 2001,[37] it did gain support from bodies such as the World Medical Association,[38] and its findings are worthy of consideration. The SIRUS Project proposed that what constituted superfluous injury and unnecessary suffering should:

[B]e determined by design-dependent, foreseeable effects of weapons when they are used against human beings and cause[d]:

- specific disease, specific abnormal physiological state, specific abnormal psychological state, specific and permanent disability or specific disfigurement; or
- field mortality of more than 25% or hospital mortality of more than 5%; or

- Grade 3 wounds as measured by the Red Cross wound classification; or
- effects for which there is no well-recognized and proven treatment.[39]

Criteria 1 and 4, in particular, appear to be of potential applicability for States considering the legality of the development or use of new ICA weapons and RCAs. Subsequently, the ICRC has specifically highlighted the importance of the SIRUS prohibition when considering the development and use of ICA weapons. For example, Peter Herby, then-Head of the Arms Unit in the Legal Division of the ICRC, argued that ICA weapons should not be assumed to "merely incapacitate by making a person sleep".[40] Instead, he has raised the potential dangers that such weapons, if they resulted in effects such as "lifelong epileptic convulsions, permanent damage to internal organs, long-term and severe vomiting, or an extended coma" could violate the SIRUS prohibition.[41]

(v) *The prohibition of deliberate attacks on civilians, the prohibition of indiscriminate weapons and of attacks that do not discriminate between civilians and military objectives:* These prohibitions are considered fundamental to IHL and are covered by both treaty law (e.g. Additional Protocol I[42]) and customary IHL.[43] These prohibitions are considered applicable to both international and non-international armed conflict.[44] The International Court of Justice has stated that the principle of distinction was one of the "cardinal principles" of IHL and one of the "intransgressible principles of international customary law".[45]

Certain commentators and/or proponents of "less lethal" weapons have argued that the rules of IHL, particularly regarding distinction, should be reviewed and applied in a different manner for "less lethal" weapons than for conventional weapons.[46] Highlighting the dangers of such an approach, Herby noted that in discussions of potential scenarios for the use of ICA weapons, "situations in which civilians are interspersed with combatants are consistently mentioned...It is likely that the use of incapacitants will lower the threshold for attacks that affect civilians and combatants without distinction, with an inherent risk that this rule [prohibiting indiscriminate attacks] will be undermined."[47] The employment of RCAs against mixed populations of civilians and combatants – as permitted under existing US policy and regulations[48] – raises related concerns.

(vi) *Requirement to respect and ensure respect of international humanitarian law:* ICA weapons are amongst a range of weapons that can degrade the cognitive ability of enemy combatants, with the consequent dangers of their committing serious breaches of IHL. Under Common Article 1 to the four Geneva Conventions, and Article 1 of Additional Protocol I, "The High Contracting Parties undertake to respect and to ensure respect" for the relevant treaties "in all circumstances".[49] The Royal Society has argued that "degrading the cognitive abilities of an adversary such that they are unable to distinguish between military targets and civilians, which often require a high degree of concentration, will undermine this requirement".[50] This is because such cognitive impairment could easily result in an unintended attack on one's own civilians or other persons or places specifically protected by law. Furthermore, the Royal Society has contended that "Such attacks could not be prosecuted because the perpetrators will have been rendered mentally incapable of being responsible for the offences."[51]

(vii) Prohibitions or restrictions based on the principles of humanity and the dictates of public conscience (the "Martens clause"): The "Martens clause" is contained in a number of IHL treaties,[52] and has been found to represent customary international law.[53] As articulated in Additional Protocol I, the clause states that: "In cases not covered by this Protocol or by other international agreements, civilians and combatants remain under the protection and authority of the principles of international law derived from established custom, from the principles of humanity and from dictates of public conscience."[54]

According to the 2006 *ICRC Guide* "to the legal review of new weapons, means and methods of warfare", a weapon which was not covered by existing rules of IHL would be considered contrary to the Martens clause if it was determined *per se* to contravene the principles of humanity or the dictates of public conscience.[55] The International Court of Justice has affirmed the importance of the Martens clause, "whose continuing existence and applicability is not to be doubted",[56] and stated that it "had proved to be an effective means of addressing rapid evolution of military technology".[57] Herby has highlighted its potential applicability to ICA weapons, and has noted that: "Public conscience is of particular relevance to the prohibition of chemical or biological incapacitants."[58]

However, although the Martens clause is considered one of the cornerstones of IHL, the interpretation and meaning are contested.[59] Consequently, the international community should first seek to employ the existing chemical and biological weapons control regimes to effectively regulate or prohibit future forms of ICA weapons and RCAs, as appropriate. However, if such avenues prove ineffective or unworkable, the application of the Martens clause in conjunction with the relevant principles of IHL highlighted in this section is worthy of further study as a potential means to constrain the malign use in armed conflict of the rapidly evolving chemical and life sciences, until more elaborated controls can be developed and implemented.

7.4. Investigation and enforcement measures

The Geneva Conventions and Additional Protocols have established a range of investigation and enforcement measures which may potentially be applicable to cases involving the alleged use of RCAs or ICA weapons in armed conflict. For example, under the four Geneva Conventions, each High

Contracting Party must enact any legislation necessary to provide effective penal sanctions for persons committing, or ordering to be committed, any of the grave breaches of the Convention;[60] search for persons alleged to have committed, or to have ordered to be committed, grave breaches; bring such persons, regardless of their nationality before its own courts or hand them over for trial to another High Contracting Party concerned, provided it has made a *prima facie* case.[61] If there is dispute about the circumstances of an alleged violation, a Party to the conflict can request an enquiry, in a manner to be decided between the interested Parties. If the enquiry procedure cannot be agreed, the Parties can choose an umpire to decide upon the procedure to be followed. Once the violation has been established, the Parties to the conflict shall put an end to it and shall repress it with the least possible delay.[62] Enhanced investigation and enforcement mechanisms have been established under Additional Protocol I, with the establishment of an independent fact-finding Commission. Following an allegation by a Party to an international armed conflict, the Commission can "enquire into any facts alleged to be a grave breach... or other serious violation of the Conventions or of this Protocol",[63] and can "facilitate... the restoration of an attitude of respect for the Conventions and this Protocol".[64] To date, however, the jurisdiction of the fact-finding Commission has never been invoked. In practice, there is very little to compel a State to enforce its obligations in relation to its own armed forces.[65]

7.5. Customary international law obligations relating to specific weapons

As well as the overarching general obligations under IHL potentially applicable to the use of all weapons, a State must also consider the prohibitions or restrictions on the use of *specific* weapons, means and methods of warfare pursuant to customary international law. These rules apply to all States, regardless of whether they are party to relevant treaties. They have been detailed in the ICRC Study on customary IHL; and include the following prohibitions that are relevant to ICA weapons, RCAs and their means of delivery:

- The use of biological weapons is prohibited.[66]
- The use of chemical weapons is prohibited.[67]
- The use of RCAs as a method of warfare is prohibited.[68]

It is important to note that, according to the ICRC Study, these rules are applicable in both international armed conflict and non-international conflict.[69] These obligations reflect those contained in the Geneva Protocol, the BTWC and the CWC discussed in detail in Chapters 5 and 6 of this publication.

7.6. Obligations to review "new" weapons under international humanitarian law

Under Article 36 of Additional Protocol I to the Geneva Conventions, all High Contracting Parties that are engaged "in the study, development, acquisition or adoption of a new weapon, means or method of warfare" are "under an obligation to determine whether its employment would, in some or all circumstances, be prohibited by this Protocol or by any other rule of international law applicable to the High Contracting Party".[70]

As of August 2015, there were 174 States party to Additional Protocol I, which are explicitly bound by this obligation. In addition, the *ICRC Guide to the Legal Review of New Weapons* has stated that this requirement is "arguably" one that applies to all States, regardless of whether or not they are party to Additional Protocol I.[71]

7.6.1. Scope of weapons to be reviewed

The *ICRC Guide* has argued that the material scope of the Article 36 legal review should be very broad, and should cover:

- weapons of all types – be they anti-personnel or anti-materiel, "lethal", "non-lethal" or "less lethal" – and weapons systems [which would therefore include any new potential ICA weapon, RCA or means of delivery intended for use in armed conflict];
- the ways in which these weapons were to be used pursuant to military doctrine, tactics, rules of engagement, operating procedures and countermeasures;
- all weapons to be acquired, be they procured further to research and development on the basis of military specifications, or purchased "off-the shelf";
- a weapon which the State was intending to acquire for the first time, without necessarily being "new" in a technical sense;
- an existing weapon that was modified in a way that altered its function, or a weapon that had already passed a legal review but that was subsequently modified;
- an existing weapon where a State had joined a new international treaty which may affect the legality of the weapon.[72]

In addition, although not specifically called for in Article 36, Daoust, Coupland and Ishoey have argued that it "would be desirable" for States also to examine the legality of weapons they intend to export. This would be in line with their obligation under Article 1, common to the four Geneva Conventions of 1949 and Additional Protocol I, "to respect and ensure respect" for these treaties.[73]

The *ICRC Guide* has stated that the temporal application of Article 36 should be very broad. It would require an assessment of the legality of new weapons at the stages of their "study, development, acquisition or adoption". This would cover all stages of the weapons procurement process, in particular the initial stages of the research phase (i.e. conception, study), the development phase (i.e. development and testing of prototypes) and the acquisition phase (including "off-the-shelf" procurement).[74] Consequently, it can be argued that States that are currently conducting research into and development of new ICA weapons, RCAs or related means of delivery should have conducted a legal review of such weapons before research commenced, and, if not, should do so at the earliest opportunity.

Similarly, those States seeking to acquire new ICA weapons, RCAs or related means of delivery, either from another State or from the commercial market, should undertake a legal review at the stage of the study of the weapon proposed for purchase, and certainly before entering into the purchasing agreement.[75] Furthermore, a range of commentators have emphasized that the purchasing State is under an obligation to conduct its own review of the weapon it is considering to acquire, and cannot simply rely on the analysis of the vendor or manufacturer as to the legality of the weapon, nor on another State's evaluation.[76]

7.6.2. Criteria for review

Although Article 36 did not establish specific criteria for States to employ in their review of new weapons, Boothby has recommended that States should incorporate the following criteria:[77]

- whether, in its normal or intended circumstances of use, the weapon was of a nature to cause SIRUS;
- consideration of the alternative weapons or methods of choice for accomplishing the military purpose intended for the weapon under review;
- whether the weapon was intended, or might be expected, to cause widespread, long-term and severe damage to the natural environment;
- whether there are any specific rules of treaty or customary law that would prohibit or restrict the use of the weapon;
- whether there are any likely future developments in the law of armed conflict that might be expected to affect the weapon.

7.6.2.1. Applicability of human rights considerations

A number of commentators have argued that States should also incorporate international human rights law (IHRL) considerations in weapons reviews. For example, Casey-Maslen has contended that the reference in Article 36 to "any other rule of international law applicable" to a State Party to the

Protocol would also encompass IHRL.[78] Kathleen Lawand, then-ICRC legal advisor and principal author of the *ICRC Guide*, has stated that:

> [In] reviewing the legality of new weapons, states may also need to consider the rules of international human rights law applicable to the use of force in situations not amounting to armed conflict. This is especially important in view of the increased involvement of armed forces in peace support operations, where troops are more likely to be involved in law enforcement than warfare.[79]

The inclusion of IHRL considerations in Article 36 reviews would be particularly critical for any new potential ICA weapons, RCAs and related means of delivery, given the proposals by some commentators for the employment of such weapons by military personnel in certain counter-insurgency, counter-terrorism, peacekeeping and other military operations other than armed conflict.

7.6.2.2. Health considerations

The *ICRC Guide* has recommended that States should take into account certain health-related considerations when reviewing any new weapon that injures "by means other than explosive or projectile force, or otherwise causes health effects that are qualitatively or quantitatively different from those of existing lawful weapons and means of warfare".[80] This approach has been supported by the 28th Conference of the Red Cross and Red Crescent, which encouraged States "to review with particular scrutiny all new weapons, means and methods of warfare that cause health effects with which medical personnel are unfamiliar".[81]

The *ICRC Guide* has outlined a range of health criteria for consideration by States; those that would be of particular relevance to the review of new ICA weapons or RCAs include:

- whether all relevant scientific evidence pertaining to the foreseeable effects on humans has been gathered;
- how the mechanism of injury was expected to impact on the health of victims;
- when used in the context of armed conflict, what was the expected field mortality and whether the later mortality (in hospital) was expected to be high;
- whether there was any predictable or expected long-term or permanent alteration to the victims' psychology or physiology;
- whether the effects would be recognized by health professionals, be manageable under field conditions and be treatable in a reasonably equipped medical facility.

- These and other health-related considerations are important to assist the reviewing authority in determining whether the weapon in question can be expected to cause *superfluous injury or unnecessary suffering* [emphasis in original].[82]

7.6.3. Restrictions on use

Article 36 requires a State to determine whether the employment of a new weapon would "in some or all circumstances" be prohibited by international law. The *ICRC Guide* has noted that a weapon or means of warfare cannot be assessed in isolation from the method of warfare by which it is to be used: "It follows that the legality of a weapon does not depend solely on its design or intended purpose, but also on the manner in which it is expected to be used on the battlefield."[83] Consequently, the *ICRC Guide* has noted that a weapon "used in one manner may 'pass' the Article 36 'test', but may fail it when used in another manner".[84] However, the ICRC's Commentary on Additional Protocol I has also noted that the scope of such considerations was limited: a State need only determine "whether the employment of a weapon for its *normal or expected use* would be prohibited under some or all circumstances. A State is *not required to foresee or analyse all possible misuses of a weapon*, for almost any weapon can be misused in a way that would be prohibited" (emphasis added).[85] Other commentators, notably Fry, have emphasized that these legal reviews should consider anticipated uses of weapons beyond those that are considered "normal".[86]

The *ICRC Guide* has recommended that, when the reviewing authority has determined that some, but not all, of the expected methods of use of the weapon were found to be unlawful, it should either place restrictions on the weapon's use which "should be incorporated into the rules of engagement or standard operating procedures", or it could request modifications to the weapon which "must be met before approval can be granted".[87]

7.6.4. Mechanism of review

Article 36 did not specify in what manner and under what authority weapons reviews should be constituted. Consequently, as the *ICRC Guide* has stated, "it is the responsibility of each State to adopt legislative, administrative, regulatory and/or other appropriate measures to effectively implement this obligation".[88] Reviews of State practice in this area that were undertaken by the Danish Red Cross,[89] and Daoust, Coupland and Ishoey,[90] have shown a variety of approaches employed in terms of the size, nature and location of the reviewing body, and whether or not it possessed effective veto power. A range of commentators have recommended that such review mechanisms should be multidisciplinary in nature, being able to draw on relevant legal, medical, military and environmental expertise.[91]

7.6.4.1. Information exchange

Although States are required to provide information, on request, to other High Contracting Parties, regarding the process of weapons review, there appears to be no obligation placed on them to disclose their conclusions as to the legality of a particular weapon.[92] However, a case can be made that reviewing States should provide information to the High Contracting Parties of any reviews they have performed that have resulted in a determination that a specific weapon, means or method of warfare would be in breach of the Additional Protocol or "any other rule of international law applicable". Such actions would appear to be in conformity with, and in fact be required by, the reviewing State's obligation under Article 1, common to the four Geneva Conventions of 1949 and Additional Protocol I, "to respect and ensure respect" for these treaties. Although the determination by one State that the employment of a particular weapon is prohibited would not be binding upon any other State,[93] it can be argued that the exchange of such information between States Parties could potentially be an important factor in "closing down" possible avenues of research that might – unchecked – lead to the development of weapons that would be contrary to international law.[94]

A number of arms control experts and relevant bodies have called for greater transparency in this area. For example the 27th Conference of the Red Cross and Red Crescent encouraged States "to promote, wherever possible, exchange of information and transparency in relation to these mechanisms, procedures and evaluations".[95] Furthermore, the 28th Conference of the Red Cross and Red Crescent "invited" States that have review procedures in place to cooperate with the ICRC, with a view to facilitating the voluntary exchange of experience on review procedures.[96]

Greater transparency with regard to the determinations of Article 36 reviews of new ICA weapons, RCAs and related means of delivery – particularly when the evaluation concluded the weapon to be prohibited – could play an important role in increasing the effectiveness of the regulation of "dual-use" chemical and life science technologies, particularly given the rapidity and scale of advances in this area and their potential military applicability.

7.6.5. Current implementation of Article 36

The obligation upon all States to review new weapons has been voiced by a number of pluri-lateral and multilateral bodies, such as the Conference of the Convention on Certain Conventional Weapons,[97] and NATO's RTO.[98] The importance of such legal reviews has also been regularly highlighted at successive International Conferences of the Red Cross and Red Crescent.[99] However, despite the widespread recognition of States' obligations to conduct legal reviews of new weapons, publicly available information indicated that, in February 2006, only ten States had instituted *formal* weapon review procedures.[100] Fry has described this number as "shockingly low" compared to the number of States that are party to Additional Protocol I. He has noted that although "more states may have informal review procedures ... there is no way of telling when the existence of these procedures is not publicized".[101] Lawand has argued that: "Establishing national mechanisms to review the legality of new weapons is *especially relevant and urgent* in view of emerging new weapons technologies such as directed energy, *incapacitants, behaviour change agents*, acoustics and nanotechnology, to name but a few" (emphasis added).[102]

7.7. Conclusions

IHL – specifically the Geneva Conventions and Additional Protocols, and corresponding international customary law – constitutes an important body of rules that places significant constraints upon the development, acquisition, stockpiling or employment of any weapon in armed conflict. It can be concluded that the employment of an ICA weapon in armed conflict would breach relevant IHL prohibitions (e.g. the prohibitions on SIRUS and deliberate attacks on civilians); or undermine relevant IHL obligations (e.g. the duty to project persons considered *hors de combat* or the duty to respect and ensure respect of IHL). Similarly, there are concerns that the use of RCAs in armed conflict, in certain proposed scenarios, would contravene the prohibitions on deliberate attacks on civilians and the prohibition on attacks that do not discriminate between civilians and military objectives. Consequently, IHL should certainly be incorporated into a comprehensive holistic arms control strategy for the regulation of these agents and related means of delivery.

Despite the potential utility of the Geneva Conventions and Additional Protocols in this area, these instruments do have important limitations. Firstly, although the Conventions and Protocols do incorporate investigation and enforcement procedures in response to allegations of grave breaches, these can only be initiated by High Contracting Parties, and normally in the first instance by the Parties to the specific conflict. If such Parties are unwilling to trigger these mechanisms, there are no routes by which individuals – such as the civilian targets of attacks involving ICA weapons or RCAs – can initiate such procedures directly. Secondly, although States would be required to undertake reviews of the legality under IHL of all new ICA weapons, RCAs and related means of delivery developed or acquired for use in armed conflict, the number of States carrying out such reviews, the nature of the review process and the results of such reviews are unknown.

Thirdly, a major limitation to the employment of IHL in the regulation of such weapons is that this body of law is only applicable to situations of armed conflict. Whilst much relevant IHL would extend to non-international armed conflicts there may well be disagreements as to whether a particular situation is a non-international armed conflict, with the relevant State instead claiming to be involved in law enforcement activities against criminals or terrorist organizations. In such circumstances, however, States would have obligations under human rights law to actively protect the right to life and prohibit torture and other cruel, inhuman or degrading treatment or punishment.

Finally, the important constraints that IHL potentially imposes on the use of ICA weapons, RCAs and related means of delivery would not be directly applicable for such weapons designed and utilized solely for law enforcement activities that fell short of armed conflict. However, in such circumstances, human rights law would, once again, be applicable.

8
Human Rights Law Applicable to ICA Weapons and Riot Control Agents

8.1. Introduction

Although human rights law does not specifically address the use of discrete arms or security equipment, it is certainly of great relevance to the employment of such weapons, as it regulates the use of force by law enforcement officials and other agents of the State. An important strength of international human rights law (IHRL) is its applicability in a broad range of circumstances where use of ICA weapons or RCAs might be considered. The International Court of Justice has affirmed that human rights law continues to apply in situations to which IHL is applicable,[1] whilst the UN Human Rights Committee has stated that, in situations of armed conflict, "both spheres of law are complementary, not mutually exclusive".[2] Thus, IHRL would be applicable to domestic policing operations, to non-international conflicts, whether or not the State recognized it as such, and to those aspects of an international conflict occurring in national territory. The importance of this breadth of coverage has been highlighted by Hampson, who has noted that "States frequently refuse to characterize an internal armed conflict as such, preferring to call it criminal or terrorist activity. In such a situation, they can hardly challenge the applicability of human rights law."[3]

While several human rights norms may be applicable to the regulation of ICA weapons and RCAs, the rights to life, to liberty and security; to freedom from torture and other cruel, inhuman or degrading treatment or punishment; to engage in "peaceful protest"; and to health; together with associated obligations on the restraint of force, are the most relevant. These rights and the attendant obligations upon States will be explored in this chapter as part of the ongoing HAC analysis, as will the consequent implications for the regulation of RCAs and ICA weapons.

8.2. Protection of the right to life and restrictions on the use of force

The "inherent" right to life is enshrined in many international,[4] and regional, human rights instruments.[5] The International Covenant on Civil and Political Rights (ICCPR), for example, stated that "every human being has the inherent right to life. This right shall be protected by law. No one shall be arbitrarily deprived of his life."[6] The UN Human Rights Committee, the body that monitors the implementation of the ICCPR, has stated that the right to life is "the supreme right from which no derogation is permitted even in time of public emergency which threatens the life of the nation ... ".[7] The Committee has further stated that States Parties to the ICCPR should take measures not only to prevent and punish deprivation of life by criminal acts, "but also to prevent arbitrary killing by their own security forces".[8] The Committee has argued that the right to life, in the case of the ICCPR at least, is a right "which should not be interpreted narrowly".[9]

An integral feature of the right to life norm is the attendant restrictions on the legitimate use of force contained in several human rights instruments. These have been further elaborated in a range of criminal justice standards adopted under the auspices of the UN that have sought to clarify those situations where the use of force may be appropriate and by whom it can be legitimately applied. Although such standards are not directly legally binding upon States, they are widely recognized as exemplifying good practice and do reflect existing international norms. Of particular relevance are two international agreements that have codified the rules by which law enforcement personnel should operate the 1990 UN Basic Principles on the Use of Force and Firearms by Law Enforcement Officials (UNBP),[10] and the 1979 UN Code of Conduct for Law Enforcement Officials (UNCoC).[11] The extent to which these standards reflect customary law has not been settled. According to Melzer, though, the UNCoC "confirms the conditions and modalities established by conventional human rights law for the resort to lethal force in law enforcement operations", while the European Court on Human Rights appears to consider the UNBP as legally binding.[12]

The UNBP has provided that: "Law enforcement officials, in carrying out their duty, shall, as far as possible, apply non-violent means before resorting to the use of force and firearms. They may use force and firearms only if other means remain ineffective or without any promise of achieving the intended result."[13] Whenever the lawful use of force has been unavoidable, the UNBP has required that law enforcement officials:

"(a) Exercise restraint in such use and act in proportion to the seriousness of the offence and the legitimate objective to be achieved;
(b) Minimize damage and injury, and respect and preserve human life;

(c) Ensure that assistance and medical aid are rendered to any injured or affected persons at the earliest possible moment..."[14]

The UNBP has also promoted the use of "non-lethal" weapons under certain circumstances:

> Governments and law enforcement agencies should develop a range of means as broad as possible and equip law enforcement officials with various types of weapons and ammunition that would allow for a differentiated use of force and firearms. These should include the development of non-lethal incapacitating weapons for use in appropriate situations, with a view to increasingly restraining the application of means capable of causing death or injury to persons...[15]

8.2.1. Application to ICA weapons

Proponents of ICA weapons have advocated their development and use in certain law enforcement scenarios where there is a need to rapidly and completely incapacitate individuals or a group without causing death or permanent disability. Although the issue is contested, certain legal experts have argued that the employment of ICA weapons may be permissible in certain extreme law enforcement situations where the authorities need to resort to potentially lethal force to resolve urgent, life-threatening situations because less violent and dangerous methods have failed, are impractical or have a low chance of success.

However, in its 2012 *Synthesis* paper, the ICRC argued against the development and use of ICA weapons. It underlined the grave dangers of the employment of these weapons in practice and the extremely limited circumstances under which the use of such "potentially lethal force" should even be considered, given the States obligations under IHRL:

> In light of the certainty that bystanders will also come to harm, the question to be asked is whether such a means is absolutely necessary to save the lives of those who are threatened, that is whether there are any other means available that would achieve the same aim while posing less of a danger to life; and whether this is an unavoidable measure of last resort, the State having exhausted all feasible less harmful means before it resorts to this means.[16]

Furthermore, even if such conditions are met, the obligations upon States under IHRL to protect the right to life still apply, with contingent constraints upon employment of the ICA weapon as well as the requirement to take appropriate remedial measures. Aceves, for example, has stated that:

[T]he right to life norm places strict limits on the use of force, which includes the use of incapacitating biochemical weapons... States must, therefore, act with due diligence in all cases involving these weapons. The use of these weapons must be carefully regulated and cannot cause indiscriminate harm. Their use must be proportionate to the perceived threat and must be justified under the circumstances.[17]

Similarly, Fidler has argued that:

The inability to control dosage or exposure environment in extreme law enforcement emergencies heightens governmental responsibility to ensure all precautions are taken to minimize harm to innocent people and to provide immediate and adequate medical attention to those exposed and perhaps adversely affected.[18]

Case law in this area is limited. To date, there has been only one well-documented instance of an ICA weapon agent employed in such extreme law enforcement situations, by the Russian Federation in 2002.[19] Two UN human rights authorities subsequently issued statements on this case.[20] In a January 2003 report, the UN Special Rapporteur on Extra-Judicial, Summary or Arbitrary Executions expressed concern "about the actions by Russian police/security forces" and stated that he "has been collecting information from various sources about the incident and plans to take the issue up in 2003 with the Government of the Russian Federation".[21] As of August 2015, however, the results of the Rapporteur's actions on this issue have not been made public.

Subsequently, in June 2003, the UN Human Rights Committee declared that: "While acknowledging the serious nature of the hostage-taking situation, the Committee cannot but be concerned at the outcome of the rescue operation in the Dubrovka theatre in Moscow on 26 October 2002."[22] The Committee further expressed "its concern that there has been no independent and impartial assessment of the circumstances, regarding medical care of the hostages after their liberation and the killing of the hostage-takers".[23] It called upon the Russian Federation to "ensure that the circumstances of the rescue operation in the Dubrovka theatre are subject to an independent, in depth investigation, the results of which are made public, and, if appropriate, prosecutions are initiated and compensation paid to the victims and their families".[24] No further reference appears to this incident in subsequent reports of the UN Human Rights Committee.

8.2.1.1. European Court of Human Rights: Finogenov and others v. Russia

A highly significant development has been a judgement in December 2011 by the European Court of Human Rights (ECtHR) on this case. In August

2003, a group of 64 former hostages and relatives filed a complaint before the Court, claiming that their right to life (protected under Article 2 of the European Convention on Human Rights [ECHR]) had been violated by the actions of the Russian authorities. The case was accepted by the Court in December 2007 and on 20th December 2011, the Court announced its ruling.[25]

The Court rejected the Russian Federation authorities' assertions that the ICA weapon had been harmless, and that, according to the official medical examinations of the bodies, no direct causal link had existed between the use of the ICA weapon and the death of the hostages.[26] The Court found that even if the ICA weapon had not been a "lethal force" but rather a "non-lethal incapacitating weapon", the ICA weapon was, "at best, potentially dangerous for an ordinary person, and potentially fatal for a weakened person",[27] so the case clearly fell within the ambit of Article 2 of the ECHR.[28] The Court declared that: "it is safe to conclude that the gas remained a primary cause of the death of a large number of the victims".[29]

Nonetheless, with regard to the decision to storm the theatre and use an ICA weapon, the Court stressed that, in situations of such a scale and complexity, it was prepared to grant the domestic authorities "a margin of appreciation,[30] even if now, with hindsight, some of the decisions taken by the authorities may appear open to doubt".[31] In this particular case, the Court believed there had been a real, serious and immediate risk of mass human losses and the authorities had every reason to believe that a forced intervention had been "the lesser evil". The Court considered that the solution, using a "dangerous and even potentially lethal" chemical agent, had put at risk the lives of hostages and hostage-takers alike; but "it was not used 'indiscriminately' as it had left the hostages a high chance of survival, which depended on the efficiency of the authorities' rescue effort".[32] Indeed, the Court believed that the use of the ICA weapon facilitated the liberation of the hostages and reduced the likelihood of an explosion. The Court therefore concluded that, in the circumstances, the authorities' decision to end the negotiations and resolve the hostage crisis by force by using an ICA weapon and storming the theatre had not been disproportionate and had not, as such, breached Article 2 of the ECHR.[33]

However, the Court found that, as a whole, the Russian authorities had not taken all feasible precautions to minimize the loss of civilian life, as the rescue operation had been inadequately prepared and carried out, in violation of Article 2.[34] In addition, the Court concluded that the investigation into the authorities' alleged negligence during the rescue operation had been neither thorough nor independent and had not therefore been effective, in further violation of Article 2.[35]

Certain academics and organizations,[36] including the ICRC,[37] have raised concerns about the Court's judgement, and there are certainly aspects of this ruling that would benefit from further legal analysis, including the inexplicable failure of the Court to assess the legality of the use of the ICA weapon under the CWC;[38] the Court's willingness to make a judgement without knowledge of the identity, nature and consequently (predicted) effects of the ICA weapon employed; and the Court's claims that it was impossible for it "to establish whether or not the gas was a 'conventional weapon' or to identify the rules for its use".[39]

However, such concerns notwithstanding, the initial decision of the Court, though subject to an appeal, has subsequently been made final.[40] Consequently, this ruling may well set important precedents and influence future judgements on the use of such weapons. It is notable that the Court considered the use of the ICA weapon as definitely falling under the ambit of the ECHR and that such use could potentially be a violation of Article 2 of the Convention. It appears that the use of such ICA weapons would only be considered consistent with the Convention in very limited, extreme situations (i.e. where there was an immediate and direct threat to life and where there was no recourse to other measures to resolve the situation with less risk of injury or death), and only when such use occurred in conjunction with proper planning and the provision of adequate medical care and remedial support. Although the Court gave the Russian Federation some "margin of appreciation" regarding its decision to use an ICA weapon with then-unknown effects, the consequences of such action have now been shown to include the deaths of a large proportion of the hostages and long-term injury for many of the survivors. Given such empirical data, it is debatable whether a court would give such a "margin of appreciation" or range of discretion to a State for the use of an ICA weapon in similar situations in the future.

Finally, and with more general application, this case has highlighted the potential utility of regional human rights mechanisms as a means that victims of human rights violations (including those inappropriately targeted with ICA weapons or RCAs) can employ to hold the relevant authorities to account (and also, in the case of the ECHR, to obtain financial compensation for the wrongs done to them).[41] Although the relevant courts[42] have a number of limitations and only have jurisdiction over those States that have ratified the relevant Conventions and Protocols, they deliver legally binding judgements.[43] Furthermore, as Hampson has noted:

> An important feature of these enforcement mechanisms is that, almost uniquely in international law, they can be directly or indirectly triggered by individuals... who are far more likely to bring such complaints than are foreign States, particularly where the claim concerns something that the State has done on its own territory.[44]

8.2.1.1.1. Employment of ICAs in judicial executions. Although the inherent right to life is enshrined and protected under IHRL and the arbitrary deprivation of life is prohibited, State-sanctioned judicial executions are not currently prohibited,[45] though they are subject to certain restrictions. For example, the ICCPR states that:

> [S]entence of death may be imposed only for the most serious crimes in accordance with the law in force at the time of the commission of the crime and not contrary to the provisions of the present Covenant and to the Convention on the Prevention and Punishment of the Crime of Genocide. This penalty can only be carried out pursuant to a final judgement rendered by a competent court.[46]

The Covenant prohibits imposition of the death penalty "for crimes committed by persons below eighteen years of age" and specifies that it "shall not be carried out on pregnant women".[47]

In China,[48] Guatemala,[49] the Maldives,[50] Papua New Guinea,[51] Taiwan,[52] Thailand,[53] Vietnam,[54] and the United States,[55] the intravenous administration of a lethal dose of certain pharmaceutical chemicals (normally including particular ICAs) – a process called "lethal injection" – is provided for as a method of execution. As of August 2015, the vast majority of lethal injection executions appear to have occurred in China and the United States.

In China, the number of death sentences issued and executions carried out is a State secret; however, human rights organizations have estimated that China has continued to execute a minimum of several thousand prisoners every year; more people than the rest of the world combined.[56] Since 1997, lethal injection and shooting have been the only methods of execution authorized by China's Criminal Procedure Law.[57] Although the composition of the pharmaceutical chemicals used for lethal injections has not been confirmed by the Chinese authorities, the mixture is reportedly similar, if not identical, to the three-drug cocktail that has been used in many States of the United States, comprising a barbiturate, muscle relaxant and potassium chloride.[58] Lethal injection has been carried out in prisons or in mobile "death vans" where prisoners were reportedly strapped to a metal bed, a needle was attached by a doctor and an automatic syringe inserted the lethal drug cocktail into the prisoner's vein.[59] In December 2003, the People's Supreme Court reportedly urged all courts throughout China to purchase mobile execution chambers "that can put to death convicted criminals immediately after sentencing".[60] The number of "death vans" now in use is not known, although a US newspaper reported in 2006 that more than 40 were deployed.[61] Executions by shooting were reportedly due to be discontinued in 2010 following a People's Supreme Court ruling in February 2009 which held that lethal injection was a more humane form of execution.[62] By December 2009, at least 15 provinces and municipalities had reportedly adopted the policy.[63]

In the United States, lethal injection was first introduced as a legal method of execution in Oklahoma and Texas, in May 1977, and the first lethal injection execution was carried out in Texas on 7th December 1982.[64] According to the Death Penalty Information Center (DPIC), lethal injection is by far the most common method of execution in the United States: as of 2nd April 2015, 88% (1,229 of the 1,404) prisoners executed, since the death penalty was reintroduced in 1977, were killed by this procedure.[65] Currently, all the 34 US States that have the death penalty authorize the use of lethal injection, and this is also the method sanctioned for use by the US military and the US Government.

In addition to the moral and human rights arguments against the death penalty *per se*, international and regional human rights bodies, non-governmental human rights organizations and medical associations have raised specific concerns about execution by lethal injection. These include the attempts of death penalty proponents to promote lethal injections as pain-free and humane, thereby masking the long-term suffering inflicted on prisoners through the entire death penalty process; the potential for this method to cause intense, sometimes prolonged physical suffering by way of botched executions and the danger that such suffering may be hidden due to the action of a paralysing agent in the lethal mixture;[66] and the involvement of health personnel, directly or indirectly, in the execution process.[67]

8.2.2. Application to riot control agents

RCAs are widely employed by law enforcement officials throughout the world for activities such as the dispersal of assemblies posing an imminent threat of serious injury, or the incapacitation of violent individuals. When used in accordance with manufacturers' instructions and in a lawful, proportionate and discriminate manner, in line with international human rights and criminal justice standards, RCAs can provide an important alternative to other applications of force more likely to result in injury or death, notably firearms. However, RCAs are also open to misuse.

As described in Chapter 3 of this publication, in order to provide a preliminary indication of the nature of the misuse of RCAs by law enforcement personnel, an analysis of documentation produced by relevant UN monitoring bodies and leading international non-governmental human rights organizations relating to reported human rights abuses over a five-year period was undertaken. The survey indicated that, from 1st January 2009 until 31st December 2013, RCAs were reportedly employed by law enforcement personnel to facilitate or carry out human rights abuses in at least 95 countries or territories.

UN human rights monitoring bodies and international non-governmental human rights organizations have regularly expressed concern regarding reports of the employment of RCAs as part of the indiscriminate, excessive or lethal use of force by law enforcement officials, particularly in crowd control situations. For example in 2003, the UN Special Rapporteur on Torture stated that:

> [C]hemical agents provided for "crowd-control" purposes are prone to abuse if used against demonstrators in an indiscriminate manner. Precise practical guidelines regarding the circumstances in which such chemical agents may be used, as well as information regarding their effects on specific categories of persons such as children, pregnant women and persons with respiratory problems, are said often to be lacking.[68]

In May 2012, in his first thematic report to the UN Human Rights Council, the UN Special Rapporteur on the rights to freedom of peaceful assembly and of association stated:

> [w]ith regard to the use of tear gas, the Special Rapporteur recalls that gas does not discriminate between demonstrators and nondemonstrators, healthy people and people with health conditions. He also warns against any modification of the chemical composition of the gas for the sole purpose of inflicting severe pain on protestors and, indirectly, bystanders.[69]

A specific recurring concern has been the employment of RCAs in excessive quantities or in confined spaces where the targeted persons cannot disperse and where the toxic properties of the agents can lead to serious injury or death, particularly to vulnerable individuals. As well as their widespread use (and misuse) in the context of crowd control and public order, RCAs are also employed as a means to subdue prisoners and maintain order in correctional centres, prisons, police stations and other places of detention. Human rights bodies have raised concerns about the appropriateness of such application. For example, the European Committee for the Prevention of Torture and Inhuman or Degrading Treatment or Punishment (CPT) has stated that tear gas is a "potentially dangerous substance" and that "only exceptional circumstances can justify [its] use...inside a place of detention – but never in a confined space such as a cell – for control purposes, and such exceptional use should be surrounded by appropriate safeguards".[70]

As well as the potential dangers to health due to the toxicity of the chemical agents employed, concerns have been raised that RCAs are used by law enforcement officials in conjunction with other "less lethal" or indeed lethal weapons, to facilitate excessive force or even enhance the application of lethal force. Between 1st January 2009 and 31st December 2013, RCAs reportedly contributed to excessive use of force in 83 countries or territories, often being employed in conjunction with firearms.

8.3. The rights to freedom of opinion and expression, of association and assembly

The rights to freedom of opinion and expression, of association and assembly are established in a wide range of regional,[71] and international,[72] human rights agreements. Indeed, the ICCPR has expressly stated that:

> The right of peaceful assembly shall be recognized. No restrictions may be placed on the exercise of this right other than those imposed in conformity with the law and which are necessary in a democratic society in the interests of national security or public safety, public order (*ordre public*), the protection of public health or morals or the protection of the rights and freedoms of others.[73]

Furthermore, the UNBP recognized that: "In the dispersal of assemblies that are unlawful but non-violent, law enforcement officials shall avoid the use of force or, where that is not practicable, shall restrict such force to the minimum extent necessary."[74]

Despite such strictures, however, RCAs have reportedly been employed to suppress freedom of opinion, expression, association or assembly in at least 75 countries or territories from 1st January 2009 to 31st December 2013. Such misuse of RCAs is just one aspect of the reported widespread inappropriate use of force by law enforcement officials in public gatherings.

The UN Human Rights Council has given increased attention to these issues, adopting a series of Resolutions, including, in 2010, 2012 and 2013, on "freedom of peaceful assembly and of association";[75] and in 2012, 2013 and 2014, on the "promotion and protection of human rights in the context of peaceful protests".[76] The most recent Resolution, 25/38 of April 2014, "calls upon" States to: "promote a safe and enabling environment" for people to exercise their rights to freedom of peaceful assembly, expression and association, including by "ensuring that their domestic legislation and procedures...clearly and explicitly establish a presumption in favour of the exercise of these rights, and that they are effectively implemented".[77] The Resolution also "urges all States to avoid using force during peaceful protests and to ensure that, where force is absolutely necessary, no one is subject to excessive or indiscriminate use of force".[78]

8.4. The rights to liberty and security

The rights to liberty and security of person are established in a range of international,[79] and regional,[80] human rights instruments. The ICCPR has stated that: "Everyone has the right to liberty and security of person. No one shall be subjected to arbitrary arrest or detention. No one shall be deprived of his liberty except on such grounds and in accordance with such procedure as are established by law."[81]

Casey-Maslen has argued that the concepts of liberty and security are "typically considered broadly, and could potentially cover the use of NKE [non-kinetic energy] weapons that prevent a person from moving.... Certain NKE weapons may cause paralysis in human beings. This paralysis may last only a few seconds or it may be more prolonged."[82] Consequently, Casey-Maslen has contended that "the use of such weapons by security officials might, in certain circumstances, be deemed to violate a person's right to liberty and security of person in addition to other human rights they may infringe".[83]

The potential relevance of such considerations to a range of ICA weapons that can immobilize or otherwise incapacitate their targets for extended time periods is underscored by NATO's definition of an ICA weapon as: "A chemical agent which produces temporary disabling conditions which (unlike those caused by riot control agents) can be physical or mental and *persist for hours or days after exposure to the agent has ceased...*" (emphasis added).[84] Although such considerations are worthy of further analysis, they do not appear to have been applied by any international human rights body, as of August 2015, with reference to toxic chemical agents.

8.5. The prohibition against torture and other cruel, inhuman or degrading treatment or punishment

The prohibition against torture and other cruel, inhuman or degrading treatment or punishment (CIDTP) is recognized in a wide range of international,[85] and regional,[86] human rights agreements, and is a customary norm, applicable at all times and in all circumstances, including in armed conflict. It is one of the few human rights for which no derogation has been permitted.[87] The UNCoC, for example, has stated that:

> No law enforcement official may inflict, instigate or tolerate any act of torture or other cruel, inhuman or degrading treatment or punishment, nor may any law enforcement official invoke superior orders or exceptional circumstances such as a state of war or a threat of war, a threat to national security, internal political instability or any other public emergency as a justification of torture or other cruel, inhuman or degrading treatment or punishment.[88]

Torture is defined under Article 1 of the UN Convention against Torture as:

> [A]ny act by which severe pain or suffering, whether physical or mental, is intentionally inflicted on a person for such purposes as obtaining from him or a third person information or a confession, punishing him

for an act he or a third person has committed or is suspected of having committed, or intimidating or coercing him or a third person, or for any reason based on discrimination of any kind, when such pain or suffering is inflicted by or at the instigation of or with the consent or acquiescence of a public official or other person acting in an official capacity. It does not include pain or suffering arising only from, inherent in or incidental to lawful sanctions.[89]

Elements of CIDTP have been defined in other relevant legal texts. For example, "inhuman treatment", in the context of war crimes, has been defined in the Elements of Crimes for the International Criminal Court as the infliction of "severe physical or mental pain or suffering".[90] The notion of "degrading treatment" has been defined by the European Commission of Human Rights as treatment or punishment that "grossly humiliates the victim before others or drives the detainee to act against his/her will or conscience".[91] The commentary to Principle 6 of the UN Body of Principles for the Protection of All Persons under Any Form of Detention or Imprisonment (which prohibits torture and other forms of CIDTP) has stated that:

> The term "cruel, inhuman or degrading treatment or punishment" should be interpreted so as to extend the widest possible protection against abuses, whether physical or mental, including the holding of a detained or imprisoned person in conditions which deprive him, temporarily or permanently, of the use of any of his natural senses, such as sight or hearing, or of his awareness of place and the passing of time.[92]

The UN Special Rapporteur on Torture has indicated that the prohibition on CIDTP places limits on the lawful use of force. The UN Special Rapporteur has emphasized that the use of force must be regulated by the principles of proportionality and that the disproportionate exercise of police powers might constitute CIDTP in certain circumstances.[93]

8.5.1. Application to riot control agents

Despite the absolute prohibition on torture and CIDTP, the misuse of RCAs for such purposes by law enforcement officials in countries has been reported by UN human rights bodies and international human rights non-governmental organisations. Presenting the results of a 2003 study on the trade in security equipment that could be used for torture and CIDTP,[94] the UN Special Rapporteur on Torture, Theo van Boven, stated that:

> [T]he allegations of torture that he has received from all regions of the world have involved instruments such as [*inter alia*] ... chemical control substances (e.g. tear gas and pepper spray). While some of the cases

have involved the use of equipment which is inherently cruel, inhuman or degrading, and would per se breach the prohibition of torture, the vast majority have involved the misuse of those instruments, legitimate in appropriate circumstances, to inflict torture or other forms of ill-treatment.[95]

Between 1st January 2009 and 31st December 2013, RCAs were reportedly employed for torture or ill-treatment in at least 18 countries or territories. In certain cases RCAs have reportedly been employed as a means of inflicting "collective punishment" upon groups of individuals or crowds. Other cases of concern involved the use of hand-held irritant sprays against individual prisoners and detainees in a targeted fashion.

8.5.1.1. European Court of Human Rights: Ali Güneş v. Turkey

A potentially significant development in relevant human rights case law has been the July 2012 ECtHR final judgement regarding the employment of tear gas against Ali Güneş, a high-school teacher who took part in a demonstration on 28th June 2004 against a NATO summit meeting being held in Istanbul on that date. According to Mr Güneş, although he was unarmed and participating peacefully with colleagues in the demonstration at a site sanctioned by the authorities, police grabbed him by the arms, sprayed him with tear gas and beat him up, following which they took him to a police station in which they kept him for 11 hours. A prosecutor saw him after that and ordered his release.[96]

In its final judgement, the Court considered "that the unwarranted spraying of the applicant's face in the circumstances described...must have subjected him to intense physical and mental suffering and was such as to arouse in him feelings of fear, anguish and inferiority capable of humiliating and debasing him".[97] The Court, therefore, concluded "that by spraying the applicant in such circumstances the police officers subjected him to inhuman and degrading treatment within the meaning of Article 3 of the Convention".[98] Consequently, the Court awarded "the applicant the sum claimed by him in full, that is EUR 10,000, in respect of non-pecuniary damage".[99] The Court's ruling is important, given the bearing it may have on future judgements on the use of RCAs in such circumstances.

8.5.2. Application to ICA weapons

Important constraints upon the use of ICA weapons arise from obligations to prevent torture and CIDTP. Such obligations are of particular relevance where ICA weapons are being considered for use against prisoners or detainees. After a review of the relevant law, Aceves concluded that:

> [T]he prohibition against cruel, inhuman or degrading treatment places significant restrictions on the use of incapacitating biochemical weapons.

These weapons are designed to impair the physical and mental integrity of the individual. Depending on the nature, duration and long-term effects of this impairment, the use of incapacitating biochemical weapons can give rise to a claim of cruel, inhuman, or degrading treatment.[100]

In its report exploring the potential use and misuse of neuroscience, the Royal Society highlighted the European Commission of Human Rights' definition of degrading treatment and considered it to be of "particular importance in considering the potential applications of neuroscience that could, for example, manipulate behaviour or thought processes".[101] A supporting note observed that: "the use of potential militarised agents including noradrenaline antagonists such as propranolol to cause selective memory loss, cholecystokinin B agonists to cause panic attacks, and substance P agonists to induce depression could all be considered violations of the prohibition against degrading treatment".[102]

Fidler has stated that: "non-consensual, non-therapeutic use of any chemical or biochemical against detained individuals would constitute degrading treatment and could, constitute cruel or inhumane treatment and perhaps even torture".[103] He did, however, believe that there may be situations where use of ICA weapons might be compatible with IHRL: where the detained person posed an immediate, violent threat to himself (e.g. attempting suicide) or to safety and order in the detention facility (e.g. attacking guards or participating in riots).[104]

In addition to the prohibitions against torture and CIDTP, further important potential constraints upon the non-consensual application of ICA weapons to detainees relate to obligations to ensure respect for the detainee's right to freedom of opinion. For example, Article 19 of the ICCPR has declared that "everyone shall have the right to hold opinions without interference".[105] In his legal commentary to the Convention, Nowak stated that this provision consequently "obligates the States Parties to refrain from any interference with freedom of opinion (by indoctrination, 'brainwashing', influencing the conscious or unconscious mind with psychoactive drugs or other means of manipulation) and to prevent private parties from doing so".[106]

Despite the prohibitions on such actions, evidence of the detention of large numbers of political prisoners and other sane detainees in mental institutions, at some stage during a period from the 1960s until the 1990s, has been documented in China,[107] and the Soviet Union.[108] In certain institutions, forcible administration of psychotropic substances for non-therapeutic reasons was alleged.[109] There are indications that such practices may still continue – albeit at a far lower level. A review of relevant documentation produced by UN human rights bodies and international human rights NGOs from 1st January 2004 to 31st December 2013 indicates that cases of non-consensual, non-therapeutic application of psychoactive chemicals against political prisoners, detainees or patients were reported in the Russian Federation,[110] Vietnam[111] and Uzbekistan,[112] apparently to intimidate or punish the victims.

8.5.2.1. Use of incapacitating chemical agents as "truth serums" in interrogation

Certain States have reportedly employed psychoactive incapacitating chemical agents (such as sodium thiopental, sodium amytal or scopolamine) as so-called "truth drugs" or "truth serums" against detainees without their consent, for the purposes of interrogation by law enforcement officials. Dando and Furmanski,[113] and Perry Robinson,[114] have documented the attempts by the United States and the Soviet Union during the Cold War to develop such chemical aids to interrogation. More recently, such methods have reportedly been employed and/or threatened in India,[115] Turkmenistan,[116] Turkey[117] and in the US detention centre at Guantanamo Bay.[118]

Such practices are contrary to IHRL prohibiting CIDTP. In addition, the UN Body of Principles for the Protection of All Persons under Any Form of Detention or Imprisonment (UN Principles for the Protection of Detainees), addressing the issue of interrogation, stated that: "No detained person while being interrogated shall be subject to violence, threats or methods of interrogation which impair his capacity of decision or his judgement."[119]

The involvement of health professionals in any form of torture or CIDTP (including the non-consensual application of "truth drugs") is prohibited under the UN Principles of Medical Ethics.[120] Under these standards it is deemed a contravention of medical ethics for health personnel, particularly physicians: "to be involved in any professional relationship with prisoners or detainees the purpose of which is not solely to evaluate, protect or improve their physical and mental health",[121] or "to apply their knowledge and skills in order to assist in the interrogation of prisoners and detainees in a manner that may adversely affect the physical or mental health or condition of such prisoners or detainees".[122] The UN Principles of Medical Ethics have also stated that "there may be no derogation from the foregoing principles on any ground whatsoever, including public emergency".[123]

Further constraints or outright prohibitions upon the use of ICAs in interrogation have been established at the national level in certain States. In May 2010, the Indian Supreme Court ruled that the administration of ICAs to detainees – a practice known in India as "narcoanalysis" – without their consent during interrogation violated the Indian Constitution and was illegal as it constituted cruel, inhuman or degrading treatment.[124] The Court stated that:

> Even though "the right against cruel, inhuman and degrading punishment" cannot be asserted in an absolute sense, there is a sufficient basis to show that Article 21 [of the Indian Constitution which protects

"personal liberty"] can be invoked to protect the "bodily integrity and dignity" of persons who are in custodial environments. This protection extends not only to prisoners who are convicts and under trials, but also to those persons who may be arrested or detained in the course of investigations in criminal cases.[125]

Concerns have also been raised by jurists in the United States, and the use of "truth serums" or "truth drugs" is not recognized as an authorized method of interrogation by US courts. Under US law, confessions made under the influence of "truth serums" are not considered as "voluntary" and are consequently inadmissible as evidence.[126] The employment of ICAs to facilitate interrogations also appears to be contrary to the Inter-American Convention to Prevent and Punish Torture. Article 2 of this Convention defines torture as including: "the use of methods upon a person intended to obliterate the personality of the victim or to diminish his physical or mental capacities, even if they do not cause physical pain or mental anguish".[127]

Further constraints upon the non-consensual use of ICAs to facilitate the interrogation of prisoners may arise from obligations under IHL that may be applicable where law enforcement activities and military operations are perceived to overlap. For example, the reported application by US military personnel of certain ICAs for the interrogation of detainees at Guantanamo Bay would appear to be in contravention of regulations laid out in the *US Army Field Manual*:

> The psychological techniques and principles in this manual should neither be confused with, nor construed to be synonymous with, unauthorized techniques such as brainwashing, physical or mental torture, including drugs that may induce lasting or permanent mental alteration or damage. Physical or mental torture and coercion revolve around eliminating the source's free will, and are expressly prohibited by [the Geneva Conventions].[128]

8.6. Health considerations and the right to health

The most comprehensive statement of the right to health is enunciated in the International Covenant on Economic, Social and Cultural Rights (ICESCR):[129]

1. "The States Parties to the present Covenant recognize the right of everyone to the enjoyment of the highest attainable standard of physical and mental health.
2. The steps to be taken by the States Parties to the present Covenant to achieve the full realization of this right shall include those necessary for:

... d. The creation of conditions which would assure to all medical service and medical attention in the event of sickness."[130]

The Committee on Economic, Social and Cultural Rights (CESCR), which monitors the IESCR, has stated that violations of the obligation to respect the right to health "are State actions, policies or laws that contravene standards set out in Article 12 of the Covenant and are likely to result in bodily harm, unnecessary morbidity and preventable mortality".[131] Furthermore, the Committee stated that violations of the obligation to protect the right to health "follow from the failure of a State to take all necessary measures to safeguard persons within their jurisdiction from infringements of the right to health by third parties...".[132]

Casey-Maslen has noted that the application of the right to health "to the use of weapons does not appear to have been tested",[133] however, he has contended that "a challenge to certain NKE weapons on the basis of their health effects, including by prisoners or patients at a mental health institution, merits consideration".[134] This area appears to be of potential relevance to the use of "less lethal" weapons employing chemical agents, given the long-standing concerns voiced by some in the medical community regarding the immediate and long-term effects of certain RCAs,[135] and ICA weapons.[136]

8.6.1. Application to RCAs

With regard to RCAs, such concerns have been acknowledged by human rights bodies. For example, in 2003, the UN Special Rapporteur on Torture stated that:

> [C]hemical agents, such as tear gas/irritant ammunition and pepper spray weapons, are said to be promoted as providing effective control without the risk to life, i.e. as "humane alternatives" to lethal force. However, according to information received, insufficient research has been undertaken into their potential effects on targeted persons.[137]

Similarly, the ECtHR recognized that the use of pepper spray: "can produce effects such as respiratory problems, nausea, vomiting, irritation of the respiratory tract, irritation of the tear ducts and eyes, spasms, chest pain, dermatitis and allergies". Furthermore, the Court noted that: "In strong doses it may cause necrosis of the tissue in the respiratory or digestive tract, pulmonary oedema or internal haemorrhaging (haemorrhaging of the suprarenal gland)."[138]

In addition, human rights monitoring bodies have highlighted the deleterious effects to health resulting from the inappropriate employment of RCAs, particularly in enclosed or confined spaces. For example, the CPT has raised concerns about the use of RCAs during the forcible removal of immigration detainees:

The CPT also has very serious reservations about the use of incapacitating or irritant gases to bring recalcitrant detainees under control in order to remove them from their cells and transfer them to the aircraft. The use of such gases in very confined spaces, such as cells, entails manifest risks to the health of both the detainee and the staff concerned.[139]

Although more discrete and targetable in nature than tear gas, concerns have been raised about the inappropriate employment of hand-held irritant sprays, in particular, the deleterious consequences to health of the use of pepper spray and its analogues against prisoners or detainees. For example, in 2003, AI stated that over 100 people were reported to have died in custody in the United States over the previous decade, after being subjected to pepper spray. While most of the deaths had officially been attributed to other factors such as drug intoxication or positional asphyxia, pepper spray was thought to have been a contributory factor in a number of cases.[140]

In its 2009 reports, following visits to the Czech Republic and Bosnia and Herzegovina (BiH), the CPT highlighted cases of inappropriate use of pepper spray against detainees, and stated that it considered that: "Pepper spray is a potentially dangerous substance and should not be used in confined spaces.... Pepper spray should never be deployed against a prisoner who has already been brought under control."[141] In its July 2012 judgement in the case of *Ali Güneş v. Turkey*, the ECtHR stated that it shared the CPT's concerns with regard to pepper spray and concurred with the CPT's recommendations. The Court stressed, "in particular, that there can be no justification for the use of such gases against an individual who has already been taken under the control of the law enforcement authorities...".[142]

In a June 2013 Resolution, the Parliamentary Assembly of the Council of Europe highlighted recent cases of the reported misuse of tear gas and/or pepper spray in public order incidents.[143] Referring to both "the position" of the CPT and "the case law of the European Court of Human Rights", the Resolution "underlines the serious health consequences of the use of tear gas".[144] Consequently, the Assembly urged the Council of Europe Member States, "to take the necessary measures to bring their legislation into line with Council of Europe standards and the case law of the European Court of Human Rights",[145] and specifically invited States "to draw up clear instructions concerning the use of tear gas (pepper spray) and prohibit its use in confined spaces".[146]

8.6.2. Application to ICA weapons and inappropriate administration of ICAs

The right to health appears to be of relevance to the employment of potentially lethal toxic chemical agents, given the long-standing concerns raised by some in the medical community regarding the immediate and long-term effects of ICA weapons.[147] Disquiet has also been voiced by medical professionals and human rights organizations regarding the potential involvement of medical personnel in the development or application of ICA weapons,[148] as well as the failure of the State to provide adequate medical treatment following the use of such weapons.[149]

In addition, it is notable that a UN Human Rights Commission report into conditions in Guantanamo Bay prepared by five UN Special Rapporteurs (including the Special Rapporteur on the right of everyone to the enjoyment of the highest attainable standard of physical and mental health) raised concerns relating to the "right to health" regarding the forcible application of drugs to detainees.[150] The Special Rapporteur on the right to health stated that he "received serious and credible reports of violations of the right to health – both health care and the underlying determinants of health – at Guantánamo Bay".[151] The reports alleged, *inter alia*, that "detainees have been subjected to non-consensual treatment, including drugging and force-feeding".[152] The Special Rapporteur also raised concerns that reports indicated that "some health professionals have been complicit in abusive treatment of detainees detrimental to their health. Such unethical conduct violates the detainees' right to health, as well as the duties of health professionals arising from the right to health."[153]

8.7. Considerations regarding means of delivery and dispersal of RCAs

Although this chapter has explored the constraints imposed by IHRL and related standards on the use of RCAs and ICA weapons *per se*, a further dimension needs to be considered, namely the mechanisms employed to deliver or disperse such agents to their targets, as such mechanisms raise important additional considerations regarding safety, lethality and discrimination. This section will focus upon RCA means of delivery, as these are widely and regularly employed by law enforcement officials throughout the world.

A range of projectiles and delivery systems which have a narrow dispersal area and emit a limited quantity of agent – such as hand-thrown or weapons-launched RCA canisters and grenades or hand-held RCA spray disseminators – are widely employed by law enforcement officials, for example in public order situations. If such devices have been properly tested and trailed, their use should not raise undue concerns as long as it is in strict accordance with the relevant criminal justice standards, specifically, UNBP, UNCoC, and UN Standard Minimum Rules for the Treatment of Prisoners,[154] and in conformity with, national deployment guidelines.

However, the kinetic impact safety of certain weapons-launched pyrotechnic RCA grenades and canisters may be of concern given limitations in their accuracy and their potential to cause trauma; furthermore, direct firing of such devices at individuals, particularly at short ranges, has led to a number of serious injuries and deaths. Consequently, this issue has been regularly highlighted by UN and regional human rights bodies as well as by international non-governmental human rights organizations.

For example, in his September 2011 report on Israeli practices affecting the human rights of the Palestinian people in the Occupied Palestinian Territory, the UN Secretary General expressed concern regarding frequent and excessive use of force against unarmed demonstrators by Israeli security forces, including live ammunition and "tear gas canisters being fired as projectiles at protesters, resulting in severe injuries".[155] In his June 2012 report,[156] the UN Special Rapporteur on the promotion and protection of the right to freedom of opinion and expression highlighted the case of Mustafa Tamimi, a young man participating in a demonstration in the village of Nabi Saleh in the West Bank on 9th December 2011, who was taken to hospital suffering from injuries caused "by a tear gas canister fired from a short range directly into his face". And who subsequently died in hospital as a result of his injuries.[157] The Special Rapporteur stated that: "while the use of tear gas to disperse a crowd may be legitimate under certain circumstances, tear gas canisters should never be used at short range or aimed directly at protesters. While it has been alleged that IDF open-fire regulations prohibit such use, the Special Rapporteur has been informed of repeated infractions by members of IDF, who are rarely sanctioned or criminally held to account."[158]

8.7.1. European Court of Human Rights: *Abdullah Yaşa and Others v. Turkey; Ataykaya v. Turkey*

Two important judgements relating to the use of RCA means of delivery have been made in recent years by the ECtHR. The first, which was made final on 16th October 2013, related to the case of *Abdullah Yaşa and Others v. Turkey*.[159] On 29th March 2006 Abdullah Yaşa, who was 13 at the time, was at the scene of an unlawful and violent demonstration in Diyarbakır, Turkey. He was injured in the head by a tear-gas grenade fired by a police officer using a tear-gas grenade launcher. He was taken to hospital the same day and underwent an operation. Uncontested video footage of the demonstration showed Abdullah Yasa being hit by a tear-gas grenade and the Court considered that "the impact suggested that the grenade had been fired directly and in a straight line rather than at an upward angle".[160]

The Court considered that the use of tear gas *per se* to disperse the violent gathering did not in itself give rise to a particular issue under Article 3 of the Convention (prohibition of torture and inhuman or degrading treatment). However, the court considered that "firing a tear-gas grenade along a direct, flat trajectory by means of a launcher cannot be regarded as an appropriate police action as it could potentially cause serious, or indeed fatal injuries".[161] The Court considered that "the severity of the injuries noted to the applicant's head could not have been commensurate with the strict use by the police officers of the force necessitated by his behaviour",[162] and was in fact a violation of Article 3 of the Convention.[163]

The second Court judgement of note was made on 22nd July 2014 concerning the case of *Ataykaya v. Turkey* (application number 50275/08).[164] On 29th March 2006, as Tarık Ataykaya was leaving the workshop where he worked, he unexpectedly found himself in the middle of a demonstration and was hit in the head by one of the tear-gas grenades fired by the police in order to disperse the demonstrators. Tarık died of his injuries a few minutes later. The autopsy performed on the following day, and an expert examination conducted a few days later, formally established that death had been caused by a type no. 12 tear-gas grenade, as used by the police, which had struck the deceased's head. As the cartridge extracted from the dead man's head had no distinguishing marks, it was impossible to identify the weapon from which it had been fired.[165]

The Court determined that "there was nothing to indicate that the use of lethal force against Mr Ataykaya's son had been absolutely necessary and proportionate, or that the police had taken the appropriate care to ensure that any risk to life was minimised".[166] Consequently, the Court ruled that there had been a violation of Article 2 (the right to life) of the ECHR. Furthermore, the Court "emphasised the need to reinforce, without further delay, the safeguards surrounding the proper use of tear-gas grenades, so as to minimise the risks of death and injury stemming from their use".[167] The Court also emphasized that, "so long as the Turkish system did not comply with the requirements of the European Convention, the inappropriate use of potentially fatal weapons during demonstrations was likely to give rise to violations similar to that in the present case".[168]

In contrast to the foregoing (and as discussed in Chapter 4), a range of delivery mechanisms have been developed for crowd control and dispersal that deliver far larger amounts of RCAs over wider areas than could be delivered by hand-held sprays and the like. Certain forms of such "wide area" RCA means of delivery may have utility in large-scale law enforcement situations provided they are employed in strict conformity with human rights standards. However, a number of these devices raise questions concerning proportionality and/or about the feasibility of their discriminate use, with the consequent danger of affecting bystanders, in contravention of criminal justice standards and human rights law.

In addition, a further range of "wide area" RCA means of delivery have been developed and promoted which are inherently unacceptable for use in law enforcement activities. These include munitions containing RCAs which have military utility, such as cluster munitions, aerial bombs, mortar rounds and artillery shells. As discussed in Chapter 5, such RCA means of delivery should be considered as chemical weapons under the CWC; all holdings of such weapons should be reported to the OPCW and verifiably destroyed.

8.8. Obligations to review and monitor the use of "less lethal" weapons

Although States are required under Article 36 of Additional Protocol I to the Geneva Conventions to carry out reviews of all new weapons intended for use in warfare,[169] IHRL does not explicitly require that States carry out comparable reviews of new weapons developed or acquired solely for the purpose of law enforcement. However, the norm-setting UNBP has provided that: "The development and deployment of 'non-lethal' incapacitating weapons should be carefully evaluated in order to minimize the risk of endangering uninvolved persons, and the use of such weapons should be carefully controlled."[170]

In addition, the UN Human Rights Council in its Resolution 25/38 of April 2014: "underline[d] the importance of thorough, independent and scientific testing of non-lethal weapons prior to deployment to establish their lethality and the extent of likely injury, and of monitoring appropriate training and use of such weapons".[171] The Resolution also "encourage[d] States to make protective equipment and non-lethal weapons available to their officials exercising law enforcement duties, while pursuing international efforts to regulate and establish protocols for the training and use of non-lethal weapons".[172]

The argument can be made that, in order to effectively fulfil their obligations to protect life, prevent torture and CIDTP, and ensure the responsible use of force by law enforcement officials, States will need to implement review mechanisms to ensure that any new weapons developed or otherwise acquired are consistent with such principles. Consequently, it can be argued that States should, as a minimum,

- ensure that new weapons are not inherently of a nature to violate relevant IHRL and standards; and,
- identify whether there are specific circumstances in which use of new weapons may breach IHRL and criminal justice standards (for example, use of RCAs in confined spaces), and restrict such use accordingly.

In order to do this effectively, it can be argued that a thorough multi-disciplinary review of the prospective weapons is needed, comparable to that recommended by the ICRC for fulfilment of Article 36 of Additional Protocol I.[173]

In addition to requiring that States "carefully evaluate" development and deployment of "non-lethal" or "less lethal" weapons, the UNBP also incorporated obligations upon States to ensure effective control of the use of force and firearms through the introduction of rules and regulations, which would certainly be applicable to such "less lethal" weapons.[174] These provisions have been supplemented by requirements relating to the training of law enforcement officials "with a view to limiting the use of force and firearms", and the review of "training programmes and operational procedures in the light of particular incidents".[175] Furthermore, the UNBP required that: "Governments shall ensure that arbitrary or abusive use of force and firearms by law enforcement officials is punished as a criminal offence under their law."[176] The Human Rights Council, in its 2014 Resolution on the "promotion and protection of human rights in the context of peaceful protests", has "call[ed] upon States to investigate any death or significant injury committed during protests, including those resulting from the discharge of firearms or the use of nonlethal weapons by officials exercising law enforcement duties".[177]

However, despite continuing widespread reports of the utilization of a range of "less lethal" weapons and security equipment in human rights violations – sometimes resulting in serious injury or death – documented by UN and regional human rights monitoring bodies and by international human rights NGOs, there are currently no internationally accepted procedures for evaluating new "less lethal" weapons and for monitoring their subsequent employment.

8.9. Conclusions

As part of the second stage of the HAC analytical process, this chapter explored the constraints upon the use of RCAs and ICA weapons imposed by human rights law derived from regional and international agreements, as well as that arising from customary international law. Human rights law was found to impose significant constraints upon the use of force by the State. Such limitations consequently have important implications as to whether, in what circumstances, by whom and how RCAs and ICA weapons could be lawfully employed.

Human rights law is particularly relevant to the discussion of the regulation of RCAs and ICA weapons as it potentially covers the full "use of force" spectrum from law enforcement activities through to armed conflict, including "grey areas" such as counter-terrorist, counter-insurgency and military operations outside of armed conflict where use of these chemical agents has been proposed. While several human rights norms may be applicable to the regulation of RCAs and ICA weapons, the rights to life; to liberty and security; to freedom from torture and other cruel, inhuman or degrading treatment or punishment; to engage in "peaceful protest"; and to health, together with attendant obligations on the restraint of force, are the most relevant.

An important aspect of human rights law is that there are a number of international and regional mechanisms to monitor adherence to relevant treaties and *jus cogens* norms. Hampson noted that two of these mechanisms – the UN Special Rapporteur on Torture and the Special Rapporteur on Extrajudicial, Summary or Arbitrary Executions – "can scrutinize relevant conduct of any UN member. They are not limited to those that have ratified a particular treaty."[178] Furthermore, Hampson has noted that the two Special Rapporteurs and two of the international treaty bodies (the Human Rights Committee and the Committee against Torture): "could monitor the use of unlawful weapons or the unlawful use of potentially lawful weapons".[179] In addition to the monitoring mechanisms, certain treaties have established enforcement mechanisms that can be accessed by individuals and which result in legally binding judgements for those States party to the relevant instrument (as was seen in the ECtHR rulings on the use of an ICA weapon by the Russian Federation, and of tear gas and related means of delivery by Turkey).

However, it must be recognized that these monitoring and enforcement mechanisms have limited preventative value (though they may have some deterrent effect), as they would only be initiated after a potential misuse of RCAs or ICA weapons has occurred. Furthermore, there are currently no internationally accepted procedures, under IHRL, for evaluating new RCAs or ICA weapons, or for monitoring their subsequent use at a national level. Despite the limitations highlighted, it is clear that IHRL imposes significant constraints upon the use of RCAs and ICA weapons, and should certainly be incorporated into a comprehensive HAC strategy for the regulation of these agents and associated means of delivery.

9
International Criminal Law Applicable to ICA Weapons and Riot Control Agents

9.1. Introduction

The preceding analysis of potential mechanisms to regulate ICA weapons, RCAs and related means of delivery has concentrated upon agreements, treaties and international law intended to constrain and influence the activities of States. Another important alternative approach is to address individual responsibility and culpability through the employment of international criminal law. According to Cassese: "International crimes are breaches of international rules entailing the personal criminal liability of the individuals concerned."[1] Some of the crimes under international law have been considered to be violations of *jus cogens*, peremptory norms that "have a rank and status superior to those of all the other rules of the international community" and which cannot be set aside by States, through for example a treaty.[2] According to Oñate, Exterkate, Tabassi and van der Borght, although there was not necessarily a consensus on a definite list of such crimes under international law it has been widely accepted that war crimes and genocide would certainly be included.[3] In addition, Hampson has added both aggression and crimes against humanity to this list.[4] In certain situations, international criminal law could be applied by national courts or international courts to the use of, and possibly the transfer of, chemical weapons. As part of the HAC stage-two analysis, this chapter explores the potential applicability of international criminal law and judicial mechanisms to cases involving the serious misuse of ICA weapons and RCAs.

9.2. Rome Statute of the International Criminal Court

The International Criminal Court (ICC), governed by the Rome Statute,[5] is the first permanent, treaty based, independent international criminal court established to help end impunity for the perpetrators of the most serious crimes of concern to the international community.[6] The ICC came into being on the 1st July 2002 when the Rome Statute entered into force, following its ratification by 60 States Parties. As of August 2015, there were 123 States Parties to the ICC, with a further 16 having signed but not ratified the Rome Statute.[7] Pursuant to the Rome Statute, the ICC Prosecutor can initiate an investigation on the basis of a referral from any State Party or from the UNSC. In addition, the Prosecutor can initiate investigations *proprio motu* on the basis of information on crimes within the jurisdiction of the Court received from individuals or organizations.

The Rome Statute has asserted jurisdiction over war crimes, crimes against humanity and genocide.[8] The Statute's definition of "war crimes" has included:

"(xvii) Employing poison or poisoned weapons;
(xviii) Employing asphyxiating, poisonous or other gases, and all analogous liquids, materials or devices..."[9]

It should be noted that whilst these two prohibitions have referred to chemical and biological weapons (CBW), they have done so implicitly rather than explicitly. The first has appeared to enunciate a norm first codified in the Second Hague Convention, whilst the second has derived from the Geneva Protocol. Certain legal commentators have stated "that it remains unclear whether all chemical weapons are included, and whether biological weapons are included at all".[10]

In September 2002, the Assembly of States Parties adopted the "Elements of Crimes", which was intended to "assist the Court in the interpretation and application of articles 6, 7 and 8, consistent with the Statute".[11] The Elements of Article 8(2) (b) (xviii) relating to employment of "asphyxiating, poisonous or other gases, and all analogous liquids, materials or devices" were specified as:

- The perpetrator employed a gas or other analogous substance or device.
- The substance was such that it causes death or serious damage to health in the ordinary course of events, through its toxic properties.
- The conduct took place in the context of and was associated with an international armed conflict.
- The perpetrator was aware of factual circumstances that established the existence of an armed conflict.[12]

The Elements of Article 8(2) (b) (xvii) relating to the employment of "poison or poisoned weapons" were essentially the same, except for the first element, which reads: "The perpetrator employed a substance or a weapon that releases a substance as a result of its employment."[13]

Given the information publicly available regarding the characteristics of the range of agents previously explored or developed as ICA weapons, coupled with the experience of the use of an ICA weapon during the Moscow siege, it can be argued that ICA weapons are likely to cause "death or serious damage to health", at least to a certain proportion of the target population "in the ordinary course of events". Consequently, the use of ICA weapons in an armed conflict appears to fall within the scope of Article 8(2) (b) (xvii) and/or 8(2) (b) (xviii), and would be considered a war crime, with the possibility that those responsible for such acts might be tried before the Court.[14]

The possible applicability of the Statute with regard to RCAs, however, has been questioned. Tabassi, for example, has argued that the use of RCAs as a method of warfare did not come under the jurisdiction of the Court.[15] Furthermore, Allen has noted "though 'asphyxiating, poisonous or other gases'... has widely been interpreted to include some chemical weapons, other chemical agents, such as irritants, may not be included since they are not poisons".[16] Indeed, it is clear that the use of RCAs will not "in the ordinary course of events" cause "death or serious damage to health" and would therefore fail to fulfil one of the "Elements of Article 8(2)". However, although the employment *per se* of RCAs in an armed conflict arguably may not fall within the scope of a crime under the Rome Statue, their use in certain specific circumstances may do so, for example, if such use facilitated other actions covered by the Rome Statute, namely "torture or inhuman treatment",[17] or was employed to "wilfully caus[e] great suffering, or serious injury to body or health".[18]

Whilst the use of ICA weapons (and potentially RCAs in certain circumstances) in armed conflict appears to be considered a crime under the ICC, the scope of ICC applicability in this area is restricted. For example, the Court's jurisdiction is limited to nationals of States Parties that have ratified the Statute, and it can only prosecute crimes committed on or after the date of its establishment.[19] Furthermore, under the Rome Statute the Court will not admit cases that are being, or have already been, investigated or prosecuted by a State which has jurisdiction over the case, unless the State is genuinely unwilling or unable to carry out the investigation or prosecution.[20] In addition, according to Tabassi, the ICC would only have jurisdiction over cases involving the use of chemical weapons, but not cases solely involving the development, production, acquisition, stockpiling or transfer of such agents *per se*.[21]

Furthermore, until recently, the ICC jurisdiction only covered the use of chemical weapons in an international armed conflict.[22] Cases involving the use of chemical weapons in internal armed conflicts (or for law enforcement operations) were not covered – except potentially in cases of genocide or crimes against humanity.[23] Tabassi and van der Borght highlighted the anachronism of such a position, which "dates back to the 1925

Geneva Protocol", particularly given the near universal adherence to the CWC, which has made no distinction between internal and international armed conflict.[24] Furthermore, the authors contended that both the jurisprudence of the International Tribunal for the former Yugoslavia (ICTY) and the ICRC study on IHL, "concluded that the customary international law prohibition, binding on all States, on the use of chemical weapons in international armed conflict now extends to the use in non-international armed conflict as well".[25]

In order to address this limitation, Belgium,[26] together with 11 co-sponsoring States,[27] submitted a proposal to the First Rome Statute Review Conference in 2010,[28] recommending the amendment of Article 8 of the Statute so as to extend the criminalization of the use of poison, poisoned weapons, asphyxiating, poisonous or other gases (as well as expanding bullets) to armed conflicts not of an international character.

The Resolution to amend Article 8 was subsequently adopted by consensus at the First Review Conference,[29] and will enter into force for individual States one year after the deposit of their instruments of ratification or acceptance.[30] As well as extending the scope of the Rome Statute prohibitions in this area, the process of review and amendment conducted by the Review Conference has been important in helping to ensure that the Rome Statute remains a "living document" capable of adapting to changing circumstances and needs.

9.3. International, hybrid and internationalized-domestic courts and tribunals

From the 1990s, a small number of international bodies have been established to deal with international crimes,[31] including the ICTY,[32] and the International Criminal Tribunal for Rwanda (ICTR).[33] In addition, four "hybrid (national/international) tribunals"[34] were created, namely the Special Panel for Serious Crimes of the Dili District Court in East Timor (and its Court of Appeal),[35] the Special Court for Sierra Leone (with Trial Chambers and an Appeal Chamber),[36] the Extraordinary Chambers of the Courts of Cambodia,[37] and the Special Tribunal for Lebanon.[38] These courts have a mixed membership of local and international judges. There are also "internationalised-domestic courts or tribunals" established in BiH,[39] Kosovo[40] and Iraq,[41] with jurisdiction for certain international crimes.

The courts highlighted in this section are limited by the terms of their constituent instruments with regard to the crimes they can try, and their jurisdiction may be limited geographically and to certain time periods.[42] As of August 2015, one case – the Anfal trial – involving the use of chemical weapons has been prosecuted in such a court (see Section 9.3.1). It is possible that the alleged use of an ICA weapon (or less likely an RCA) to facilitate war crimes could fall within the jurisdiction of such courts, but this would be dependent on the specific nature of the alleged crimes as well as that of the court itself.

9.3.1. Case study: Anfal trial of Iraqi officials

On 21st August 2006, the Anfal trial opened before the Iraqi High Tribunal in Baghdad.[43] Seven former Iraqi high officials were charged with crimes against humanity and war crimes,[44] and two of them (former Iraqi president Saddam Hussein and his cousin and former military commander in the region, Ali Hassan Al-Majid) with the additional charge of genocide.[45] The charges related to the defendants' involvement in the Anfal military operation against the Iraqi Kurdish population in 1988. During this operation, an estimated 182,000 people lost their lives, entire villages were razed, mass executions were carried out with the dead buried in mass graves, survivors were forcibly relocated to detention camps and chemical weapons were used repeatedly. The testimony of numerous witnesses and victims of the chemical attacks were heard, and documentary evidence and film footage linking the defendants to the chemical weapons attacks was presented.

On 24th June 2007, the Court sentenced three of the accused – Ali Hassan al-Majid, Sultan Hashim Ahmed and Hussein Rashid al-Tikriti – to death by hanging; Farhan al-Jibouri and Saber Abdul Aziz were sentenced to life in prison; all charges were dropped for lack of evidence against Taher Muhammad al-Ani. Previously, the trial against Saddam Hussein had been discontinued following his execution on 30th December 2006. Whilst the trial was important in documenting the use of chemical weapons in activities the Tribunal considered as amounting to war crimes, crimes against humanity and genocide, there were serious concerns expressed about the fairness of the trial and the lack of an adequate appellate process.[46]

9.4. National courts

In the case of international crimes, any State is free to try a suspected perpetrator providing for requisite jurisdiction and subject to sovereign immunity.[47] Individuals responsible for crimes under international law can be prosecuted once the jurisdiction of the State, which initiates the prosecution, has been asserted. A number of principles exist, under international law, to determine the legal grounds for jurisdiction. These include the principle of territoriality (the crime has been committed on the territory of the State which intends to prosecute); the principle of active nationality (the perpetrator of the crime is a national of the State which is initiating prosecution); the principle of passive nationality (the victim of the crime is a national of the State which is initiating prosecution); the principle of protection (fundamental [i.e. security] interests of the State are affected); and the universality principle (no specific connection exists between the State which is initiating the prosecution and the offender).[48]

As Rikhof has noted, while it had previously been possible in the domestic context to initiate criminal prosecutions for genocide and war crimes as a result of ratifying the 1948 Genocide Convention,[49] and the 1949 Geneva Conventions,[50] the coming into force of the Rome Statute has:

> [P]rovided an important impetus for a large number of countries to not only examine their domestic legislation dealing with the regulation of war crimes, crimes against humanity and genocide but also to introduce changes to their laws to ensure that they were in compliance with international obligations and the tenets of the Rome Statute.[51]

Legal scholars such as Hankin,[52] and Rikhof,[53] have analysed the divergent mechanisms by which countries have incorporated international criminal law into their domestic legislation. Whether and how such domestic legislation could be applied to the use of ICA weapons or RCAs in armed conflict would have to be determined on a case-by-case basis. In a number of States, national law dealing with the international crimes of genocide, crimes against humanity and war crimes has incorporated,[54] or made reference,[55] to the relevant definitions set out in the Rome Statute. In such countries, at least, a case could be made that the use of ICA weapons or RCAs to commit or facilitate genocide, crimes against humanity or war crimes would fall within the national court's jurisdiction, as would the use *per se* of ICA weapons in armed conflict.

To date (August 2015), there has been one case – the van Anraat trial (see Section 9.4.1) – where a national court has employed international criminal law to try an individual accused of complicity in international crimes involving the use of chemical weapons. In addition, the Republic of South Africa unsuccessfully prosecuted Wouter Basson on 61 charges of domestic crimes relating to activities whilst he was head of Apartheid South Africa's chemical weapons programme (Project Coast) – including murder, conspiracy to murder, possession of drugs of addiction and fraud.[56] Other related cases of interest include attempts to prosecute individuals accused of facilitating provision of conventional arms to those engaged in international crimes.[57]

9.4.1. Case study: Van Anraat trial – Complicity in war crimes

On 23rd December 2005, the District Court of The Hague, Criminal Law Section, found Dutch businessman Frans van Anraat guilty, under Dutch law, of complicity in war crimes committed by the former Iraqi regime.[58] The Court sentenced van Anraat to the maximum penalty of 15 years' imprisonment, while concluding that the sentence was insufficient due to the severe repercussions of the chemical attacks and the nature of the crime.[59] During the 1980s, Mr Van Anraat was Saddam Hussein's most important supplier of chemicals used for the production of mustard gas. According to the Court, van Anraat's involvement in supplying chemicals to Iraq was an essential contribution to the chemical weapons programme of Saddam Hussein's regime.[60]

The Court determined that Mr van Anraat knowingly and intentionally supplied chemicals to the former Iraqi regime, which were used to produce chemical weapons employed by Iraq in Iraqi Kurdistan and in the Islamic Republic of Iran during the period 1984–1988.[61] Mr van Anraat was convicted of complicity in war crimes since his deliveries facilitated the attacks on the Kurdish population and made the carrying out of the regime's ambitions considerably easier.[62] Mr van Anraat was acquitted of a second charge: complicity in genocide. Although the Court determined that the attacks in Iraqi Kurdistan (including the use of chemical weapons) formed part of a genocidal campaign in the period 1985–1988, the Court was not convinced that the accused had actual knowledge of the genocidal intention of the perpetrators – which was a necessary element for a conviction of "complicity to genocide".[63]

9.5. Harvard Sussex Draft Convention

The Harvard Sussex Program (HSP) has proposed an alternative approach to employing international criminal law to address individual responsibility and culpability in the development or misuse of CBWs, through the development of a stand-alone convention.

Meselson and Robinson have argued that: "Any development, production, acquisition, or use of biological and chemical weapons is the result of decisions and actions of individual persons, whether they are government officials, commercial suppliers, weapons experts, or terrorists."[64] However, they have contended that the BTWC and CWC "are directed primarily to the actions of states, and address the matter of individual responsibility to only a limited degree".[65]

Although both the BTWC and the CWC have required that States Parties enact domestic legislation which criminalized the prohibitions contained in these conventions, these are limited in scope. For example, the BTWC and the CWC have stopped short of requiring a State Party to establish criminal jurisdiction applicable to foreign nationals on its territory who have committed biological or chemical weapons offences elsewhere.

Furthermore, Meselson and Robinson have noted that: "Purely national statutes present daunting problems of harmonizing their various provisions regarding the definition of crimes, rights of the accused, dispute resolution, and judicial assistance, among others."[66] Consequently, Meselson and Robinson have argued that: "What is needed is a new treaty, one that defines specific acts involving biological or chemical weapons as international crimes, like piracy or aircraft hijacking."[67]

To this end, HSP, with advice from an international group of legal authorities, developed a *Draft Convention to Prohibit Biological and Chemical Weapons under International Criminal Law*. If enacted, the Harvard Sussex Draft Convention would make it a crime under international law for any person: to knowingly develop, produce, acquire, retain, transfer or use biological or chemical weapons; to order, direct or knowingly render substantial assistance to those activities; or to threaten to use biological or chemical weapons.[68]

Under the Harvard Sussex Draft Convention, each State Party would be required to "establish jurisdiction with respect to such crimes according to established principles of judicial law", and where the State had jurisdiction and was satisfied that the facts warranted such action, "to submit those cases to competent authorities for the purpose of extradition or prosecution".[69] Furthermore, with respect to the actual use of chemical or biological weapons, each State Party would be required to "establish jurisdiction over all persons found on its territory regardless of their nationality or place of the offence".[70]

The Harvard Sussex Draft Convention has defined biological and chemical weapons as they were defined in the BTWC and the CWC, on the basis of a general purpose criterion worded so as to prohibit activities undertaken with hostile intent, while not prohibiting those intended for protective, prophylactic or other peaceful purposes.[71] Consequently, the Draft Convention appears to cover ICA weapons, RCAs and associated means of delivery within its scope when used for prohibited purposes.

The Draft Convention has garnered support from academics concerned with CBWs.[72] Dando has stated his belief that:

> Widespread adoption of a convention such as this draft [HSP] treaty, or a similar development of the Rome Statute, would add to the web of preventive policies that help minimize the potential for the development and use of these weapons in the future. Individuals who carried out such activities would do so knowing their actions had international, legal ramifications.[73]

HSP has explored a number of potential routes by which the Draft Convention could be adopted and implemented by the international governmental community. Firstly, a group of States could submit the proposed Draft Convention or a similar text in the form of a resolution for consideration by the UNGA, seeking its referral to the UNGA Sixth (legal) Committee for negotiation of an agreed instrument.[74] Alternatively, a regional or other grouping of States might convene a diplomatic conference to produce an agreed instrument that could then be opened for signature and ratification by any State wishing to do so.[75] Although the current strength and depth of support for this initiative amongst States is unclear, the UK Foreign and Commonwealth Office has previously indicated its support,[76] and "a number of other European governments ha[d] the convention under consideration".[77]

In 2011, Robinson explained that, "Once we are satisfied that the political environment is favourable, our plan is to convene an international conference that will bring together policy makers, jurists and exponents of the Draft Convention."[78]

9.6. Conclusions

The progressive development of international criminal law and attendant judicial bodies, notably the ICC, are important mechanisms in combating "the most serious crimes of concern to the international community".[79] As the International Military Tribunal at Nuremberg in 1948 recognized: "[C]rimes against international law are committed by men, not abstract entities, and only by punishing individuals who commit such crimes can the provisions of international law be enforced."[80]

The application of international criminal law and attendant judicial mechanisms to crimes involving chemical weapons can play an important role in strengthening and enforcing the prohibition against the use of such weapons, allowing those directly or indirectly responsible for such crimes – be they government officials, military commanders, scientists, chemical suppliers – to be brought to justice, even if their own States were unwilling to do so. Such actions would potentially have important deterrent effects. For, as a group of eminent prosecutors have stated, "Ending impunity by perpetrators of crimes of concern to the international community is a necessary part of preventing the recurrence of atrocities."[81]

To date (August 2015), whilst two cases involving direct or indirect complicity in the use of chemical weapons have been tried, no case involving ICA weapons or RCAs has been brought before the relevant courts. Indeed, the applicability of international criminal law to RCAs is unclear; the case being stronger for ICA weapons. Of the existing judicial bodies, the ICC – with its attendant review and amendment mechanisms – appears to afford the best route for clarification in this area and should be explored as part of a HAC regulatory strategy.

10
Mechanisms to Regulate the Transfer of ICA Weapons, Riot Control Agents and Related Means of Delivery

10.1. Introduction

This chapter will explore the potential application of transfer controls as part of a HAC approach to the regulation of ICA weapons, RCAs and related means of delivery. The transfer of these chemical agents and related means of delivery currently appears to fall within the scope of three types of control regimes, which cover chemical and biological weapons (and so-called weapons of mass destruction); conventional arms, (para)-military equipment and related technology; and security equipment utilized in torture and ill-treatment. Finally, the employment of certain pluri-lateral and international arms and equipment embargoes to halt transfers of these agents and means of delivery in certain limited circumstances is assessed.

10.2. Regimes addressing chemical weapons, biological weapons and weapons of mass destruction

The Chemical Weapons Convention (CWC) and the Biological and Toxin Weapons Convention (BTWC) contain legally binding provisions requiring States Parties to introduce controls over the transfer of (chemical or biological) agents and related goods that could be utilized in the development, production and employment of weapons prohibited under the relevant conventions. Such measures have subsequently been reinforced by an additional politically binding pluri-lateral control regime, the Australia Group. Although each of the three regimes contains provisions which appear to regulate the transfer of at least certain ICAs, RCAs and related means of delivery in certain circumstances, there are a number of limitations in the application of such provisions in practice.

10.2.1. Chemical Weapons Convention
The CWC has attempted to strike a balance between the rights and desires of States Parties to "participate in, the fullest possible exchange of chemicals,

equipment and scientific and technical information relating to the development and application of chemistry for purposes not prohibited under this Convention",[1] and their obligations "never, under any circumstances, to... transfer, directly or indirectly, chemical weapons to anyone",[2] or "to assist, encourage or induce, in any way, anyone to engage in any activity prohibited under the Convention".[3]

Consequently, the CWC has included provisions explicitly covering the transfer of chemicals listed in three Schedules. Schedule 1 chemicals, which are precursors and agents that have been developed as chemical weapons and which have little or no significant legitimate commercial use, may only be acquired on the territory of a State Party and can only be transferred to other States Parties for "research, medical, pharmaceutical or protective purposes" in quantities of up to a total for all Schedule 1 chemicals of 1 tonne per year.[4] All transfers are subject to advance notification and annual declaration. Re-export is not permitted. These restrictions apply irrespective of the amount to be transferred or the concentration of the chemical if transferred in a mixture.[5] Transfer to any non-State Party is forbidden under any circumstances.[6]

Schedule 2 chemicals are those considered to pose a significant risk to the object and purpose of the Convention, but which also have legitimate commercial uses in quantities in excess of 1 tonne per year. Since 29th April 2000, Schedule 2 chemicals may only be exported or imported to/from other States Parties.[7] Although trade in such chemicals between States Parties is not specifically regulated under the Convention, all States Parties must annually declare quantities held, and, if relevant, production sites, with the potential for onsite monitoring over certain thresholds.[8]

Schedule 3 chemicals are considered to pose a risk to the object and purpose of the Convention but are manufactured in very large quantities for legitimate commercial purposes. Although trade of such chemicals between States Parties has not been specifically regulated under the CWC, they may only be exported to non-State Parties if such recipient States issue an end-use certificate confirming they will only be used for purposes not prohibited under the Convention and will not be retransferred.[9]

As of August 2015, the major factor that has determined (and therefore limited) whether transfer controls should be introduced for a particular chemical agent has been whether that agent was included in one of the CWC Schedules. However, although the Convention has provided for the Schedules to be updated under a simplified technical change procedure that can be initiated by a State Party,[10] the process has appeared to be heavily politicized and the Schedules have not been revised on a regular basis. Consequently, at present the only ICA to have been listed under a CWC Schedule and thus to have triggered, albeit limited, transfer controls under the Convention has been BZ[11] (and two of its immediate precursors, 3-Quinuclidinol and Benzilic Acid[12]). All other chemical agents explored as potential ICA weapons as of

2015, as well as all RCAs,[13] must be considered as non-Scheduled chemicals with no specific CWC provisions restricting their transfer.

Although the Convention does not contain *specific* provisions relating to the transfer of non-Scheduled chemicals, there are *general obligations* under the Convention that are applicable to the transfer of all toxic chemicals. Each State Party is required to "adopt the necessary measures to ensure that toxic chemicals and their precursors are only developed, produced, otherwise acquired, retained, transferred, or used within its territory or in any other place under its jurisdiction or control for purposes not prohibited under this Convention".[14]

Furthermore, as the OPCW has acknowledged "...unilateral legislation may be adopted by Member States establishing additional restrictions on the trade of non-Scheduled chemicals, equipment and technologies which will have an impact on any country, particularly developing countries that are not States Parties of the Chemical Weapons Convention".[15]

A number of CWC States Parties have introduced "additional restrictions" which go beyond those explicitly stipulated in the Convention that are directly applicable to the transfer of ICAs, RCAs and related means of delivery. Of particular importance in this regard has been the establishment and activities of the Australia Group (see Section 10.2.3).

10.2.2. Biological and Toxin Weapons Convention

Under Article III of the BWTC, each State Party to the Convention has undertaken:

> [N]ot to transfer to any recipient whatsoever, directly or indirectly, and not in any way to assist, encourage, or induce any State, group of States or international organizations to manufacture or otherwise acquire any of the agents, toxins, weapons, equipment or means of delivery specified in Article I of this Convention.[16]

However, in provisions similar to the CWC, BTWC States Parties have also undertaken to "facilitate, and have the right to participate in, the fullest possible exchange of equipment, materials and scientific and technological information for the use of bacteriological (biological) agents and toxins for peaceful purposes...".[17] Furthermore, Article X requires that the Convention should be implemented so as to "avoid hampering the economic or technological development" of BTWC States Parties or international cooperation in "peaceful bacteriological (biological) activities", including the international exchange of biological agents, toxins and related equipment for such purposes.[18]

There is thus a tension in the regime and a balance that must be struck between the non-proliferation aspect of the Convention and that promoting the peaceful use of biological agents and toxins. This tension continues to be expressed, but not resolved, at BTWC Review Conferences. For example, the 7th BTWC Review Conference called for "appropriate measures, including effective national export controls... to ensure that direct and indirect transfers relevant to the Convention, to any recipient whatsoever, are authorized only when the intended use is for purposes not prohibited under the Convention",[19] but also reiterated that States Parties "should not use the provisions of... Article [III] to impose restrictions and/or limitations on transfers for purposes consistent with the objectives and provisions of the Convention...".[20]

As discussed in Chapter 6, the scope of materials covered by the BTWC includes RCAs and ICAs of biological origin, together with related means of delivery. Consequently, BTWC States Parties would be prohibited from transferring or otherwise assisting in the transfer or acquisition of such agents and materials that were not to be utilized for "peaceful purposes",[21] but were in fact intended to be employed for "hostile purposes or in armed conflict".[22] However, the application of the BTWC (as with the CWC) in this area suffers from a "double ambiguity", firstly with regard to the characterization of the agents and their treatment under the Convention (explored in Chapter 6), and secondly from the unresolved free trade/non-proliferation tension in the regime, as described. Consequently, it is likely that there will be divergent interpretation and implementation of the Convention in this area by States Parties.

10.2.3. The Australia Group

The principal objective of the Australia Group (AG) has been to employ licensing measures to ensure that "exports of certain chemicals, biological agents, and dual-use chemical and biological manufacturing facilities and equipment, do not contribute to the spread of CBW [chemical and biological weapons]".[23] The Group has attempted to achieve this by harmonizing participating countries' national export licensing measures.[24] The Group meets annually to discuss ways of increasing the effectiveness of participating countries' national export licensing measures to prevent would-be proliferators from obtaining materials for CBW programmes.[25] The 42 members of the AG[26] who are all party to the Geneva Protocol, BTWC and CWC have not undertaken any legally binding obligations. Instead, "the effectiveness of their cooperation depends solely on a shared commitment to CBW non-proliferation goals and the strength of their respective national measures".[27]

Key considerations in the formulation of participants' export licensing measures are that they should be effective in impeding the production of CBWs; practical, and reasonably easy to implement; and not impede the normal trade of materials and equipment used for legitimate purposes.[28] Under these measures, exports are denied "only if there is a well-founded concern about potential diversion for CBW purposes".[29]

10.2.3.1. Range of items controlled

The AG has developed Common Control Lists (reflected in national export control regimes) covering chemical weapons precursors;[30] dual-use chemical manufacturing facilities and equipment and related technology and software;[31] dual-use biological equipment and related technology and software;[32] human and animal pathogens and toxins for export control.[33]

Since the AG's objectives have not explicitly differentiated between "lethal" and "less lethal" chemical or biological weapons, the control regime would appear, potentially, to cover biological agents and toxic chemical agents (together with their precursors) that could be utilized as RCAs or ICA weapons. However, no specific RCAs or their precursors have been included in the AG Control Lists. Although no ICAs have been explicitly listed, the AG Chemical Weapons Precursors Control List (which was last updated in September 2014)[34] contained five precursors of BZ, two of which – Benzilic Acid and 3-Quinuclidinol – are also included in Schedule 2.b of the CWC, whilst the remaining three – Methyl Benzilate, 3-Hydroxy-1-methylpiperidine and 3-Quinuclidone – did not appear in any of the Convention's Schedules.

In addition, the list of human and animal pathogens and toxins for export control (last updated January 2014) has included "*Stapholoccous aureus* toxins",[35] and thus would cover SEB, which was previously explored as a potential ICA weapon by the United States in the 1960s.[36] As of August 2015, no AG Control Lists have included specific munitions for delivery of chemical or biological agents within their scope of coverage. However, the dual-use biological equipment and related technology and software list (last updated September 2014)[37] has included "spraying and fogging systems", "Spray booms or arrays of aerosol generating units" and "aerosol generating units", which are "specially designed or modified for fitting to aircraft, lighter than air vehicles or UAVs",[38] all of which could be applicable to devices intended for dispersal of RCAs and ICA weapons of biological origin.

10.2.3.2. Transfer control criteria

Attempting to improve the effectiveness and consistency of application of the Control Lists, the AG agreed "Guidelines for Transfers of Sensitive Chemical or Biological Items" in June 2007, and subsequently revised these in January 2009, and then again in June 2012.[39] These were intended to "form the basis for controlling transfers to any destination beyond the Government's national jurisdiction or control of materials, equipment, technology and software that could contribute to CBW activities".[40] AG States committed themselves to implementing the Guidelines in accordance with their national legislation, and to ensuring that the Guidelines were "applied to each transfer of any item in the AG control lists".[41]

When evaluating export applications for goods covered by the AG Control Lists, Participating States agreed to "take into account" a "non-exhaustive list of factors", including[42] "information about proliferation and terrorism involving CBW, including any proliferation or terrorism-related activity, or about involvement in clandestine or illegal procurement activities, of the parties to the transaction",[43] and also "the capabilities and objectives of the chemical and biological activities of the recipient state".[44] As of August 2015, however, neither respect for international humanitarian law nor respect for international human rights law and standards have been included in the list of factors.

The Guidelines incorporated two important innovations: a "catch-all" clause and a "no-undercutting" policy. Under the "catch-all" clause, Participating States committed themselves to include a requirement in their regulations for: " ... an authorisation for the transfer of non-listed items where the exporter is informed by the competent authorities... that the items... may be intended, in their entirety or part, for use in connection with chemical or biological weapons activities".[45] Furthermore, if the exporter was "aware that non-listed items are intended to contribute to such activities it must notify the [relevant] authorities... which will decide whether or not it is expedient to make the export concerned subject to authorisation".[46]

Under the "no-undercut" policy, a licence for an export that was "essentially identical" to one denied by another AG participant would only be granted after consultations with that participant, provided it had not expired or been rescinded.[47] An "essentially identical" good was defined as "being the same biological agent or chemical or, in the case of dual-use equipment, equipment which has the same or similar specifications and performance being sold to the same consignee".[48]

Following the introduction of these provisions, it can be argued that the transfer of all types of RCAs and incapacitating agents, as well as dual-use manufacturing equipment and facilities, would potentially fall within the AG's area of concern, when such agents may be intended "for use in connection with chemical or biological weapons activities".[49] Furthermore, if one AG State considered that a specific requested transfer of an RCA or incapacitating agent (or their precursors) was intended for use in a chemical or biological weapons programme and consequently prohibited its export, there appears to be a presumption of export denial for similar transfers from all other AG States, pending consultation with the AG State which initiated the denial.

10.3. Regimes addressing conventional arms, related (para)-military equipment and technology

The term "conventional arms" or "conventional weapons" has not been formally defined under international law. However, according to Brehm,[50] these weapons can be distinguished from WMD as defined by the UN, namely, "atomic explosive weapons, radioactive material weapons, lethal CBWs, and any weapons developed in the future which have characteristics comparable in destructive effect to those of the atomic bomb or other weapons mentioned above".[51] The issue of whether ICA weapons, RCAs and related means of delivery should be considered as conventional arms does not appear to have been specifically addressed collectively by the international governmental community and remains open. Examination of the instruments regulating conventional arms transfers has shown divergence in the scope of items covered by such agreements and how they are described. Consequently, whilst many mechanisms regulating the transfer of conventional arms do not currently appear to include ICA weapons, RCAs or related means of delivery within their scope, a number of mechanisms do.

Similar ambiguities arise with regard to the criteria to be applied to determine whether specific conventional arms transfers should be permitted or prohibited. According to the 1996 UN Disarmament Commission Guidelines on International Arms Transfers, "Limitations on arms transfers can be found in international treaties, binding decisions adopted by the Security Council under Chapter VII of the Charter of the United Nations and the principles and purposes of the Charter."[52] Furthermore, according to the Commission, the scope of "Illicit arms trafficking is understood to cover that international trade in conventional arms, which is contrary to the laws of States and/or international law."[53]

From the 1990s onwards, a number of regional and pluri-lateral bodies began to develop instruments or other measures regulating the trade in conventional arms and associated goods. Two of those encompassing ICAs, RCAs and related means of delivery – EU Common Position 2008/944/CFSP and the Wassenaar Arrangement – are examined in Sections 10.3.1 and 10.3.2.

Building upon such regional and pluri-lateral mechanisms, in 1996 a coalition of States, arms control and human rights organizations and Nobel Peace Laureates began to advocate for the development of a set of explicit standards based on relevant international law which UN Member States should apply to the international transfer of conventional arms. A process to develop such standards through the UN-mandated negotiation of an international Arms Trade Treaty (ATT) was agreed in December 2009 and commenced in 2012.[54] In April 2013, the UNGA adopted the ATT,[55] and the Treaty came into force on 24th December 2014. The ATT was signed by 130 States and, as of October 2015, 76 have become State Parties to the Treaty.[56] Although certain States and leading civil society organisations had called for the ATT to cover a broad range of conventional arms and security equipment, unfortunately, during the negotiations the scope of the instrument was narrowed so that it would only include battle tanks; armoured combat vehicles; large-calibre artillery systems; combat aircraft; attack helicopters; warships; missiles and missile launchers; small arms and light weapons; and related ammunition/munitions.[57] Consequently, the transfer of RCAs and ICA weapons *per se* would not currently be covered by its provisions, though certain relevant means of delivery may fall within its scope.

10.3.1. The Wassenaar Arrangement

The Wassenaar Arrangement (WA) was established in July 1996 with the expressed intention of contributing to "regional and international security and stability, by promoting transparency and greater responsibility in transfers of conventional arms and dual-use goods and technologies, thus preventing destabilising accumulations".[58] Consequently, WA Participating States have sought, through their national policies, to "ensure that transfers of these items do not contribute to the development or enhancement of military capabilities which undermine these goals, and are not diverted to support such capabilities".[59]

The WA became operational in September 1996 and, as of August 2015, comprised 41 Participating States.[60] Its decision-making body, the Plenary, normally meets once a year in December, and its subsidiary bodies meet periodically. All decisions are taken by consensus and the deliberations are kept in confidence.[61]

WA Participating States have agreed to maintain national export controls on a range of commonly agreed listed items; report on transfers and denials of specified controlled items to destinations outside the Arrangement; exchange information on sensitive dual-use goods and technologies; and be guided by agreed Best Practices, Guidelines or Elements.[62]

10.3.1.1. Range of items controlled

The range of materials controlled by the Wassenaar Arrangement has been established in the WA Dual Use Goods and Technologies List and the WA Munitions List (last updated 25th March 2015).[63] RCAs are contained within the WA Munitions List, under the ML7 category (chemical or biological toxic agents, "riot control agents", radioactive materials, related equipment, components and materials) as follows:

> d. 'Riot control agents', active constituent chemicals and combinations thereof, including: 1. α-Bromobenzeneacetonitrile...(CA); 2. [(2-chlorophenyl) methylene] propanedinitrile...(CS); 3. 2-Chloro-1-phenylethanone (CN); 4. Dibenz-(b,f)-1,4-oxazephine, (CR); 5. 10-Chloro-5,10-dihydrophenarsazine (DM); 6. N-Nonanoylmorpholine (MPA).[64]

Although ML7 has listed a number of widely used RCAs, it does not include all RCAs identified by the OPCW, in May 2014, as being currently marketed and/or declared by CWC States Parties as being held for riot control purposes.[65] Two particularly important exceptions not specifically covered in this list are PAVA and OC, which are held by a number of States for law enforcement. In contrast, DM is listed, even though the OPCW has recommended its use as an RCA be discontinued.[66]

The WA Munitions List has provided a definition of RCAs as: "Substances which, *under the expected conditions of use for riot control purposes*, produce rapidly in humans sensory irritation or disabling physical effects which disappear within a short time following termination of exposure. (Tear gases are a subset of 'riot control agents'.)" (emphasis added).[67] Although this definition was based upon Article II.7 of the CWC, the highlighted additional text recognized that the effects of such chemical agents might alter if not employed under the "expected conditions of use for riot control purposes". However, exactly what these "expected conditions" were, has not been further defined.

ICAs have been included in the WA Munitions List under ML7.b. as: "b.3. CW incapacitating agents, *such as*: 3-Quinuclidinyl benzilate (BZ) (CAS 6581-06-2)" (emphasis added).[68] Although only BZ has been specifically mentioned in the control list to date, the wording of ML7 has implied that other chemical incapacitating agents could potentially be added to this list, but were not specifically controlled at present. Despite the reported use of an ICA weapon employing two fentanyl derivatives by Russian Federation security forces in October 2002, and contemporary research by a number of States into a range of ICAs with potential weapons utility, ML7.b. has not been updated to include such agents. Ambiguities regarding the scope of this category and the range of agents controlled have been further exacerbated as the term "CW incapacitating agents" has not been defined in the WA Munitions List (or indeed elsewhere).

ML7.a. of the WA Munitions List covers: "Biological agents or radioactive materials, 'adapted for use in war' to produce casualties in humans or animals, degrade equipment or damage crops or the environment."[69] The term "biological agents" has not been defined and, unlike the chemical agents section, there has been no list of specific agents covered. Although this issue has not been clarified by WA Participating States, ML7.a. does appear to cover "less lethal" biological agents (potentially including RCAs [such as OC] and incapacitating agents [such as SEB] of biological origin) as long as they were "adapted for use in war" to produce human casualties.

As well as including a range of chemical and biological agents, the WA Munitions List, under ML7.e., also covers:

> Equipment, specially designed or modified for military use, designed or modified for the dissemination of any of the following, and specially designed components therefor:...Materials or agents specified by ML7.a., ML7.b. or ML7.d.; or...CW agents made up of precursors specified by ML7.c.[70]

It should be noted that ML7.e. only covers equipment "specially designed or modified for military use", consequently law enforcement dissemination mechanisms – even if identical to military devices – do not currently appear to fall within the scope of these controls.

Finally, the WA Munitions list, under ML18, also covers "specially designed or modified 'production' equipment for the 'production' of products specified by the Munitions List, and specially designed components thereof"; as well as "specially designed environmental test facilities and specially designed equipment thereof, for the certification, qualification or testing of products specified by the Munitions List".[71]

10.3.1.2. Transfer control criteria

All measures with respect to the Wassenaar Arrangement have been taken in accordance with national legislation and policies, and have been "implemented on the basis of national discretion".[72] Consequently, although the scope of items covered in Participating States' export controls has been determined by the WA lists, the practical implementation has varied from country to country in accordance with national procedures.[73] As the WA has stated, "the decision to transfer or deny transfer of any item will be the sole responsibility of each Participating State".[74]

Participating States have, however, agreed certain guidelines, elements and procedures as a basis for decision-making through the application of their own national legislation and policies. These included the advisory, non-binding "Elements for Objective Analysis and Advice Concerning Potentially Destabilising Accumulations of Conventional Weapons".[75] This recommended that WA States consider a range of factors when making decisions on relevant transfers, including the risk that weapons would be used to commit or facilitate the violation and suppression of human rights and fundamental freedoms or the laws of armed conflict, or be used offensively against another country or in a manner inconsistent with the UN Charter.[76]

10.3.2. EU Common Position 2008/944/CFSP

In December 2008, the EU Member States adopted an EU Council Common Position 2008/944/CFSP, "defining common rules governing control of exports of military technology and equipment".[77] The Common Position replaced and built upon the politically binding EU Code of Conduct on Arms Exports, adopted in 1998.[78] In addition to the 28 EU Member States legally bound by the Common Position, BiH, Canada, the Former Yugoslav Republic of Macedonia, Iceland, Montenegro and Norway "officially aligned themselves" with its "criteria and principles".[79] Furthermore, Lichtenstein, Switzerland and Turkey had previously aligned themselves to the principles of the EU Code.[80]

10.3.2.1. Transfer control criteria

In the Common Position pre-ambulatory paragraphs, Member States declared themselves "determined to set high common standards which shall be regarded as the minimum for the management of, and restraint in, transfers of military technology and equipment by all Member States",[81] and to "prevent the export of military technology and equipment which might be used for internal repression or international aggression or contribute to regional instability".[82]

The Common Position required that each Member State "assess the export licence applications made to it for items on the EU Common Military List ... on a case-by-case basis against the criteria of Article 2",[83] namely, international obligations and commitments; human rights and international humanitarian law; the internal situation of the country; regional peace, security and stability; national security; terrorism and respect for international law; the risk of diversion or re-export under undesirable conditions; and the compatibility of the exports with the technical and economic capacity of the recipient country.[84]

Article 2, Criterion 2, regarding human rights and international humanitarian law, is of particular relevance. Under this Criterion, Member States shall "deny an export licence if there is a clear risk that the military technology or equipment to be exported might be used for internal repression".[85] They are also required to "exercise special caution and vigilance" in issuing licences "on a case-by-case basis" and taking account of the nature of the military technology or equipment, to countries where serious violations of human rights have been established by competent bodies of the UN, EU or Council of Europe.[86]

Furthermore, in an important advance on the EU Code of Conduct text, the Common Position introduced a specific requirement upon Member States to assess "the recipient country's attitude towards relevant principles established by instruments of international humanitarian law",[87] and to "deny an export licence if there is a clear risk that the military technology or equipment to be exported might be used in the commission of serious violations of international humanitarian law".[88]

In addition to these binding export criteria, the Common Position also contained innovative operative provisions to combat "undercutting". Member States are required to notify each other of arms export licences they have refused when a proposed arms export has failed to meet the Common Position criteria. Furthermore, before any Member State can grant a licence that has been denied by another Member State (for an essentially identical transaction in the preceding three years), it is required to consult the Member State that denied the original licence. Although the power to take the final decision remains with individual Member States, if a licence is granted in these circumstances, the licensing Member State will have to provide a detailed explanation of its reasoning.[89]

Further operative provisions imposed annual reporting obligations on Member States,[90] and required Member States to possess national legislation which enabled them to control the export of the technology and equipment of concern,[91] which extended beyond physical exports of goods to include transit, transhipment and brokering activities, as well as intangible transfers of technology.[92]

10.3.2.2. Range of items controlled

Whilst the Common Position instituted relevant operative provisions and the criteria by which export applications will be judged, the EU Common Military List,[93] and EC Regulation 1334/2000,[94] established the recommended range of items covered.[95] Examination of the EU Common Military List has shown that the relevant text effectively replicated the Wassenaar Control List wording with regard to ICA weapons, RCAs and their means of delivery. Although the EU Common Military List has been in place since July 2000, and has been reviewed regularly,[96] with the latest version adopted by the Council on 9th February 2015, the relevant text (regarding ML7.a., b., d. and e.) has remained unchanged.[97] Consequently, the direct "import" of WA language without adaptation at the time of drafting the EU Code of Conduct, and during subsequent revision of the text at the time of its transformation into a legally binding Common Position, appears to have "imported" limitations and ambiguities existing from the WA control regime into the EU control regime.[98]

10.3.2.3. Mechanisms to facilitate implementation and transparency

Because implementation of the Common Position has been mediated at the national level, the decisions on whether to deny or grant individual export licences have lain with the relevant Member States. Consequently, there have been concerns that the Common Position criteria would be interpreted differently by different Member States. To offset this, the Council developed a publicly accessible User's Guide in 2003 (which was last updated in 2015) containing advice regarding the interpretation of the EU Code/Common Position criteria.[99]

There have also been concerns about the limited public transparency mechanisms contained within the Common Position. Although the Common Position has required each Member State to produce a national annual report of "its exports of military technology and equipment",[100] such reports have not been harmonized; consequently, the organization of information and level of detail provided by different Member States has been extremely variable.[101] The Common Position has also required Member States to provide the Commission with "information on their implementation" of the Common Position, which would form the basis of an annual compilation report of Member State licences, produced by the Commission.[102] However, this Annual Report (the latest of which was released on 25th March 2015) has only recorded licences granted by each Member State under the broad ML categories (e.g. ML7) and not by the subcategories (e.g. ML7.a.–i.).[103] Thus, it has not been possible to ascertain whether licences have been granted by a particular Member State for RCAs (ML7.d.) or ICA weapons (ML7.b.), or indeed protective and decontamination equipment (ML7. f.), let alone whether a specific RCA had been exported.

Despite the increased harmonization of arms transfer controls across the region, facilitated by the introduction of the EU Common Position, and the EU Code before it, irresponsible transfers of RCAs from EU Member States to military or law enforcement agencies that have persistently abused such agents have been reported.[104]

10.4. Regimes addressing security equipment that can be utilized in torture and cruel, inhuman, degrading treatment or punishment

Although the prohibition on torture and CIDTP is absolute, these practices are still perpetrated in countries in all regions of the world. UN independent experts, UN human rights bodies and international non-governmental human rights organizations have documented the use of different types of equipment, including RCAs and ICAs, to commit such human rights violations.[105] Consequently, previous UN Special Rapporteurs on Torture,[106] as well as human rights NGOs, particularly AI and the Omega Research Foundation (ORF),[107] have promoted the development of mechanisms to prohibit or severely restrict the transfer of a range of policing and security equipment that could be utilized for torture and CIDTP. In 2006, the EU introduced the world's first multilateral trade controls in this area.

10.4.1. EC Regulation 1236/2005

EC Regulation 1236/2005,[108] which entered into force in July 2006, is a legally binding instrument applicable to all EU Member States. The EC Regulation prohibited imports and exports to or from the EU of certain goods "which have no practical use other than...for the purpose of torture and other cruel, inhuman or degrading treatment or punishment", so-called Annex II goods.[109] The EC Regulation also required national export authorizations for exports of certain items, "that could be used for the purpose of torture and other cruel, inhuman or degrading treatment or punishment", so-called Annex III goods.[110]

Article 6 of the EC Regulation obliged Member States to regulate the export of controlled items, and to deny authorizations for exports of such items "when there are reasonable grounds to believe that goods listed in Annex III might be used for torture" or other ill-treatment.[111] Under Article 6.1, decisions on export applications for Annex III goods shall be taken by the competent authority "on a case by case basis", and taking into account "all relevant considerations, including in particular, whether an application for authorisation of an essentially identical export has been dismissed by another Member State in the preceding three years".[112]

Further operative provisions required States to introduce rules imposing penalties on violators of the EC Regulation;[113] fulfil notification and consultation requirements;[114] implement information exchange mechanisms to combat "undercutting";[115] and produce public annual activity reports.[116]

10.4.1.1. Range of items controlled

In order that the EC Regulation and consequent control regime could "take into account new data and technological developments, the list of goods covered by this Regulation should be kept under review and provision should be made for a specific procedure to amend these lists".[117] Consequently, Article 12 of the EC Regulation empowered the Commission to amend the lists of prohibited (Annex II) and controlled (Annex III) goods.[118]

As of August 2015, no RCAs or ICAs have been included within the Annex II list of prohibited goods. However, the range of Annex III goods whose trade is controlled by the EC Regulation has included[119]

> 3. Substances for the purpose of riot control or self-protection and related portable dissemination equipment, as follows:
> 3.1. Portable devices for the purpose of riot control or self-protection by the administration or dissemination of an incapacitating chemical substance[120]
> 3.2. Pelargonic acid vanillylamide (PAVA)
> 3.3. Oleoresin capsicum (OC).

10.4.1.1.1. Introduction of export controls on ICAs employed in death penalty.

Lethal injection is the practice of executing a person using a lethal dose of drugs (normally including an ICA) administered intravenously. It is currently provided for as a method of execution in China, Guatemala, the Maldives, Papua New Guinea, Taiwan, Thailand, Vietnam and the United States.[121] Lethal injection is the most common method of execution in the United States.[122] Until 2010–2011, the majority of US death-penalty States employed a three-drug protocol, comprising an anaesthetic to put the prisoner to sleep (either sodium thiopental or pentobarbital), a paralytic to halt breathing (pancuronium bromide) and potassium chloride to stop the heart. Death normally results from anaesthetic overdose and respiratory and cardiac arrest while the condemned person is unconscious. An alternative protocol favoured by a small number of US States utilized one large dose of a barbiturate, normally either sodium thiopental or pentobarbital; death results from respiratory arrest.

In 2010, Hospira, the sole US manufacturer of sodium thiopental, suspended production of the drug. Subsequently, certain US death-penalty

States whose stockpiles of sodium thiopental were running low sought to acquire further stocks from overseas suppliers, including from the EU. In December 2011, to ensure that EU-sourced ICAs were not employed in lethal injection executions, the European Commission amended Annex III of the EC Regulation to include a new category of controlled goods: "Products which could be used for the execution of human beings by means of lethal injection... Short and intermediate acting barbiturate anaesthetic agents."[123] This amendment specifically included, but was not limited to, amobarbital, pentobarbital, secobarbital and thiopental and their respective sodium salts.[124] These measures subsequently appear to have been reflected in the licensing practices of at least certain Member States. For example, according to its annual reports of relevant export licensing activities, whilst Germany granted 66 licences in 2012, and 49 in 2013, for the export of such dual-use drugs to States with no record of "lethal injection", it refused two licence applications for these drugs in 2013, to China and Vietnam.[125]

Although this amendment was introduced with the intention of halting the transfer of such drugs for lethal injection, it brought within the EC Regulation's scope "barbiturate anaesthetic agents", some of which have reportedly been employed in torture and ill-treatment, including as "truth serums" in interrogations, or have been explored (though apparently discounted) as potential ICA weapons by certain States.[126] In the future, if Member States were so minded, the EC Regulation could potentially be amended to specifically cover ICAs intended for law enforcement purposes beyond the death penalty, with a prohibition on the transfer of such agents to end users likely to misuse them for torture and ill-treatment.[127] If such a measure were to be agreed, additional relevant ICAs could be added to Annex III.

10.4.1.1.2. Further expansion of control lists and ongoing process to strength EC Regulation 1236/2005. On 16th July 2014, the European Commission introduced a new Commission Implementing Regulation (EU) No. 775/2014, further expanding the lists of prohibited (Annex II) goods and controlled (Annex III) goods covered by the Regulation.[128] The European Commission adopted the legal changes to the lists after consultation with Member State officials in the Committee on common rules for exports of products, and following a year-long review of the Regulation by a Commission appointed group of independent experts. The new lists, which entered into force on 20th July 2014, are legally binding and directly applicable now in all 28 EU Member States. Among the items added were:

> Fixed equipment for the dissemination of incapacitating or irritating chemical substances, which can be attached to a wall or to a ceiling inside a building, comprises a canister of irritating or incapacitating chemical agents and is activated using a remote control system;[129]

Fixed or mountable equipment for the dissemination of incapacitating or irritating chemical agents that covers a wide area and is not designed to be attached to a wall or to a ceiling inside a building; ...

Notes: 1. This item does not control equipment controlled by item ML7(e) of the Common Military List of the European Union 2. This item also controls water cannons...[130]

The inclusion of fixed-installation RCA dissemination devices and "wide area" RCA dissemination devices in the Regulation is a highly important development. Firstly, such inclusion on the control list of an anti-torture instrument recognizes the potential for such devices to be misused for torture and ill-treatment, and the consequent obligation of Member States to ensure that they are not exported to end users likely to employ them for such violations. In addition, the definition of "wide area" dissemination devices under paragraph 3.5 is very broad and potentially captures many items of concern (as raised in Chapters 4 and 5 of this publication).

The introduction of the Implementing Regulation by the Commission in July 2014, expanding the list of prohibited and controlled goods, was the first part of a wide-ranging overhaul of EC Regulation 1236/2005. The Commission, Council and Parliament are now engaged in a substantive review of the 2005 Regulation's operative mechanisms. In January 2014, the Commission presented proposals to the Council of Member States and the European Parliament for strengthening the Regulation.[131] Human rights and arms control NGOs have also proposed that the EU consider additional measures to strengthen the Regulation, including a targeted end-use clause.[132] The development and introduction of such a clause was originally proposed by the United Kingdom,[133] and such measures have previously been supported by the European Parliament.[134] Further discussions on the potential benefits and difficulties of applying such a mechanism have taken place both within relevant Committees of the European Parliament and amongst Member States in consultation with the European Commission. If agreed, this targeted end-use clause would allow individual Member States to prohibit specific transfers of items not currently listed in the Regulation Annexes where they had intelligence that the specific transfers of these items would be used for the death penalty, torture or ill-treatment.[135] If such a mechanism were to be introduced, and depending upon its scope and nature, it could potentially cover RCAs, ICAs and related means of delivery not currently listed on the EC Regulation Annexes.

10.4.1.2. Potential development of controls in other countries, regions or internationally

Growing international concern about the supply to law enforcement agencies of equipment used in torture and ill-treatment has also been reflected in the UN General Assembly, where, since 2001, a bi-annual resolution on torture and CIDTP has incorporated a paragraph on transfer controls. The resolution adopted by UNGA initially in November 2011 (and subsequently in December 2012 and December 2013), contained the following strengthened language calling upon all States to:

> [T]ake appropriate effective legislative, administrative, judicial and other measures to prevent and prohibit the production, trade, export, import and use of equipment that have no practical use other than for the purpose of torture or other cruel, inhuman or degrading treatment or punishment.[136]

Whilst recognizing the limitations of EC Regulation 1236/2005 in certain areas and the inadequate implementation of the instrument in some European countries,[137] a number of observers, including the UN Special Rapporteur on Torture,[138] and the European Parliament,[139] have highlighted the potential for the EC Regulation to act as a model for similar controls in other regions or as the basis for international controls in this area.

10.5. Embargoes on the transfer of arms and (para)-military equipment

Embargoes on the supply of arms and (para)-military equipment to particular States or non-State actors have been increasingly deployed since the end of the Cold War by a wide range of individual States, regional, pluri-lateral and international organizations. Where such embargoes have covered ICA weapons, RCAs and related means of delivery, they have provided a potentially powerful mechanism for halting or at least restricting the supply of such items to specific end users that have seriously misused or were likely to misuse them (normally in concert with other arms and (para)-military equipment) for grave human rights abuses, breaches of international humanitarian law or for armed aggression.

10.5.1. United Nations embargoes

Under Chapter 7 of the UN Charter, the UN Security Council "may call upon the Members of the United Nations to apply... complete or partial interruption of economic relations... and the severance of diplomatic relations",[140] when it has determined "the existence of any threat to the peace, breach of the peace, or act of aggression".[141] Amongst the sanctions available to the UNSC are arms embargoes, which can be either voluntary or mandatory (and legally binding) upon all UN Member States. The UNSC Sanctions Committee website explains that the use of mandatory sanctions was intended to "apply pressure" on a State or entity to comply with the objectives set by the

UNSC "without resorting to the use of force", and thus "offer the Security Council an important instrument to enforce its decisions".[142]

Since the end of the Cold War, the use of UN arms embargoes and other targeted sanctions has increased, and the range of purposes for which they were intended has also widened. The UNSC has used them as a tool to repel aggression, restore democracy (and in certain cases protect human rights), and apply pressure to regimes supporting terrorist activities and others charged with international crimes.[143] Whereas sanctions have traditionally been used against States, the UNSC also has increasingly imposed sanctions against a range of non-State actors. For example, of the 20 mandatory UNSC embargoes established from 1st January 1990 till 31st December 2009, 10 included provisions against non-State actors.[144]

10.5.1.1. Range of items controlled

Analysis of the relevant UNSCRs, which established or amended embargoes adopted from 1st January 1990 to 31st December 2009, has shown that none of them specifically cited ICA weapons, RCAs or related means of delivery as items contained within their scope of coverage. However, the majority (14) of the embargoes contained the following language in the description of goods covered:[145] "Arms and related materiel of all types including weapons and ammunition, military vehicles and equipment, paramilitary equipment, and spare parts for the aforementioned" (emphasis added).[146]

A further four UNSCRs which established embargoes contained variants on the terms "arms and related materiel" or "arms and munitions", but did not specifically list "paramilitary equipment".[147] In addition, a number of UNSC embargo Resolutions have explicitly allowed the transfer of "non-lethal military equipment" to certain end users in the embargoed country, such as peacekeeping troops[148] thus indicating that the supply of such equipment to non-exempted end users was covered (and prohibited) under the scope of the embargo. Since ICA weapons, RCAs and related means of delivery appear to fall within the scope of all these terms, it can be intimated that 18 of the 20 mandatory embargoes established during the period studied covered these items.

As no UNSC embargo Resolution contained an annexed list detailing the specific range of arms and related equipment included within its scope, there is considerable ambiguity over precisely which ICA weapons, RCAs and related means of delivery would be covered by each of the 18 embargoes. This ambiguity has been compounded by the fact that there have been no agreed definitions of key terms such as "paramilitary equipment", "arms and related materiel" or "non-lethal military equipment", nor have there been any indicative lists of materiel covered by these terms. Furthermore, although a number of UN Member States and regional or pluri-lateral organizations (such as the EU and the WA) have agreed and published lists of arms, and military equipment whose export is regulated,[149] there were and are still no international military or paramilitary lists accepted by all UN Member States and applicable to UN embargoes generally. Instead, the scope of coverage for each embargo has and will be determined by national interpretation of the relevant descriptive terms in the UNSCR (and may also be informed by recommendations of relevant UN bodies such as UNSC Sanctions Committees). This of course presents the danger that different UN Member States will interpret the scope of arms and equipment to be embargoed differently, leading to inconsistent implementation and the potential weakening of the embargo regime.[150]

Although the majority of UN embargoes in force (in the time period studied) covered conventional arms and related military and paramilitary equipment, an additional small numbers have included WMD within their scope. For example, certain previous embargoes against Iraq prohibited the transfer of "[a]ll chemical and biological weapons and all stocks of agents and all related subsystems and components and all research, development, support and manufacturing facilities".[151] Furthermore, the current embargo in force against North Korea has prohibited the transfer of "items, materials, equipment, goods and technology, determined by the Security Council or the Committee, which could contribute to DPRK's nuclear-related, ballistic missile-related or other weapons of mass destruction related programmes".[152] Although it appears that these embargoes would not cover RCAs, they do appear to potentially cover materiel utilized in the production of certain incapacitating chemical or biological weapons. However, once again the specific goods covered by these terms, and, consequently, the embargoes' scope have not been detailed and are determined by national implementation.

10.5.1.2. Monitoring and verification of embargoes

Mandatory UN arms embargoes currently in force contain provisions for monitoring their implementation. As a minimum, a UNSC Sanctions Committee is normally established to undertake a range of monitoring tasks, which can include seeking information from all UN Member States on embargo implementation measures; considering information concerning violations; periodic reporting on violations and violators to the Security Council; considering requests for humanitarian exceptions; providing guidelines for implementation; and making information publicly available through the media.[153] UNSC Sanctions Committees often face resource constraints and normally work within highly politicized environments. They have a varied record of effectiveness which is often dependent on the personnel and personalities driving each one.[154]

The work of UNSC Sanctions Committees has been increasingly aided by Panels of Experts or Monitoring Groups tasked with investigating serious suspected violations or allegations of non-compliance, some with notable success. For example, the Fowler Report on violations of sanctions against the União Nacional para a Independência Total de Angola (UNITA)[155] was, according to Shields, "ground-breaking in its scope and candour",[156] in its exposure of the direct involvement of Burkina Faso, Togo and Zaire in sanctions-busting. However, such examples notwithstanding, Fruchart and colleagues have noted that, in cases where the UN has mandated panels of experts or monitoring teams to investigate the implementation of arms embargoes, "they have reported that they lack the authority and powers to explore the ways in which the embargoes were breached".[157]

10.5.1.3. The effectiveness of UN arms embargoes

UN bodies, as well as academic and NGO researchers, have highlighted the failings and limitations of UN arms embargoes.[158] Factors believed to contribute to the limited effectiveness of UN embargoes include widespread lack of political will and insufficient resources committed to support enforcement, ineffective border controls between the embargoed State and its neighbours, and the failure to incorporate arms embargoes into national legislation. Often, ineffective national implementation happens simply because many UN Member States do not possess the required technical or financial resources.

In addition, leading human rights organizations have highlighted the delays in imposition of certain UNSC embargoes which have allowed protagonists to a conflict or those engaged in human rights abuses to procure sufficient arms to continue their activities after the embargo is finally imposed.[159] Furthermore, there have been a number of cases where a humanitarian or human rights crisis has developed and the UNSC has been unable or unwilling to impose an arms embargo on the offending State or non-State actor.[160]

10.5.2. Arms embargoes established by regional and pluri-lateral organizations

As the UN Sanctions Committee has noted, the "universal character" of the UN makes it an "especially appropriate body to establish and monitor [sanctions including arms embargoes]".[161] However, it is not the only body that has the authority to introduce such measures. Indeed, there are a range of regional and pluri-lateral organizations and groupings of States that potentially can institute sanctions in this area, including the African Union, the Commonwealth, Economic Community Of West African States (ECOWAS), the EU, the League of Arab States, the Organization of American States (OAS) and the Organization for Security and Cooperation in Europe (OSCE).

However, apart from the EU, which will be dealt with in detail in Section 10.5.3, no other regional body, pluri-lateral organization or grouping of States have established systematic mechanisms for introducing and implementing embargoes on arms and security equipment. Whilst certain additional bodies have, on occasion, introduced embargoes, these have generally been poorly defined, non-mandatory sanctions, without any form of monitoring or enforcement mechanism. Furthermore, given the limited information available on such embargoes, it is unclear which, if any, contain ICA weapons, RCAs and related means of delivery within their scope of coverage.

10.5.3. EU arms embargoes

According to the European Commission, EU arms embargoes may be applied to "stop the flow of arms and military equipment to conflict areas or to regimes that are likely to use them for internal repression or aggression against a foreign country".[162] EU embargoes can take two forms, the first, as a minimum, directly implements the corresponding UN arms embargo established under a UNSCR. Such EU embargoes, of which 20 have been introduced in the period from January 1990 till December 2009, have often contained the same language and have covered the same arms and equipment as the UNSCR they have duplicated. In certain cases, however, the coverage of these EU arms embargoes has been wider than the corresponding UN arms embargo in terms of geographic scope and entities embargoed, and/or the relevant EU embargo has been adopted long before the corresponding UNSCR.[163] For example, the EU embargo against the entire Democratic Republic of the Congo (DRC) was established in April 1993,[164] whilst the corresponding UNSC embargo (which was limited to the Ituri and the North and South Kivu districts of the DRC) was not introduced until July 2003.[165] Certain EU arms embargoes have also had a wider scope of coverage in terms of goods embargoed. For example, the EU embargo against the Federal Republic of Yugoslavia contained an explicit ban on the transfer of "equipment which might be used for internal repression or terrorism",[166] which was not included in the corresponding UNSC embargo.[167]

In addition to such UNSCR-implementing embargoes, the EU has instituted autonomous embargoes in situations where a human security, human rights or humanitarian crisis have developed and the UNSC has not been able, for reasons of *realpolitik*, to adopt a UNSCR. From January 1990 till December 2009, the EU instituted autonomous embargoes on Guinea, Indonesia, Myanmar, Nigeria, Uzbekistan and Zimbabwe.[168]

Article 346 of the Treaty on the Functioning of the European Union (previously Article 296 of the Treaty establishing the European Community) has allowed for an embargo relating to military goods to be implemented by EU Member States using national measures.[169] It is, therefore, common practice that arms embargoes are imposed by a politically binding Common Position or Council Decision, and enforced on the basis of export control legislation of EU Member States.[170] EU arms embargoes have generally comprised, at a minimum, a prohibition on the "sale, supply, transfer or export of arms and related materiel of all types, including weapons and ammunition, military vehicles and equipment, paramilitary equipment and spare parts".[171]

In addition, these "core arms" embargoes have often been accompanied by an additional prohibition on the "provision of financing and financial assistance and technical assistance, brokering services and other services related to military activities and to the provision, manufacture, maintenance and use of arms and related materiel of all types".[172] Unlike the "core arms" embargoes, these additional prohibitions have been imposed through Council Regulations which were directly applicable and had direct effect in the EU Member States, creating obligations and rights for those subject to them (including EU citizens and economic operators). Their application and enforcement has been a task attributed to the competent authorities of the EU Member States and the Commission.[173]

A number of analysts have highlighted the potential difficulties with this dual mechanism. For example, Shields has argued that "[T]his divergence in the way specific elements of embargoes are implemented is a fatal flaw because sanctions regimes are on uneven legal footing at their inception. This can lead to insufficient implementation and a multitude of interpretations, resulting in disjointed and ineffective policy."[174]

10.5.3.1. Range of items controlled

Prior to the end of 2003, there had been a certain ambiguity regarding the range of arms and equipment covered by EU arms embargoes. This was caused by a lack of harmonization of wording in texts authorizing EU embargoes, including failures to define key terms used, such as "arms and related materiel", and to explicitly describe the list of goods covered. In 2003, the European Council embarked upon an exercise of examination of its sanctions practice and policy. This resulted in the European Council's adoption of "Guidelines on Sanctions", which established standardized wording for EU embargoes, and declared that "unless otherwise specified, [EU] arms embargoes should be interpreted as covering at least all those goods and technologies included in the 2000 Common List of Military Equipment".[175] Consequently, all EU embargoes, be they autonomous or those implementing UNSCRs, cover the range of ICA weapons, RCAs and related means of delivery listed in the EU Common Military List but would not include goods such as the commonly used chemical irritants OC and PAVA.

In addition, the 2003 "Guidelines on Sanctions" recognized that "If a policy of internal repression is at the basis of the imposition of restrictive [EU] measures, a ban on exports of certain [security] equipment is appropriate."[176] In such circumstances, "EU legal instruments could refer to or use an agreed list when deciding an embargo on exports of items that could be used for internal repression".[177] A "list of equipment which might be used for internal repression" was annexed to the "Guidelines on Sanctions" and included: "Portable devices designed or modified for the purpose of riot control or self-protection by the administration of an incapacitating substance (such as tear gas or pepper sprays), and specially designed components therefore."[178]

Subsequently, the autonomous EU embargoes for Cote d'Ivoire, Indonesia, Myanmar, Uzbekistan and Zimbabwe, which were imposed on human rights grounds, were further strengthened by Council Regulations covering equipment that could be used for internal repression (these Regulations incorporated the text mentioned referring to tear gas and pepper sprays).[179] Consequently, these five embargoes, at least, did appear to cover OC and PAVA within their scope. In 2009, however, the revised "Guidelines on Sanctions" contained a shortened "list of repression equipment", which no longer incorporated text on tear gas and pepper sprays.[180] It thus appears that subsequent EU sanctions introduced on human rights grounds did not explicitly cover those tear gases and pepper sprays (such as OC and PAVA) that were not specifically identified in the EU Military List.[181]

10.5.3.2. Monitoring and verification of embargoes

In 2004, pursuant to the EU's Sanctions Guidelines, a "Sanctions Formation" of the Foreign Relations Counsellors Working Group (RELEX) was created and mandated with the development of best practices in the implementation and application of restrictive measures through the exchange of information and experiences.[182] Analysts have, however, highlighted the limited effectiveness of such mechanisms to monitor embargo implementation. For example, in 2005 Shields highlighted the "notable, even critical, lack of monitoring and enforcement provisions in EU arms embargoes", and specifically criticized the lack of "mandatory reporting or information exchange concerning implementation, violations or decisions handed down by national courts".[183]

In 2010, Holton and Bromley reviewed EU arms embargo implementation and concluded that there were no "independent monitoring mechanisms to ensure implementation and to assess the positive and negative impacts" of such sanctions.[184] The researchers highlighted the limited mandate of RELEX/Sanctions, which "only exchanges information on alleged violations" and cannot "investigate or commission investigators" to look into such alleged violations. In addition, RELEX/Sanctions had no mechanisms to publicly report on the implementation and enforcement of EU arms embargoes or on investigations of alleged violations.[185] These concerns were exacerbated by the lack of adequate oversight by the Council of the EU. In 2005, the Council stated that there should be "regular reporting on the implementing measures and enforcement actions taken by Member States to give effect to restrictive measures";[186] however, according to Holton and Bromley, the Council did not subsequently "call...for [any] reports on investigations into alleged and actual violations".[187]

10.6. Conclusions

Transfer control regimes could form an important element of an HAC approach to the regulation of ICA weapons, RCAs and related means of delivery, providing mechanisms for combating proliferation and misuse of these agents and goods by prohibiting their transfer, at least in circumstances where their use is deemed inappropriate, for example, as a method of warfare or to facilitate human rights abuses.

However, existing transfer control regimes have important limitations. Apart from the BTWC and the CWC, which can be regarded as global disarmament and non-proliferation regimes incorporating transfer control mechanisms, the agreements highlighted are either politically binding or, if legally binding, are of limited membership. Furthermore, the resulting control regimes have divergent transfer control criteria and cover differing ranges of agents and goods within their scope. Although the international community negotiated and introduced an Arms Trade Treaty in 2013 incorporating global transfer controls on conventional arms, unfortunately this instrument did not include ICA weapons or RCAs within its scope (though certain relevant means of delivery may be covered).

In addition, the fairness and legitimacy of certain existing regional and pluri-lateral transfer controls such as the Australia Group has been questioned. If such mechanisms are employed, particularly in isolation, there is a danger that they will be seen by those States outside the regimes as discriminatory or as attempts by certain States to keep the perceived benefits of certain agents to themselves. Furthermore, the reliance upon such transfer controls as the principal or indeed only mechanism to address ICA weapons, RCAs and related means of delivery would prove ineffective as it would fail to sufficiently address the dangers of existing agent production as well as future indigenous development in new States. Such concerns could, at least partially, be overcome if transfer control mechanisms were to be employed as part of an HAC approach to the regulation of such agents (i.e. if such mechanisms were coupled with stringent regulation of RCA use for law enforcement purposes and a moratorium or prohibition on development and use of ICA weapons for law enforcement purposes in potential supplier States).

Legally binding embargoes (particularly those introduced by the UN) on the supply of arms and security equipment are potentially important tools for halting, or at least restricting, the supply of ICA weapons, RCAs and related means of delivery to specific end users. The employment of such embargoes could become a powerful adjunct in an HAC approach to address the misuse of these agents; to be utilized, as a measure of last resort, in cases of persistent, widespread and/or serious abuse (e.g. in gross human rights violations) or where there was a significant danger to international peace and security. However, in practice such embargoes are currently difficult and time-consuming to introduce, relying on strong and widespread inter-governmental support (particularly from the P-5 members of the UNSC). Even when such embargoes are adopted, they often have limited effectiveness in practice, due to ambiguities in their scope of coverage, delayed and patchy implementation, and poor or non-existing monitoring and enforcement measures.

11
Application of the United Nations Drug Control Conventions to ICA Weapons

11.1. Introduction

As discussed in Chapter 2 of this publication, the range of pharmaceutical chemicals investigated as potential ICA weapons has included a range of narcotic and psychotropic drugs. Many of these substances have legitimate medical, veterinary or other scientific uses, whilst others have been utilized by the civil population as recreational drugs, often resulting in drug dependency and addiction. Since the signing of the International Opium Convention in 1912,[1] the international governmental community has developed a number of agreements and other mechanisms to combat the illicit trade and misuse of narcotic and psychotropic drugs whilst safeguarding the legitimate employment of such substances. As part of the second stage of the HAC analytical process, this chapter surveys the potential applicability of two such agreements – the Single Convention on Narcotic Drugs (SCND) and the UN Convention on Psychotropic Substances (CPS) – to the regulation of ICA weapons intended for law enforcement or military purposes.[2]

11.2. Single Convention on Narcotic Drugs

The SCND was signed in 1961, a Protocol amending the text was agreed in 1972 and the revised text came into force in August 1975.[3] As of August 2015, it had 185 States Parties.[4] The Single Convention codified all existing multilateral treaties on narcotic drug control and extended the existing control systems to include the cultivation of plants that were grown as the raw material of narcotic drugs.[5] The principal objectives of this Convention have been to restrict the possession, use, trade in, distribution, import, export, manufacture and production of narcotic drugs exclusively to medical and scientific purposes,[6] and also to address narcotic drug trafficking through international cooperation.

The narcotic drugs covered by the Convention have been elaborated in four Scheduled lists, which can be updated following recommendations by States Parties or the World Health Organisation (WHO) to the UN Secretary General.[7] The current Schedules list a number of narcotic drugs that have been explored as potential ICA weapons, including cocaine,[8] etorphine,[9] fentanyl,[10] morphine[11] and sufentanil.[12]

Article 4 of the Convention, which sets out of the "General Obligations", states that:

> The parties shall take such legislative and administrative measures as may be necessary: (c) Subject to the provisions of this Convention, to limit exclusively to medical and scientific purposes the production, manufacture, export, import, distribution of, trade in, use and possession of drugs.[13]

The explicit requirement under Article 4 to "limit" narcotic drugs "exclusively to medical and scientific purposes"[14] appears to question the legitimacy of the development, stockpiling, transfer and use of such drugs by States Parties for employment as weapons in armed conflict, military operations other than war, or in law enforcement operations. However, the applicability of the Convention to such activities is uncertain as the SCND was developed as a crime control instrument and not as a means of arms disarmament or arms control, and, as of August 2015, there is no public record of the Convention being employed for such purposes. If the Convention could be applied to the weaponization of narcotic drugs, it is uncertain which activities would be considered permissible and which prohibited.

The SCND does allow for the retention by the State of so-called "special stocks" of narcotic drugs. These "special stocks" are defined under Article 1 as: "drugs held in a country or territory by the Government of such country or territory for special government purposes and to meet exceptional circumstances".[15] Although neither "special government purposes" nor "exceptional circumstances" are defined under the Convention, their meaning is discussed in the Commentary to the Convention.[16] The Commentary states that " 'Exceptional circumstances' is meant to cover such catastrophic events as large-scale epidemics and major earthquakes."[17] Furthermore, the Commentary states that: " 'Special government purposes' can cover all purposes other than the usual needs of the civilian population or 'exceptional circumstances'. In practice the phrase is generally synonymous with 'military purposes'."[18] However, according to Rabbat, although " 'Special government purposes' includes use by the military ... this does not imply that the use is to be military in nature (such non-military uses may for example include medical treatment of military personnel)."[19] Furthermore, Rabbat has stated that: " 'Special stocks' remain subject to the obligation to limit use of controlled substances to medical and scientific purposes."[20]

Although it is currently unclear whether and how the Convention's limitations on use could be applied to the weaponization of narcotic drugs, the Convention has included certain reporting and investigatory provisions that could still be utilized by a State Party to obtain information of potential benefit in the regulation of certain potential ICA weapons. The Convention has obliged States Parties to provide the International Narcotics Control Board (INCB) – the independent and quasi-judicial monitoring body for the implementation of the UN international drug control conventions[21] – with annual estimates of drug requirements,[22] and drug production,[23] for Scheduled chemicals. The Convention has not, however, required States Parties to furnish statistical returns respecting "special stocks".[24] The statistical information provided by States Parties has subsequently been incorporated into INCB annual reports which have been made publicly available.[25] A review of all published INCB annual reports could find no details of research, development or use of weapons employing ICAs with narcotic properties intended for law enforcement or other purposes, and it is unknown whether States Parties have provided details of such activities to the INCB.[26]

Although these issues have yet to be publicly addressed by the States Parties, in 2003 the INCB, in its Annual Report, did raise the following concerns:

> The Board is aware that drugs scheduled under the 1961 Convention or the 1971 Convention, mainly drugs of the amphetamine-type group, continue to be used by some military forces, for example during armed conflict, and that research into further possible uses is taking place. The Board is of the opinion that this type of drug use may not be in line with the international drug control conventions, which require Governments to limit the use of narcotic drugs to medical and scientific purposes only. The Board appeals to Governments to ensure that the military and law enforcement sectors follow the principles of sound medical practice in their use of internationally controlled substances and that the international drug control conventions are respected in those sectors.[27]

Whilst it does not appear that the INCB was referring to the use of Scheduled drugs as ICA weapons, the Board statement is very important in confirming that law enforcement and military forces are not exempt from the restriction on States to restrict the use of drugs covered by either Convention to scientific and medical purposes.

The Convention includes a range of procedures that allow the INCB to investigate potential circumstances of concern.[28] If the aims of the Convention were being "seriously endangered" and it had not been possible to "resolve the matter satisfactorily in any other way" – and particularly where "the situation is serious" and required "cooperative action at the international level" – the INCB could alert the States Parties, the Council and the Commission to the matter. Consequently, if required, the Council could "draw the attention of the UN General Assembly to the matter".[29] It is not known whether the INCB has received information or investigated any reports of research, development, stockpiling or use of ICA weapons employing scheduled chemicals by States Parties for military or law enforcement purposes. For example, as of August 2015, no information is publicly available concerning INCB deliberations or actions regarding the Russian Federation's use of an ICA weapon employing fentanyl (or its derivatives) to end the Moscow theatre siege, and its presumptive manufacture and stockpiling of such potentially Scheduled chemicals.[30]

11.3. The UN Convention on Psychotropic Substances

The UN CPS was agreed in 1971 and came into force in August 1976. As of August 2015, it had 183 States Parties and 34 Signatory States.[31] This Convention established an international control system for psychotropic substances. The Convention was developed in response to the diversification and expansion of the spectrum of drugs of abuse and it introduced controls over a number of synthetic drugs according to their abuse potential on the one hand and their therapeutic value on the other.[32]

As with the SCND, the drugs covered by the CPS have been elaborated in four Scheduled lists which could be updated following recommendations by States Parties or the WHO to the UN Secretary General.[33] The current Schedules list a number of psychotropic drugs that have been explored as potential ICA weapons for use in law enforcement or military activities, including amphetamine,[34] dexamphetamine,[35] LSD,[36] mescaline,[37] methamphetamine,[38] midazolam,[39] phencyclidine[40] and psilocybine.[41] In addition, the Schedules also include amobarbital,[42] pentobarbital,[43] secobarbital[44] and temazepam,[45] which have been explored or utilized in judicially sanctioned "lethal injection" executions and/or narcoanalysis.[46]

Under the Convention, each State Party has been required to "limit by such measures as it considers appropriate the manufacture, export, import, distribution and stocks of, trade in, and use and possession of, substances in Schedules II, III and IV to medical and scientific purposes".[47] The restrictions for Schedule I chemicals are even stricter, with States Parties required to "prohibit all use except for scientific and very limited medical purposes by duly authorized persons, in medical or scientific establishments which are directly under the control of their Governments or specifically approved by them",[48] and to "require that manufacture, trade, distribution and possession be under a special licence or prior authorization".[49]

Unlike the SCND, the CPS has contained no ambiguously worded provision allowing the State Parties to hold "special stocks" of drugs to be used for "special government purposes" or in "exceptional circumstances".

Instead, the Convention has allowed the use of Schedule II, III and IV psychotropic substances for three specific activities – the carrying by international travellers of small quantities of preparations for personal use, during the manufacture of non-psychotropic substances and for the capture of animals.[50] There has been no specific exemption detailed for the use of such substances in law enforcement activities or by the military.

Although the applicability of the CPS to the employment of psychotropic substances as ICA weapons in military or law enforcement operations is uncertain, the restrictions elaborated under the Convention do appear to put into question the legitimacy of the development, stockpiling, transfer and use of certain psychotropic drugs by States Parties for such activities. Similarly, the development, stockpiling, transfer or use of psychotropic substances in either "lethal injection" executions or for narcoanalysis are not specifically provided for in any of the exemptions and therefore appear to be prohibited under the Convention. As of August 2015, there are no documents publicly available of the States Parties or relevant organs of the CPS clarifying these issues.[51]

As with the SCND, the CPS has included a number of reporting and investigation provisions that could potentially be utilized to provide information of benefit in the regulation of certain potential ICA weapons. Under the Convention, States Parties have been required to provide the INCB with an annual report of quantities of each Schedule I, II, III and IV drug manufactured, imported and exported, as well as details of stocks held of Schedule I and II chemicals – though there has been no requirement to provide information on the intended use of such chemicals.[52] The statistical information provided by States Parties has subsequently been incorporated into INCB annual reports, which have been made publicly available.[53] A review of all published INCB annual reports could find no details of research, development or use of ICA weapons employing psychotropic substances for law enforcement or military activities, and it is unknown whether States Parties have provided details of such activities to the INCB.

If, on the basis of its examination of information submitted by Governments to the INCB or of information communicated by UN organs,[54] the INCB had reason to believe that the aims of this Convention were "being seriously endangered" by the actions of a country or region, it could seek "explanations from the Government of the country or region in question".[55] If the matter was resolved at this stage then such requests and the information provided would be treated as confidential.[56] If unsatisfied by the State's response, the INCB could bring the matter to the States Parties, the Council and the Commission, and could recommend that the States Parties "stop the export, import, or both, of particular psychotropic substances, from or to the country or region concerned" for a "designated period" or until the INCB was "satisfied as to the situation in that country or region".[57] As of

August 2015, it is not known whether the INCB has formally received information or investigated any reports of research, development, stockpiling or use of ICA weapons employing Scheduled chemicals by States Parties for law enforcement or military purposes.[58]

11.4. Conclusion

From the HAC stage-two analysis summarized in this chapter, it is clear that, although both the SCND and the UN CPS were developed as crime control instruments and not as mechanisms for disarmament or arms control, these treaties do appear to restrict the use of a range of narcotic and psychotropic drugs previously investigated as potential ICA weapons to "medical and scientific purposes". Further research is required to establish the implications of these restrictions upon current or future attempts by States to develop, stockpile, transfer or use such drugs as ICA weapons.

Although no States have yet made their positions publicly known on these matters, as of August 2015; the ICRC in September 2012 highlighted the relevance of both Conventions to State activity in these areas. The ICRC noted that the Conventions "place strict controls" on "some of the toxic chemicals that have been considered as weapons for law enforcement" and require that "production, manufacture, export, import, distribution of, trade in, use and possession of controlled drugs must be limited exclusively to 'medical and scientific purposes' ".[59]

Even if the limitations on use established by one or both treaties were not deemed, by the States Parties, to be applicable to drugs employed as ICA weapons; the reporting (and potentially the investigatory) mechanisms of both Conventions should be considered as potential routes for obtaining information relevant to the regulation of such agents, as part of a broader HAC regulatory strategy.

12
The Role of Civil Society in Combating the Misuse of Incapacitating Chemical Agents and Riot Control Agents

12.1. Introduction

The previous chapters have explored the State-centric nexus of arms control and disarmament treaties and agreements, relevant international law and other regulatory regimes that are potentially applicable to ICA weapons, RCAs and related means of delivery. It is clear from such analysis that many of these instruments and regimes have ambiguities, weaknesses and limitations which the relevant Member States have been unable or unwilling to address. Furthermore, a number suffer from inadequate and patchy national implementation and a failure of States Parties and relevant regime organizations to challenge reported treaty violations by certain Member States.

In the light of such failings, and as part of the second stage of the HAC analytic process, this chapter will explore the potential roles that civil society, particularly the scientific and medical communities, can play in combating the misuse of RCAs and ICAs. The chapter will focus in particular upon the potential application of "societal verification" as a complement to the existing official verification mechanisms; the development of a "culture of responsibility" amongst the scientific and medical communities; and finally the possible roles that scientists, health professionals, academics and other civil society actors can play in informing and influencing the policy and practice of States in this area, and holding them to account for their actions and inaction.

12.2. Societal monitoring and verification

The concept of "societal verification" has been discussed in the peace and disarmament literature for many years under terms such as "citizens reporting", "inspection by the people", "knowledge detection" or "social monitoring". The idea was originally introduced in the late 1950s by Bohn[1] and Melman[2]; and subsequently utilized in the 1960s by Clark and Sohn,[3] and by Portnoy in the 1970s.[4] In the 1990s, Rotblat and others, principally around the Pugwash group, "... took up these old ideas and applied them to the concept of a treaty on the complete elimination of nuclear weapons".[5] In the following decades, scholars, including Deiseroth[6] and Falter[7] continued to develop and promote the concept of "societal verification" and have sought to apply its concepts to other arms control and disarmament treaties.

Although there is no agreed formal definition, Deiseroth has described societal verification as "connot[ing] the involvement of civil society in monitoring national compliance with, and overall implementation of, international treaties or agreements. One important element is citizens' reporting of violations or attempted violations of agreements by their own government or others in their own country."[8] Dieseroth noted that, unlike the "official verification organisations employing professional experts, societal verification may involve the whole of society or groups within it".[9]

Similarly, Rotblat previously contended that the main form of societal verification should involve "inducing the citizens of the countries signing the treaty to report to an appropriate international authority any information about attempted violation going on in their countries".[10] Rotblat believed that, for this system of verification to be effective, it was "vital that all such reporting becomes the right and the civic duty of the citizen".[11] Consequently, he argued that: "This right and duty will have to become part of the treaty [and] of the national codes of law in the countries party to the treaty."[12]

Rotblat also highlighted the potential role of the scientific community in this process: "organizations of scientists and technologists could be set up for the specific purpose of acting as a watchdog of compliance with treaties, by monitoring the activities of individuals likely to become involved in illegal projects".[13] He argued that such monitoring could be done "by keeping a register of scientists and technologists, and by noting changes of place of work or pattern of publications (or their absence)".[14] Other potential "giveaways" of attempted clandestine activities that could be monitored included the "start of new projects at academic institutions without proper justification; the recruitment of young scientists and engineers in numbers not warranted by the declared purpose of the project; or the large scale procurement of certain types of apparatus, materials, and equipment".[15]

However, Rotblat acknowledged that:

> [E]ven if governments were persuaded to pass laws to make reporting legitimate, [which itself would be a revolutionary development and counter to the existing arms control and disarmament policy and practice of many States] this goes so much against traditional loyalties that it would require a considerable educational effort to induce people to act on it voluntarily.[16]

Although the original vision of a global societal verification network – involving large numbers of civil society actors resident in all States party to relevant treaties, who are able to monitor their own State's implementation of such treaty obligations – appears idealistic, if not utopian, in the near to medium term, the idea remains a very powerful one.

An adapted form of societal monitoring and verification could be envisaged, comprised of a smaller number of activist researchers, who have access to the relevant technical expertise and can, at the very least, undertake open source monitoring and analysis of the activities of relevant State and non-State actors. Due to resource, personnel, political and security constraints, such groups are likely to be limited in terms of the countries from which they can operate and, consequently, the quantity and quality of information they are able to receive, particularly from inaccessible regions and closed or semi-closed authoritarian countries. Despite such constraints, variant forms of such limited societal monitoring are presently carried out for certain environmental and human rights agreements, and have been taken up by certain arms control and human security communities to monitor agreements prohibiting landmines and cluster munitions, and regulate small arms and light weapons.[17]

Societal monitoring relating to ICA weapons, RCAs and means of delivery appears, as of August 2015, to be at a very rudimentary level, with *ad hoc*, limited and uncoordinated research activities being carried out by a small number of researchers. Some illustrative examples of research methodologies that have been or could be utilized for such work will be explored in Section 12.2.1.

12.2.1. Open source monitoring and analysis

Certain academics and NGOs have undertaken the monitoring and analysis of open source information relating to the research, development, marketing, transfer and utilization of RCAs and related means of delivery, or ICA weapons, by States and non-State actors. Some of these groups, particularly human rights and civil liberties organizations, have concentrated upon the supply of RCAs and means of delivery to and/or misuse by military, security or police officials.

Although R&D programmes for ICA weapons, RCAs and related means of delivery are often considered to be issues of national security and remain classified, information, albeit partial, can sometimes be obtained from monitoring commercial documentation (e.g. company annual reports, product documentation, industry publications); scientific papers and conferences; and Government policy papers and tender solicitations for relevant new technologies. Non-governmental researchers, particularly in North America and Europe, have successfully obtained information on these activities or on relevant Government policies regarding such weapons through utilizing national governmental oversight, transparency and freedom of information mechanisms.

Open source monitoring and analysis is often time-consuming and resource intensive, and the information obtained can be limited and heavily censored (or redacted) – as a result of national security restrictions and commercial confidentiality considerations.[18] Furthermore, as the US/German NGO, the Sunshine Project, has discovered and described, sometimes Government officials or agencies can attempt to conceal potentially controversial material by exaggerating exemptions allowed for under relevant legislation, thereby "trying to keep secrets that they are not legally entitled to maintain".[19] Occasionally, however, the documents obtained do provide important information on the policy and practices of States in this area. Furthermore, although uncovering and highlighting new information and analysis on Government programmes is often the primary purpose of employing freedom of information legislation or other transparency mechanisms, another very important benefit lies in "asserting and maintaining the public's right to this information".[20]

Despite the methodological difficulties and the limitations in the information obtained, such work is vital to the formation of an informed public discourse on the existing threats and potential dangers of the proliferation and misuse of ICA weapons, RCAs and related means of delivery. It can also help in the development of timely and realistic publicly available threat assessments relating to R&D, deployment or utilization of such weapons in specific countries. Furthermore, information derived from civil society research can be presented to relevant inter-governmental organizations, most notably the OPCW. At present, such evidence would not be *formally* recognized by the OPCW, unless it was submitted by a State Party. However, if reliable, the information may serve to alert relevant Member State and/or Technical Secretariat officials to an issue of potential concern and potentially lead to the OPCW Director General raising the issue informally with relevant States; or alternatively may result in a State Party utilizing the relevant consultation, clarification and fact-finding mechanisms available under the CWC.[21] In some circumstances, the information provided can be so compelling that the State Party directly implicated decides unilaterally to take remedial action to address failures of its implementation (as illustrated in the case study in Section 12.2.1.1).

12.2.1.1. Investigations by civil society organizations leading to the destruction of Turkey's stockpile of 120 mm mortar munitions containing RCAs

In November 2003, *Jane's Defence Weekly* reported that the Turkish (State-owned) arms manufacturer Makina ve Kimya Endustrisi Kurumu (MKEK) had developed a 120 mm mortar round – the CS MKE MOD 251 – filled with CS.[22] (Full details of the munition are described in Chapter 4). It was promoted by MKEK on their website,[23] and at international security exhibitions, including the 7th International Defense Industry Fair (IDEF) held in Ankara, Turkey, in September 2005.

Given the nature of the RCA munition, it was clear that its use for riot control or other domestic law enforcement operations would have been inappropriate. Consequently, any manufacture, stockpiling and deployment of such munitions apparently breached Article I.1.(a), I.5 and II.1.(a) of the CWC. In addition, any promotion and transfer of such munitions would have potentially breached Article I.1.(a) and Article I.1.(d) of the Convention.

Following correspondence from Bradford Non-Lethal Weapons Research Project (BNLWRP) to the Turkish Government and MKEK highlighting concerns about this munition, all information concerning the CS MKE MOD 251 mortar round was subsequently removed from the MKEK website.[24]

However, a collaborative investigation by BNLWRP, the Institute for Security Studies (ISS) and the Omega Research Foundation (ORF), uncovered and documented the continued promotion of the CS MKE MOD 251 mortar round by MKEK at the AAD exhibition held in Cape Town, South Africa, in September 2010. BNLWRP, ISS and ORF wrote to the Turkish Government, the South African Government and MKEK, raising concerns about the promotion of these munitions.

Two additional Turkish companies – Furkan Defense Industry (FDI) and ASCIM Defense Industry (DI) – were subsequently found to be promoting these munitions on their websites.[25] Following written notification sent by BNLWRP, ISS and ORF to the Turkish Government and FDI and ASCIM DI, all relevant promotional information was subsequently removed from both company websites.

BNLWRP, ORF and ISS continued correspondence with representatives of MKEK and the Turkish Ministry of Foreign Affairs regarding the development and promotion of CS MKE MOD 251 munitions, and sought to raise this issue with the OPCW. On 29th November 2010, the author highlighted these issues during a presentation given at the OPCW Open Forum, attended by a number of CWC States Party delegations, representatives from the OPCW Technical Secretariat and civil society organizations.[26] During the subsequent question and answer session, a representative of the Turkish delegation stated that Turkey was investigating the issue and would publish its results in a transparent manner.

In February 2011, in correspondence with BNLWRP, ISS and ORF, the Turkish OPCW Ambassador stated that 1,000 CS MKE MOD 251 munitions had been produced in 1996, prior to Turkey's ratification of the Convention, and that "around 150 of the said ammunitions were used for testing purposes during the initial R&D phase in 1997".[27] In July 2011 correspondence, the Turkish Counsellor to the OPCW stated that: "At the time of ratification, there remained 850 pieces of CS MKE MOD 251 type munitions in the inventory of the Turkish Armed Forces. The facility for their production was discontinued after 1997."[28] From the Counsellor's statement it appeared that Turkey had not formally declared the existence of the outstanding

850 munitions or provided details of the relevant production facilities to the OPCW Technical Secretariat. Such a declaration would appear to have been required under Article III, paragraphs 1(a) and (c) of the CWC.[29] It is unclear whether Turkey subsequently submitted a declaration.

The Turkish OPCW Ambassador explained that: "The remaining 850 [munitions], whose dates of expiry have passed, are stored at the Turkish Armed Forces ammunition destruction facility awaiting disposal."[30] Subsequently, in his July correspondence, the Turkish Counsellor reported that: "The destruction of CS containing canisters of the remaining CS MKE MOD 251 munitions has now been completed at our state-of-the-art munitions disposal facility located near Ankara."[31] In addition to destroying all remaining 120 mm RCA mortar munitions, Turkey also put in place measures to halt promotion and brokering of such munitions by Turkish companies or individuals.

Evidence obtained by civil society, for example concerning the misuse of RCAs in a certain country, can be utilized to help inform and influence the policies and practices of the wider international governmental community towards that country. In certain circumstances such evidence has included information identifying the range of RCA means of delivery misused by a State, and/or indications that a State misusing RCAs and related means of delivery is seeking to obtain additional (sometimes large) quantities of such items. This information has been employed successfully by concerned civil society organizations to ensure that a State misusing such RCA delivery mechanisms does not receive fresh supplies of such goods from key exporting States (as illustrated in the case study in Section 12.2.1.2).

12.2.1.2. Investigative research, legal advocacy and public campaigning by civil society organizations lead to suspension of requested transfer of a large quantity of tear-gas shells and canisters from South Korea to Bahrain

Human rights organizations have documented the widespread misuse of RCAs and related means of delivery by Bahrani law enforcement officials to facilitate serious human rights violations in that country (see, for example, Section 12.2.2.2).

On 16th October 2013, Bahrain Watch publicly leaked a tender document it had obtained, dated 16th June 2013, issued by Bahrain's Ministry of Interior, soliciting proposals to supply the Ministry with a large shipment of tear-gas grenades and shells.[32] The document, signed by "Assistant Undersecretary Abdulla Bin Ahmed Al-Khalifa", called for all proposals to be submitted "not later than 16th July 2013".[33] The tender called for arms companies to supply Bahrain with the following items: 800,000 CS tear-gas long-range shells 37/40 mm; 400,000 CS tear-gas short-range shells 37/40 mm; 400,000 CS tear-gas shell, multiple sub-munition (five-way) 37/40 mm; 45,000 CS hand grenades (one-way); 45,000 tear-gas hand grenades (five-way) and 145,000 sound & flash grenades.[34]

On 16th October 2013, an international campaign of arms control and human rights organizations – Stop the Shipment – was initiated to halt the fulfilment of this order. Members of affiliated organizations around the world conducted public campaigning and advocacy activities. On 17th January 2013, the European Parliament called for sanctions against "individuals directly responsible [for] human rights abuses" in Bahrain, and for "restrictions on EU exports of surveillance technology, tear gas and crowd-control material".[35] Additionally, both the United States[36] and the United Kingdom[37] suspended shipments of tear gas to Bahrain because of its disproportionate and improper use by security forces.

Further research by the Stop the Shipment Campaign indicated that Bahrain was seeking to obtain the tear-gas munitions from South Korea. Since late 2011, unmarked tear-gas canisters, as well as tear-gas grenades, which were visually similar if not identical to those originally manufactured by South Korean company, Dae-Kwang Chemical Company Ltd, and reportedly exported by a second South Korean company, CNO Tech Ltd, were documented in Bahrain. The Korean Defense Acquisition Program Administration (DAPA), which regulates the export of tear gas, subsequently informed AI Korea that Dae-Kwang had previously sought its permission to export 1,546,680 canisters of tear gas to Bahrain in 2011 and 2012,[38] including 1,249,680 DK-38S CS cartridges. A *Financial Times* article published on 21st October 2013 quoted a senior executive at Dae-Kwang as confirming that his company had exported approximately 1 million rounds of tear gas in 2011 and 2012.[39] The executive stated that, at the time of the interview, no contract had been signed for any subsequent shipment.[40] In December 2013 Bahrain Watch stated that they had indications that DAPA was "currently considering a request from an unnamed Korean Company to export tear gas to Bahrain."[41]

The Campaign continued its lobbying and public advocacy activities. Its legal team – headed by Michael Mansfield QC – submitted a formal complaint to the Organisation for Economic Co-operation and Development (OECD) requesting they prohibit the requested transfer from South Korea.[42] A separate complaint was sent to four UN Special Rapporteurs,[43] asking them to take urgent action to halt further shipments of tear gas to Bahrain, and investigate the legality of previous shipments of South Korean tear gas. Participants in the Campaign reportedly placed calls and sent over 390,000 emails to the Korean Government asking them to prohibit the requested transfer.[44]

On 9th January 2014, the *Financial Times* reported that the South Korean Defence Acquisition Program Administration had informed two companies that had sought approval to export to Bahrain in October and November "to suspend shipments".[45] The article quoted Lee Jung-geun, a DAPA spokesman, who said that the decision had been made because of the "unstable politics in the country [Bahrain], people's death due to tear gas and complaints from human rights groups".[46]

12.2.2. Field missions and witness testimony

Independent scientists, health professionals and NGOs can sometimes collect their own information, first hand, from onsite investigations, or may be able to utilize information (e.g. witness testimony) and analyse materials (e.g. spent cartridges, munition shell fragments, clothing, blood, soil samples, etc.) obtained from other civil society actors operating in the field (e.g. journalists, national NGOs, etc.). There are several potential constraints upon such evidence-gathering, including access, logistics and translation; safety considerations for researchers and witnesses; difficulties ensuring "chain of custody" of the evidence, as well as establishing the representativeness of the information obtained. Despite such constraints, material collected during field missions can provide information that could not be obtained by any other means, potentially allowing identification of toxic chemical agents utilized during a military or law enforcement operation,[47] or, conversely, casting doubt on allegations of such use.[48]

12.2.2.1. Analysis by UK scientists identifies new riot control agent used in the West Bank

In July 2005, the Israeli Army reportedly employed a new RCA against Palestinian and Israeli civilians protesting about the erection of the "Separation Wall" on the West Bank. The RCA, which was delivered via small plastic projectiles fired from launchers, caused severe skin injuries in the targets. The Israeli authorities subsequently refused to identify the agent employed.

Scientists based in the United Kingdom, led by Professor Alastair Hay, obtained one of the RCA projectiles utilized. They extracted the enclosed RCA, which was subsequently subjected to physical and chemical analysis at two independent UK laboratories,[49] and was found to contain capsaicin with an inert carrier and a dispersal agent.[50] Comparing the resulting data against the material safety data sheets of commercially available RCA ammunition, the results were found to correspond with "Pepperball Tactical Powder" marketed by the US company Pepperball Technologies, Inc.

The UK researches also obtained signed affidavits from Palestinians affected by exposure to the RCA. One individual (whose skin injuries were documented and photographed) was participating in a peaceful demonstration against the Separation Wall on 28th April 2005 when he was "shot at from a distance of not more than three metres". He described the effect of the RCA projectile as "being like an electric shock" and of becoming "extremely agitated".[51]

The authors noted that, although "Pepperball tactical powder pellets are clearly designed to irritate the skin", the skin injuries that occurred in the two individuals detailed in their paper were "far more severe than the effects that material safety sheets for the product suggest may occur".[52] Consequently, they argued that, for "civilians injured in a peaceful march or demonstration and where treating clinicians have no knowledge of the effects of exposure to the capsaicin pellets and how to ameliorate these, far more severe injuries, such as those we describe, are inevitable".[53]

Such research clearly has utility in alerting clinicians (and demonstrators) to the nature and effects of new chemical agents they may have to face and treat in the future. The research methodology described can also be employed to identify specific RCAs and means of delivery used, and, consequently, potential companies and international transfers of such chemical agents and devices. It can also sensitize the international community to potentially inappropriate use of such agents. For example, a 2012 Royal Society report, citing the research by Hay and colleagues, concluded that: "While the use of RCAs for domestic policing is permitted by the CWC, in the context of the conflict between Israel and Palestinian territories the legality of using capsaicin-type munitions against demonstrators is a matter of concern."[54]

12.2.2.2. US NGO fact-finding mission documents serious and widespread misuse of RCAs by Bahraini law enforcement officials

A two-person team comprising a medical doctor and lawyer from US-based non-governmental human rights organization, Physicians for Human Rights (PHR), conducted field research in Bahrain from 7th to 12th April 2012. The mission team investigated reported violations of medical neutrality and excessive use of force by law enforcement officials alleged to have taken place in Bahrain during the previous year, and published its findings in August 2012.[55]

The team conducted a total of 102 semi-structured interviews with: Bahrainis who reported human rights violations, corroborating eyewitnesses to these alleged events, civil society leaders and also Government officials.[56] As well as locating and independently interviewing corroborating witnesses to certain alleged events, PHR conducted or utilized the following means to establish the veracity of witness testimony: (a) physical examinations by the PHR physician of victims of reported human rights violations; (b) evaluation of medical records of victims of reported human rights violations; (c) visual verification of locations of reported human rights violations; and (d) review of photographic, radiographic and video evidence. This on-sight evidence-gathering was subsequently analysed and supplemented by further extensive desk research employing medical and scientific literature databases.[57]

The PHR report highlighted the failure of Bahrani law enforcement officials to exercise restraint before resorting to force; their employment of disproportionate force when responding to protesters; and their failure to minimize damage and injury to demonstrators. The report documented cases of injured protesters, examined by PHR investigators, who had suffered from blunt force trauma and lacerations to the head, torso and limbs due to the impact of metal tear-gas canisters being fired at them by law enforcement officials at close range.[58]

The PHR report also detailed numerous cases of misuse of tear gas to "methodically...attack Shi'a civilians inside their homes and cars", which, it argued, constituted the "transformation of toxic chemical agents into weapons". PHR contended that "such unprovoked and flagrant assaults on families – who pose no threat to the safety of others – flout international human rights law and constitute torture, cruel, and inhuman treatment". On the basis of the evidence gathered, PHR concluded that the "weaponized toxic chemical agent attacks against Bahraini civilians are intentional – and may be official policy – because of the frequency of the attacks by officials throughout the police force and the lack of accountability for those who perpetrate the attacks".[59] The report also warned that the "persistent targeting" by law enforcement officials of civilians with tear gas in enclosed spaces "may lead to serious long-term health consequences", which could include "miscarriages, severe respiratory distress resulting in premature death, and a projected rise in asthma among the population".[60]

Prior to and following its field mission, PHR conducted research from outside Bahrain, concerning the misuse of tear gas by Bahrani law enforcement officials. For example, in March 2012 PHR released information of 34 "tear gas related deaths" that had taken place since the beginning of the uprising in 2011.[61] The information was "based on interviews with local physicians and analysis of news reports".[62] Following further research, in October 2013 PHR released an expanded list of 39 "confirmed deaths" related to the police employment of tear gas.[63]

The PHR field mission and desk-based study of RCA-related deaths in Bahrain have been highly influential. The PHR research has been cited by the US State Department,[64] and has been widely covered in the international media.[65] The research was subsequently utilized in the international campaign to halt the transfer of RCAs from South Korea to Bahrain (see Section 12.2.1.2). The work of PHR illustrates the importance of field-based evidence-gathering, and of objectively recording and independently corroborating witness testimonies, supported by medical and other documentary evidence, and highlights the critical role that medical professionals in particular can play in documenting and analysing evidence obtained through such research missions.

12.3. Utilizing civil law in cases of alleged involvement in misuse of certain weapons

In certain countries, attempts have been made to utilize civil law (i.e. the law determining private rights and liabilities) in attempts to hold individuals and companies accountable for their alleged involvement in the misuse of certain weapons (including RCAs), and to obtain damages or other redress for those allegedly injured by such weapons. Such legal mechanisms have been employed in the United States, including through the use of "class action" civil law suits and the attempted employment of the Alien Tort Statute.

12.3.1. Alien Tort Statute

The Alien Tort Statute (ATS) (also known as the Alien Tort Claims Act) which is established under the United States Code determines that US district courts "shall have original jurisdiction of any civil action by an alien for a tort[66] only, committed in violation of the law of nations or a treaty of the United States."[67] Originally instituted in 1789, the ATS has, since the 1980 landmark judgement by the US Second Circuit Court of Appeals in the case of *Fidartiga v. Pedna-Irala*,[68] been applied by those attempting to obtain civil redress for the victims of human rights abuse when the abuser was found in the United States.[69] Although initially defendants were individuals including police officers, military commanders or even former heads of State, from the 1990s ATS cases were filed against multinational companies for alleged participation in, or responsibility for, human rights violations or breaches of other international law.[70]

In 1991, the case of *Abu-Zeineh vs. Federal Laboratories Inc.* was brought to the court of the Western District of Pennsylvania under the Alien Tort Statute. The plaintiffs were relatives of Palestinians who died after Israeli troops reportedly fired canisters of US manufactured CS gas into a confined space where they were working. The plaintiffs sued Federal Laboratories for wrongful death, claiming that the company "allegedly sold gas to Israel with the knowledge that [the] gas caused many civilian deaths".[71] The case was dismissed by the District court on the grounds that the plaintiffs – as Palestinians – were Stateless and therefore had no "foreign citizenship for the purposes of diversity jurisdiction" in US federal courts.[72] Although the federal case was dismissed, the plaintiffs reportedly refiled the case in US State court, where the case was eventually settled.[73]

The current utility of the ATS as a viable mechanism for determining the human rights liability of multinational companies is currently uncertain and contested. In 2010 in *Kiobel v. Royal Dutch Petroleum Co.*, the US Court of Appeals for the 2nd Circuit determined that corporations may not be sued under the ATS; by reaching this conclusion, they created a "circuit split".[74] The decision was reviewed in 2013 by the US Supreme Court, however they ultimately decided the case on different grounds (as described below). To date, the issue of corporate liability under the ATS is still open for interpretation; however according to Bradley "most lower courts that had addressed the question had held in favour of corporate liability."[75]

The April 2013 decision of the US Supreme Court in the case of *Kiobel v. Royal Dutch Petroleum Co.*[76] rather than determining corporate liability, instead held that there is a presumption against "extraterritoriality" in ATS cases unless the "claims touch and concern the territory of the United States with sufficient force," in which case the presumption can be displaced.[77] In *Kiobel*, the Supreme Court decided that the "mere corporate presence" of the defendant in the United States did not overcome the presumption.[78] The implications of the ruling have subsequently been interpreted in differing ways by lower US courts and it is at present unclear whether (and if so what) consensus will emerge regarding the factors required for a determination of extra-territorial liability under the ATS.[79]

12.3.2. Civil law class actions

According to the Legal Information Institute at Cornel University Law School, a class action is a "procedural device that permits one or more plaintiffs to file and prosecute a lawsuit on behalf of a larger group, or 'class' ".[80] As well as seeking to obtain financial restitution for those injured by the allegedly inappropriate employment of RCAs, class actions have also been brought with the intention of curtailing the employment of RCAs against certain classes of individuals or in certain circumstances.

12.3.2.1. Case study: J.W. ex rel. Williams v. A.C. Roper

In 2010, the Southern Poverty Law Centre of Alabama (SPLC) filed a federal lawsuit in a US District Court against the Birmingham Police Department on behalf of former students claiming that police stationed in Birmingham high schools as school resource officers (SROs) had inappropriately employed pepper spray against children in violation of their constitutional rights.[81] The SPLC filed the federal lawsuit "to end the practice and other abusive and unconstitutional behaviour" and it was filed "after the Birmingham School Board refused to address the issue" once the SPLC brought it to their attention. The SPLC subsequently uncovered additional pepper spray incidents, including several involving female students and, in one instance, a pregnant student.[82] After discovering these incidents, the SPLC filed a motion, which was subsequently granted in August 2012, to certify the complaint as a class-action lawsuit on behalf of all current and future students in the district.[83]

According to the SPLC, from 2006 to 2011, police in Birmingham public schools – whose students are predominantly African-American – used pepper spray on about 300 students. And more than 1,000 others were allegedly exposed to the spray during these incidents. The SPLC suit alleged that many students were already restrained or posed no safety threat when they were pepper-sprayed.[84] The Court heard arguments in January and February 2015, and on 30 September 2015 gave its ruling.[85] Although the Court recognized that pepper spray could be used in some circumstances, such as when students were engaged in physical fights or other violent behaviour, the Judge ruled that certain Birmingham Police officers had violated the constitutional rights of certain students by using excessive force for minor discipline problems. In his ruling, the Judge stated that "the court was profoundly disturbed by some of the testimony it heard at trial. The defendant SROs uniformly displayed a cavalier attitude toward the use of [the pepper spray] Freeze +P."[86] The Judge severely criticized the failure of the SROs to decontaminate children sprayed with Freeze +P – despite a legal obligation to do so – and "instead left them to suffer the effects of the chemical until they dissipated over time."[87] The Court ordered the Birmingham Police Department and the SPLC to work together to develop a training plan for SROs to ensure pepper spray is not used for basic school discipline.[88]

12.3.2.2. Case study: Baker v. Katehi

In February 2012, 21 students and alumni filed a federal lawsuit against the Regents of the University of California Davis over the University's alleged ill-treatment of protesters during a demonstration in November 2011 in which campus police were filmed employing pepper spray against seated non-violent protesters.[89] The class action law suit alleged that campus police violated the plaintiffs First, Fourth, and Fourteenth Amendment rights (freedom of speech & assembly, unreasonable search & seizure, and taking property without due process of law) under the US Constitution, as well as provisions set out under California penal, civil, and government codes of false imprisonment/arrest, and failure to provide medical care (after pepper spray was used against them).[90] Following a settlement announced in September 2012 and approved by a judge in January 2013, the case (*Baker v. Katehi*) was settled out of court. As part of the settlement, the University paid $1 million, and UC Davis Chancellor Linda Katehi issued a formal written apology to the students and recent alumni who were pepper sprayed or arrested. The University also agreed to work with the American Civil Liberties Union to develop new policies on student demonstrations, crowd management, and use of force.[91]

12.4. Building a culture of responsibility within life science and biomedical communities

In its 2004 public statement "Preventing Hostile Use of the Life Sciences", the ICRC declared: "If measures to prevent the hostile use of advances in the life sciences are to work, a culture of responsibility is necessary among individual life scientists."[92] According to the ICRC, such a culture of responsibility should apply "whether these scientists are working in industry, academia, health, defense or in related fields such as engineering and information technology", and should encompass the "institutions that employ scientists and fund research in the life sciences".[93]

Similar calls to the scientific (and medical) communities have also been made by the States Parties to the BTWC and the CWC at Review

Conferences.[94] The following sections will explore the current range of initiatives being undertaken by those in the scientific and medical communities to nurture a "culture of responsibility", beginning with the growing recognition of the "dual-use" dilemma and the consequent requirement for effective oversight of research (Section 12.4.1). This will then be followed by a discussion of the potential utility of oaths, codes and pledges (Section 12.4.2), and the parallel processes of education and awareness-raising in building the appropriate norms of behaviour for the scientific and biomedical communities and beyond (Sections 12.4.3–12.4.5). Complimentary initiatives to develop and apply relevant ethical standards by health professionals are detailed in Section 12.4.6. The practical application of ethical principles by individual scientists through such practices as whistle-blowing will be explored (Section 12.4.7), as well as the duty of individual scientists and health professionals to inform the policies and practices of Governments in this area (Section 12.5).

12.4.1. Regulating "dual-use" research

Whilst scientific discoveries and their application are intrinsic to the advance of civilization, most science, to varied degrees, is "dual use" or "multi-use" in nature[95] – it has the potential for benign and malign pathways and application. A balance must be achieved between allowing scientific progress and open source disclosure, whilst ensuring against potential misuse of that knowledge. As of August 2015, much of the discourse amongst the life science community concerning how best to combat the proliferation and misuse of chemical and biological weapons (CBWs) has concentrated on regulating the actions of individual life scientists conducting "dual-use" research of potential concern. Highly influential in this discourse have been the 2004 Fink Report[96] and the 2006 Lemon Report,[97] both produced under the auspices of the National Research Council of the US National Academies. Both reports highlighted the importance of taking a comprehensive approach to analysing "dual-use" research of potential concern, with the Lemon Report recommending the adoption of "a broadened awareness of threats beyond the classical 'select agents' and other pathogenic organisms and toxins, so as to include, for example, approaches for disrupting host homeostatic and defense systems and for creating synthetic organisms".[98] The broad threat spectrums enunciated by both reports, and particularly that of the Lemon Committee, appeared to capture ICAs (but not RCAs) within their scope.

As a result of the Fink and Lemon Reports, a range of oversight structures and processes have been established by certain Governments – notably in Europe and the United States – and scientific bodies, academic institutions, funders and publishers, to review "dual-use" research of potential concern, in order to assess the risks and benefits of such research and determine whether the proposals needed to be modified or withdrawn.[99]

However, following analysis of the application of such oversight measures in practice, Rappert has concluded that: "such procedures rarely conclude that manuscripts, grant applications or experiment proposals should not be undertaken or restricted".[100] Similarly, Van Aken and Hunger have analysed the application of biosecurity policies agreed by a group of 32 influential science journals under which manuscripts could be modified or rejected where "the potential harm of publication outweighs the potential societal benefits".[101] Despite such policies having been established in 2003, Van Aken and Hunger found, in 2009, that no manuscript had ever been rejected on security grounds.[102] In 2010, Rappert contended the "same could be said" of those funders that have established submission-oversight systems.[103] Furthermore, Rappert stated that "even more notable with these review processes is the infrequency with which they have identified items 'of concern' in the first place".[104] Whilst information relating to the research controls of Government departments (especially defence-related ones) was not readily available, Rappert argued that, in relation to universities and other publicly funded agencies, "it seems justifiable to conclude that – barring dramatic changes – oversight processes will identify little research as posing security concerns and will stop next to nothing".[105]

Others have criticized the voluntary nature of the existing controls on life science "dual-use" research. For example, commenting upon the release of a draft of the National Science Advisory Board for Biosecurity (NSABB) Proposed Framework for the Oversight of Dual Use Life Sciences Research: Strategies for Minimizing the Potential Misuse of Research Information,[106] the Sunshine Project Director, Edward Hammond, stated that the "NSABB is divorced from reality if its members believe that another set of voluntary NIH [National Institutes of Health] guidelines is sufficient, and would be remotely effective, at preventing dual-use disasters...".[107] Research conducted by the Sunshine Project from 2004 to 2007 indicated that many US organizations obliged to follow NIH guidelines did not do so.[108] A 2007 Sunshine Project survey discovered that 18 of the top 20 US biotechnology companies did not comply with existing voluntary NIH biotechnology guidelines.[109] Instead of a voluntary approach, Hammond stated: "Effective federal management of dual-use risks requires making safety and security oversight truly mandatory and subject to the sobering light of public scrutiny."[110]

As well as scepticism regarding the implementation of such voluntary oversight systems in practice, a further concern relates to the limited range of issues being considered by such bodies. The discourse and much of the previous and current activity appears to be concentrated upon preventing the diffusion of "dual-use" knowledge, skills and materials to various non-State actors with malign intent, principally terrorist organizations. Insufficient attention has been given to utilizing existing "dual-use" monitoring mechanisms or adopting additional process to specifically combat the misuse of "dual-use" expertise in State programmes, even though national CBW R&D programmes arguably pose a greater danger to the CWC and BTWC than the limited activities of non-State actors in this area.

12.4.2. Oaths, codes and pledges for the life science community

One approach to building a culture of responsibility has been through the development of a range of non-binding ethical codes, codes of conduct and oaths or pledges. The development of codes of conduct became one of the priority areas for BTWC State Party activity during both the BTWC first (2003–2005) and second (2007–2010) intersessional processes. Consequently, initiatives supporting such codes were undertaken by a wide range of scientific associations and organizations, including the American Society of Microbiology,[111] National Academy of Sciences, Royal Society,[112] International Centre for Genetic Engineering and Biotechnology, International Union of Biochemistry and Molecular Biology and International Council for the Life Sciences. These activities have been complemented and stimulated by the ICRC as well as through the work of individual scientists and academics.[113]

In comparison to the time and energy expended by a diverse range of organizations in the life sciences, the chemical science community's efforts to develop codes of conduct were (initially at least) more limited and mainly focused upon the activities of the International Union of Pure and Applied Chemistry (IUPAC). In 2004, the IUPAC President and the Director General of the OPCW agreed a joint project on chemistry education, outreach and the professional conduct of chemists. This led to a joint IUPAC/OPCW international workshop in 2005, which concluded that codes of conduct were needed for all those engaged in science and technology using chemicals, so as to "protect public health and the environment and to ensure that [such] activities...are, and are perceived to be, in compliance, with international treaties, national laws and regulations such as those relating to illicit drugs, chemical and biological weapons, banned and severely restricted chemicals...".[114] The workshop also concluded that such codes were "complementary to national implementing legislation for the CWC" and would "help to achieve in-depth compliance throughout academia, industry, and government of those engaged in science and technology using chemicals". They would also "extend awareness of the general-purpose criteria of both the CWC and the BTWC and thus help ensure its effective implementation".[115] The workshop recommended that IUPAC should develop a model code of principles as well as draft elements for codes which might be promulgated to IUPAC national adhering authorities (NAOs) and associate national adhering authorities (ANAOs), urging them to review any existing codes to ensure these elements are included.[116] IUPAC and its Committee on Chemical Research Applied to World Needs (CHEMRAWN) subsequently initiated a project to develop such a code.[117] The project was completed in November 2011, and although a formal IUPAC code of conduct has not been established as of August 2015, guiding principles for a code were developed and promoted.[118]

The proponents of codes of conduct assert their potential utility in helping to sensitize chemical and life scientists to the dangers of "dual-use" research, and to reinforce the importance of, and promulgate, ethical "red lines" where the legal prohibitions or normative taboos are already clearly defined and widely accepted. However, the effectiveness of such an approach has been questioned by a range of scholars.[119] One important limitation of the majority of code-based initiatives is that the resulting instruments are aspirational and non-binding in nature, with no clearly identified penalties elaborated for those individuals who breach the prohibitions, or mechanisms established to monitor and enforce such prohibitions.

Recognizing the weaknesses of self-governance initiatives to effect change in this area, some have called for codes of conduct to become binding, with those breaching such codes facing sanction from their peers (or the State). For example, Rotblat, in a letter to a Pugwash Workshop on Science, Ethics and Society in 2004, stated his belief that: "Perhaps the time has come for a binding code of conduct, where only those who abide by the code should be entitled to be practicing scientists, something which applies now to medical practice."[120]

However, examination of the literature reveals that no such binding codes of conduct prohibiting research, development or utilization of biological or chemical agents for hostile purposes (and that specifically included ICA weapons and RCAs) have been established by any international scientific organization, as of August 2015.

Rappert has noted that "... if codes are to go beyond reiterating platitudes about the abhorrence of using modern biology toward malign ends, then they are likely to confront major issues of controversy. For instance, codes could comment on the acceptability of disputed attempts to develop 'non-lethal' incapacitating agents...".[121] However, it is those areas of dispute or controversy, such as the research and development of ICA weapons, where codes remain silent, or at best provide ambiguous guidance. Similarly, as with "dual-use" research oversight, whilst numerous codes condemn and seek to prevent the involvement of scientists in development of biological and chemical weapons by non-State actors, it is questionable whether enough energy has been devoted to targeting the more contentious issues of the involvement of life scientists in State-run weapons programmes. More recently, however, there have been some indications of an increasing awareness amongst certain life science communities of the dangers of co-option into State programmes, and initiatives undertaken to address these dangers.[122]

As well as questioning the effectiveness of self-governance measures – particularly codes of conduct – concern has also been raised over the time and resources devoted to such initiatives, particularly as other aspects of the CWC or BTWC control regimes are in need of strengthening. For example, in 2006, Perry Robinson stated his belief that the international community should "... put most effort into (a) enhancing the OPCW and (b) strengthening the norm [against biological and chemical armament]. In comparison, the various other sorts of proposal, such as codes of conduct for scientists, seem mere tinkering at the edges...."[123]

Similarly, in 2006, Corneliussen questioned why voluntary self-governance regimes (and codes of conduct in particular) focusing on individual scientists were being given so much attention when they had "significant shortcomings in practice" and the "greatest risk of misuse is at the level of national biological weapons programmes".[124] Consequently, Corneliussen contended that:

> [T]he current sole focus on codes, and the extensive investment of resources that accompanies it, might well serve to detract from other more crucial regulatory measures that target not only individual scientists but also state programmes. Without this plurality of regulatory measures in place, codes of conduct are doomed to fail.[125]

In 2010, Rappert concluded that, despite the energy and resources expended upon the development and promotion of codes, such "efforts to devise meaningful codes have largely floundered". Rappert contended that: "In no small part, this has been due to the lack of prior awareness and attention by researchers as well as science organisations to the destructive applications of the life sciences. Before codes can help teach, education is needed."[126]

12.4.3. Education, awareness-raising and promulgation of the CWC

The OPCW has previously recognized the important roles that education, promulgation and awareness-raising initiatives can play in facilitating national implementation of the CWC and in combating the development and use of chemical weapons.[127] In order to facilitate work in this area, the OPCW, in 2001, established its Ethics Project, with the aims of promoting a "development of awareness" among chemistry/engineering professionals of the "object and purpose of the CWC" and a "culture of compliance" with the requirements of the Convention; integrating ethical and scientific aspects of chemical weapons disarmament into chemistry and chemical engineering education;[128] and attempting to establish a "non-proliferation code of conduct" for professionals working with chemicals.[129]

Following analysis of preliminary surveys indicating that "very few educational institutions include ethical issues in their curricula", the Ethics Project sought to work with universities to "target students pursuing degrees in chemistry and/or chemical engineering",[130] and to engage National Authorities of States Parties and international organizations such as the United Nations Educational, Scientific and Cultural Organization (UNESCO).[131] It has also collaborated with scientific societies and professional associations – most notably IUPAC[132] – to "develop public awareness of the ethics involved in the CWC to target relevant professionals".[133]

Despite the recognition of the importance of education and awareness-raising by successive CWC Review Conferences, and the OPCW institutional machinery potentially available to assist such initiatives, it appears that activity in this area had been relatively weak and unsustained,[134] with limited engagement by both governmental and non-governmental communities. In its 2008 report for States preparing for the CWC Review Conference, IUPAC warned that: "Awareness in the scientific and technological communities in all countries about the CWC and its norms, prohibitions, and implementation requirements remains poor." IUPAC argued that this "proves the need for further efforts to incorporate [such] ethical norms...into chemistry education ... [They] should become a regular part, at an early stage, of the education of every student of chemistry and chemical engineering."[135]

In recent years, however, the OPCW has devoted renewed effort and resources to this area, often working in collaboration with the scientific community, particularly IUPAC.[136] In 2012, to facilitate and inform such activities, the Director General established a Temporary Working Group on Education and Outreach in Science and Technology Relevant to the CWC.[137] However, it appears that the educational initiatives that have been undertaken to date (August 2015) have concentrated upon the non-contentious areas of the CWC. Although there has been some (albeit limited) material produced regarding the regulation of RCAs, there does not appear to have been any significant attempts to address the contested obligations regarding ICAs.[138]

12.4.4. Education and awareness-raising amongst the life science community

Although States Parties to the BTWC have voiced support for education, promulgation and awareness-raising activities amongst the life science community regarding the Convention and measures to combat biological weapons development and use,[139] surveys undertaken from 2005 onwards have revealed low levels of awareness amongst life scientists in this area.[140] In a 2009 *Nature* article, Dando highlighted the "lack of engagement with this issue among life scientists", which he considered "alarming",[141] specifically with regard to the consequent dangers of the misuse of scientific and technological advances for the development of ICA weapons.

In a 2006 paper reviewing the effectiveness of then-existing education and awareness-raising initiatives, Rappert, Chevrier and Dando concluded:

"Many of the calls and statements surveyed earlier regarding the need for education and outreach are largely just that...Thus far, the translation of these calls into educational activities have been rather modest."[142]

Similarly, in 2010, Whitby and Dando stated that "correcting this deficiency in education and awareness-levels of life scientists will be a massive task".[143] It is one that commentators believe will require action by a broad range of constituencies involved in life science education, including Governments, bodies responsible for the administration of standards in higher education, funders of life science education, civil society groups and NGOs involved in the production of educational material, and teachers and trainers.[144] Unlike the education and promulgation activities for chemists which have been facilitated in part by the OPCW, corresponding activities amongst the life science community are currently being driven primarily by academic and NGO bodies.[145] Such life science initiatives have actively explored a range of "contested" issues, such as the development of ICA weapons, which appear to have been largely avoided in previous CWC-focused activities.

12.4.4.1. Civil society development of bioethics educational materials

Bradford Disarmament Research Centre (BDRC), together with the Australian National University and the Universities of Bath and Exeter, has undertaken a collaborative project: Building a Sustainable Capacity in Dual-Use Bioethics.[146] A core element in this programme has been the development by BDRC, the National Defence Medical College (Japan) and the Landau Network Centro Volta (Italy) of an educational module resource (EMR) designed to support life scientists and educators in learning about bio-security and "dual-use" issues, and also in building educational material for teaching students. The EMR has been designed to be "modified and tailored in order to fit the requirements of different local educational contexts". It was "intended to be a resource that can be used by a lecturer in order to develop one or more lectures, seminars, role-plays or other teaching aids suitable for the course he or she is presenting".[147] The EMR consists of 21 lectures covering a history of offensive biological weapons, the BTWC and its implementation, the web of prevention, "dual-use" dilemmas and the responsibilities of life sciences; and includes two lectures devoted to the CWC and ICA weapons.[148] As of August 2015, the EMR has been translated into French, Georgian, Japanese, Polish, Romanian-Moldovan, Russian, Spanish and Urdu.[149]

12.4.5. Public awareness initiatives about "dual-use" concerns

Whilst a number of, albeit relatively small and isolated, initiatives have been undertaken to educate and raise awareness amongst chemists and life scientists regarding "dual-use" dilemmas and the potential dangers of research being misused for chemical or biological weapons development, there do not appear to be any sustained activities designed to foster greater awareness and knowledge of such issues beyond these scientific communities. In 2010, Rappert noted that: "Scant efforts made prior to 2001 (and even since) by scientists to popularise how their work might aid the production of bioweapons indicate the historical pattern of not seeking to foster wider debate and awareness."[150]

It is, however, worth considering whether and how those scientists, academics and educators concerned about ICA weapons and RCAs, who are currently conducting CBW education and awareness-raising activities, can also engage with key civil society actors in areas such as human rights, international humanitarian law and medical ethics, who at present have limited or no knowledge of the dangers posed by the potential harnessing of advances in chemistry and the life sciences to hostile purposes. The education and engagement of such expert and/or activist communities may well enrich and inform the existing discourse concerning ICA weapons and RCAs, and potentially broaden the range of actors seeking effective restrictions of such weapons. In considering such issues, it may be worth exploring the roles of previous civil society awareness-raising initiatives and public education in helping to build successful multidisciplinary coalitions dedicated to addressing complex issues such as the prohibition of anti-personnel landmines and cluster munitions; addressing climate change; and promoting the establishment of the International Criminal Court.

12.4.6. Developing and applying ethical standards for health professionals

In addition to life and chemical scientists, the participation of physicians and other health professionals in previous State-run RCA or ICA weaponization programmes has been documented in a range of States, including, for example, South Africa, the United Kingdom and the United States.[151] More recently, a number of health professionals in certain countries appeared to have voiced support for medical involvement in the research and development of ICA weapons.[152]

The importance of medical participation for the viability of certain ICA weapons research has been discussed by Gross, who has argued that ICAs are among a limited range of "less lethal" weapons that are, in effect: " 'medicalized' in that they rely on advances in neuroscience, physiology, and pharmacology and on the active participation of physicians and other medical workers".[153]

Gross has further contended that:

> Some wonder whether it might be possible to build nonlethal weapons without the help of medical doctors and restrict weapons design, development, and testing instead to medical scientists, but there is no nonlethal

weapons program that can do this. Medical oversight, in the very least, is crucial to test devices that flirt with the limits of human endurance. Even if one might isolate medical doctors from weapons research... this becomes impossible if we consider the entire medical community, which includes health care professionals and medical scientists alike. They may be the last people you want to build weapons, but, sometimes, they are the only ones who can.[154]

In comparison to the life and chemical science communities, ethical discourse amongst certain sectors, at least, of the medical professional community, regarding the involvement in development and use of "less lethal" weapons employing chemical agents, appears to be more advanced. There are a number of ethical codes and declarations that are potentially applicable and may well constrain the involvement of health professionals in this area. Firstly, there are a range of declarations and regulations that guide health professionals in situations of conflict and unrest, and specifically prohibit their involvement in torture, ill-treatment and other forms of human rights abuse.[155]

The World Medical Association (WMA) Declaration of Tokyo has stated that: "the physician shall not countenance, condone or participate... [nor]... provide any premises, instruments, substances or knowledge to facilitate the practice of torture or other forms of cruel, inhuman or degrading treatment or to diminish the ability of the victim to resist such treatment".[156]

In addition, the WMA Regulations in Times of Armed Conflict has clearly stated that it was "deemed unethical" for physicians to:

> Give advice or perform prophylactic, diagnostic or therapeutic procedures that are not justifiable for the patient's health care... Weaken the physical or mental strength of a human being without therapeutic justification... Employ scientific knowledge to imperil health or destroy life... Condone, facilitate or participate in the practice of torture or any form of cruel, inhuman or degrading treatment.[157]

The WMA has also sought to develop ethical guidelines in the area of weapons development, for example by prohibiting biomedical involvement in the development of certain internationally outlawed weapons. In its 1990 Rancho Mirage Declaration on Chemical and Biological Weapons, the WMA stated that:

> [T]he World Medical Association considers that it would be unethical for the physician, whose mission is to provide health care, to participate in the research and development of chemical and biological weapons, and to use his or her personal and scientific knowledge in the conception and manufacture of such weapons.[158]

Furthermore, under the Declaration, the WMA: "condemns the development and use of chemical and biological weapons... asks all governments to refrain from the development and use of chemical and biological weapons... [and] asks all National Medical Associations to join WMA in actively supporting this Declaration".[159]

In its 2002 Washington Declaration on Biological Weapons, the WMA recommended:

> [T]he World Medical Association and National Medical Associations worldwide take an active role in promoting an international ethos condemning the development, production, or use of toxins and biological agents that have no justification for prophylactic, protective, or other peaceful purposes... [and that]... the World Medical Association urge all who participate in biomedical research to consider the implications and possible applications of their work and to weigh carefully in the balance the pursuit of scientific knowledge with their ethical responsibilities to society.[160]

In addition to such WMA Declarations and Regulations, guidance has been developed by national medical associations and other medical bodies on the ethical considerations surrounding the involvement of health professional in weapons development more generally.[161]

National medical bodies have also established mechanisms and structures to implement ethical standards, including ethics boards, which have the authority to suspend or disbar physicians from practising medicine in cases of extreme misconduct. In certain countries, such bodies have investigated the alleged participation of health professionals in CBW development programmes.

12.4.6.1. Implementing medical ethics – investigating the former head of Apartheid South Africa's CBW research programme

In 2007, the Health Professions Council of South Africa (HPCSA) issued *Guidelines for Good Practice in the Health Professions: Research, Development and Use of Chemical and Biological Weapons*.[162] The preamble noted that:

> [T]he South African military authorities under the apartheid government sustained a covert programme for the development of chemical and biological weapons, and recruited health care practitioners and scientists to staff the programme. The evidence that emerged in the TRC's [Truth and Reconciliation Commission's] investigations into this clandestine project has pointed to the importance of developing clear guidelines for the health professions in regard to participation in such programmes. It was particularly evident that the secrecy surrounding the apartheid government's

CBW programme enabled health professional scientists to conduct research outside of any ethical oversight.[163]

The preamble further stated "... It is completely contrary to the fundamental principles of the ethics of the health professions for a health care practitioner to participate in research activities directed at generating materials intended to cause harm to human health and well-being...."[164]

The guidelines consequently established the following ethical obligations upon healthcare professions:

> All research to develop CBWs designed to inflict harm on humans is unethical and health care practitioners should not participate in such activities. Should health care practitioners find themselves in dual loyalty situations where they are coerced or experience pressure or threats to comply from the military or other authorities, they should appeal to the HPCSA or any other appropriate professional body for support in resisting such pressures.[165]

While the guidelines recognized the potential for physicians to become involved in legitimate research to "protect military or civilian personnel against CBWs", such research "should be subject to open peer review and ethical oversight by a suitably appointed independent body...", as elaborated in the guidelines.[166]

The HPCSA considered that "the guidelines form an integral part of the standards of professional conduct against which a complaint of professional misconduct will be evaluated".[167] And, furthermore, that: "Health care practitioners who decide not to follow the guidance... must be prepared to explain and justify their actions and decisions to patients and their families, their colleagues and, if necessary, to the courts and the HPCSA."[168]

The HPCSA has been putting these ethical principles into practice through its investigation of the activities of Dr Wouter Basson, former head of Project Coast, the Apartheid South African CBW R&D programme. Dr Basson initially faced six charges of alleged professional misconduct as a result of his activities at Project Coast from 1981 to 1993.[169] The HPCSA investigation was initiated after it received a complaint submitted by over 40 doctors in 2000 of Basson's unethical conduct, but was delayed pending Basson's criminal trial on related charges.[170] The HPCSA disciplinary hearing, which followed Basson's acquittal in 2002 on all criminal charges, was blocked following claims by Dr Basson that the HPCSA was biased against him. However, on 10th May 2010, the North Gauteng High

Court of Pretoria found the claim to be unjustified, prompting the continuation of the HPCSA investigation.[171] Although two charges were dropped, the Professional Conduct Committee (PCC) of the HSPCA ruled that a hearing on the four remaining charges should continue,[172] and initiated proceedings. Three of the charges related specifically to RCAs and ICA weapons. Firstly, Dr Basson was accused of coordinating "the production of the following drugs and tear gases on a major scale...methaqualone...MDMA...BZ...CS...CR...".[173] Secondly, Dr Basson was accused of having "weaponize[d] thousands of 120mm mortars with tear gas",[174] and of overseeing the filling of "some 120 mm mortars with CR...which...were supplied by the South African Defence Force to one Savimbi in Angola for use".[175] Thirdly, Dr Basson was accused of providing "disorientation substances for over the border kidnapping ('grab') exercises, where the substances were used to tranquilise the person to be kidnapped".[176]

Although the substance of these charges was not in dispute, Dr Basson presented nine arguments as to why his conduct was not "unprofessional"; these were all rejected by the PCC.[177] For example, Dr Basson stated in his defence that the chemicals developed, manufactured and provided were specifically designed to weaken and disorientate, but not kill people. He stated that this was done to reduce fatalities.[178] However, the PCC stated:

> With regard to tear gas (CR) in charges 2.2 and 4: Evidence has been provided that people with asthma may have attacks precipitated; and in closed spaces, asphyxia may occur, especially among the very young and the elderly. These chemicals therefore are not harmless. The reality is that the substances were developed, produced and provided to be administered to people without their consent or against their own will.

> With regard to charge 5 (relating to disorientation substances), the PCC stated that "the purpose of the drugs was to capture the person and keep him or her alive so that he or she can be interrogated for information. This is hardly a good beneficence/non-maleficence argument."[179]

On 18th December 2013, the PCC gave their ruling regarding the four charges, concluding that the "conduct under discussion was in breach of established ethical rules" and that "the breaches of medical ethics amount to unprofessional conduct".[180] The Chairperson of the PCC, Professor Jannie Hugo, stated that Dr Basson contravened international protocols and conventions: "He confused ethics of a doctor with that of a soldier while discharging his duties. A doctor cannot rely on military orders to escape the consequences of his duties."[181] Dr Buyiswa Mjamba-Matshoba, Chief Executive Officer of the HPCSA, said: "As the custodian of ethics in healthcare, we are delighted at the Committee's decision. This is a watershed moment, not only for practitioners in the South African healthcare industry, but the outcome is also of critical interest to international publics."[182] Dr Basson has been working as a private practice cardiologist in Cape Town, South Africa. As of August 2015, the PCC is considering suitable sentencing; this could result in the loss of Dr Basson's licence to practise medicine.

There are thus ethical frameworks and mechanisms in place that describe and regulate the duty of health professionals to abide by and promote aspects of human rights and international humanitarian law, and specifically prohibit engagement in acts such as torture or the development of certain weapons. Whilst these and other ethical standards – particularly those concerned with medical research involving human subjects[183] – can in theory be applied to the development and utilization of ICA weapons and RCAs, this does not appear to have occurred in a consistent manner as of August 2015. There are no internationally accepted guidelines specifically determining the permissibility or non-permissibility of physician involvement in the development, testing or utilization of so-called "less lethal" weapons in general, and ICA weapons and RCAs in particular. Indeed, the issue appears, at present, to be both underexplored and contentious, with a spectrum of opinion held by health professionals and medical ethicists.

Amongst national medical associations, in addition to the important actions of the Health Professions Council of South Africa, it is the British Medical Association (BMA) that has taken the lead in the development of ethical guidance for the health community on the issue of purported "less lethal" weapons employing chemical agents, and in particular ICA weapons. In a 2010 presentation to an expert meeting convened by the ICRC on ICA weapons, the BMA's Head of Science and Ethics, Professor Nathanson, stated that: "doctors cannot develop weapons, participate in the use of weapons or of interrogation or otherwise use their medical and medico-scientific skills for anything other than the alleviation of suffering, while remaining doctors with the duties and responsibilities inherent in that title".[184]

Previously, in its 2007 publication "Drugs as Weapons", which explored the implications of ICA weapons research, development and use,[185] the BMA declared that:

> [D]octors should not knowingly use their skills and knowledge for weapons' development for the same reasons that these ethical

considerations oppose doctors' involvement in torture and the development of more effective methods of execution. In other words, the duty to avoid doing harm rises above, for instance, a duty to contribute to national security.[186]

The BMA report specifically recommended that national organizations that represent healthcare professionals should:

> Work to promote the norms prohibiting the use of poisons, and therefore the BTWC and the CWC. They should further promote understanding that the use of drugs as weapons would violate such norms… Advocate against the use of drugs as weapons and not be involved in the training of military or law enforcement personnel in the administration of drugs as weapons.[187]

As of August 2015, and with the important exceptions of the BMA and HPCSA, it is unclear whether other national medical or health professional associations have issued statements or developed guidance in these areas. Consequently, the issue does not appear to have been specifically addressed formally by the WMA during its General Assembly or in any other public WMA policy body.[188] Given the importance of medical participation to development, testing and utilization of ICA weapons and RCAs, the development of clear guidance in this area is needed from the WMA and professional associations representing other health professionals such as anaesthetists and medical toxicologists.

12.4.7. Non-participation/whistle-blowing

Any serious attempt by State or non-State actors to develop new or indeed existing chemical or biological weapons, be they considered "less lethal" or otherwise, would require the involvement of an array of scientists, engineers, technicians and other ancillary workers. Whilst such staff are essential to the development and production of such weapons, they are also potentially capable of "blowing the whistle" on such weapons programmes through public denunciations, leaking information to journalists, or by reporting concerns about potential or realized breaches of national regulations or violations of international treaties directly to the relevant national or international regulatory bodies.

In his 1995 Nobel acceptance speech, Rotblat stated that:

> The purpose of some government or industrial research is sometimes concealed, and misleading information is presented to the public. It should be the duty of scientists to expose such malfeasance. "Whistle-blowing" should become part of the scientist's ethos. This may bring

reprisals; a price to be paid for one's convictions. The price may be very heavy...[189]

Deiseroth has highlighted the particular vulnerability of whistle-blowers:

> Compared with normal citizens, employees are in a special situation because they owe their employer a certain loyalty and, by law, are normally not allowed to disclose internal or confidential information. Whistle-blowers, therefore, need protection if they make a disclosure in good faith and on the basis of reliable evidence.[190]

Similarly, Falter has noted: "it is neither realistic nor legitimate to put the full burden of whistle-blowing and potential retaliation on individual scientists and their moral sensibilities".[191]

Consequently, whilst it is the duty of individual scientists to make known their concerns about the misuse of scientific research for activities that breach ethical standards or international law, it is the responsibility of the scientific community as a whole to ensure that such whistle-blowers are fully protected. This was recognized by the ICRC in its 2004 statement, which declared that: "Those working in life sciences who voice concern and take responsible action require and deserve political and professional support and protection",[192] and the corresponding action point, which was to "ensure that adequate mechanisms exist for voicing such concerns without fear of retribution".[193]

However, although a number of States, such as South Africa,[194] the United Kingdom,[195] and the United States,[196] have legislation relating to whistle-blowing activities on their statute books, the effectiveness of such legislation and its enforcement is variable. Furthermore, Martin, who has long experience of working with, and seeking to protect, whistle-blowers in many different spheres, has argued that: "...the track record of whistle-blower protection measures – whistle-blower laws, hot-lines, ombudsmen and the like – is abysmal. In many cases, these formal processes give only an illusion of protection. Codes of ethics seem similarly impotent in the face of the problems."[197]

It is important that independent scientists, health professionals and professional bodies – in cooperation with human rights, civil liberties and whistle-blowing organizations – promote the establishment of truly effective mechanisms under international and domestic law that provide legal protection against discrimination and criminal prosecution for whistle-blowers. Furthermore, given the failings of the current systems of whistle-blower protection, the chemical and life science communities, together with those in the health professions, have a duty to support those individuals who refuse to participate in what they consider immoral R&D projects, and those who "blow the whistle" on such activities. A number of scientific associations and professional bodies have mechanisms for promoting ethical standards amongst their members, which can also be utilized to support colleagues facing reprisals for acting ethically.[198]

12.5. Engagement with States Parties to promote effective mechanisms to address ICA weapons, RCAs and related means of delivery

Scholars have discussed the responsibility that informed civil society actors – particularly from the medical and scientific communities – have to highlight the limitations and failings of the existing CBW arms control regimes, and to work to develop and promote appropriate mechanisms to strengthen these regimes. According to Dando, Pearson, Rozsa, Robinson and Wheelis: "Whatever else is needed, one crucial ingredient is clear: people with scientific and medical expertise surely have a special responsibility to alert policymakers in governments around the world to the very real dangers of inaction in regard to the BWC."[199]

Similarly, Robinson has argued that:

> When it comes to arms control, all of us...need reminding that treaties such as the CWC are engagements, not between governments, but between States Parties. Governments may represent States Parties in the [relevant regime fora]...but organs of civil society are also elements of those same states, no less responsible for proper implementation of the treaty.[200]

Previously, CWC States Parties, and the OPCW as a whole, have been reluctant to receive information from, and interact with, civil society in a systematic manner. For example, whilst representatives of civil society organizations – such as NGOs, academia, professional scientific and engineering associations, and industry – are routinely invited to address plenary sessions at Review Conferences of the BTWC and the Non-Proliferation Treaty, until recently no such opportunity existed at the CWC Review Conference. Indeed, certain international organizations were also denied such opportunities. At the First CWC Review Conference, the International Committee of the Red Cross (ICRC) was scheduled to give a presentation to delegates on its concerns regarding ICA weapons, in which it stated that:

> In an age of rapid developments in science and, in particular, in the field of chemistry and biotechnology the Convention's integrity is crucially dependent on vigilance regarding new technologies that could undermine its object and purpose. Participation of and frank debate with the scientific, industrial and medical communities on the implications of new developments are essential.[201]

However, although the ICRC was originally told that it could address the Conference as an international organization, it subsequently had this invitation rescinded, reportedly at the request of certain States Parties, including the United States.[202]

Despite these restrictions, there have been some limited opportunities for civil society organizations to interact with State Party delegates, and a number of NGOs have used these opportunities to highlight concerns about ICA weapons. For example, at each Review Conference the Technical Secretariat has hosted an "Open Forum on the Chemical Weapons Convention",[203] the first of which included a panel discussion on "The Chemical Weapons Ban and the Use of Incapacitants in Warfare and Law Enforcement",[204] and the second a presentation on "Toxic Chemicals and Law Enforcement".[205] An editorial in the *CBW Conventions Bulletin* highlighted the benefit of such initiatives: "The interest by delegations in the Open Forum...seems to indicate that their [NGO] involvement is a valuable addition to the review process, not least for the ability to highlight sensitive topics that are politically untouchable by delegations."[206] Open Forums, organized by the CWC Coalition, have been held at each of the subsequent Conference of States Parties (CSP14–CSP19), and these have included presentations by BNLWRP on ICA weapons, RCAs and their means of delivery.[207]

12.5.1. Case study: Chemical Weapons Convention Coalition

One potentially important development for civil society/OPCW interaction has been the establishment in December 2009 of the Chemical Weapons Convention Coalition (CWCC). The CWCC is an independent, international grouping of NGOs whose mission is to support the aims of the CWC with focused civil society action aimed at "achieving full membership of the CWC, the safe and timely elimination of all chemical weapons, preventing the misuse of chemicals for hostile purposes, and promoting their peaceful use".[208] The CWCC is unusual in that it has been established with the active involvement of the OPCW Technical Secretariat. Then-OPCW Director General, Ambassador Pfirter, was very supportive of this initiative, and this support has been continued by the current Director General.[209] The Technical Secretariat, particularly the Media and Public Affairs Department, has given valuable assistance and support to the CWCC and has tried to foster wider engagement of the OPCW with civil society. For example, relevant articles, reports and other publications produced by academics and NGOs can now be posted on the OPCW website[210] (though any such document would be removed immediately if a State Party raised an objection).[211]

Michael Luhan, then-head of the OPCW Media and Public Affairs Department, has stated that both the OPCW and its Member States can "benefit from outside voices that can point out faults in the system... There's a lot of things that go unsaid in more formal venues."[212] Angela Woodward, a programme director at VERTIC, one of the CWCC founding organizations, has noted that: "There remain certain significant problems with the convention which states parties have utterly failed to deal with, such as noncompliance issues (like 'nonlethal weapons')... When states parties... cannot deal with these problems, it is civil society's responsibility to air these problems and constructively work towards finding solutions to them."[213]

After six years, the CWCC is still in a formative period. As of August 2015, it had 36 members (predominately national organizations) in 24 countries. It has a functioning website and has helped to increase the numbers of civil society organizations attending and participating in OPCW meetings, most notably the Conferences of States Parties. It does provide important (and sometimes the only) opportunities for civil society to interact with CWC States Parties and the Technical Secretariat in semi-official fora most notably the Open Forum meetings that have been organized in the margins of the last six Conferences of States Parties. The CWCC has also played a role in promoting the CWC and the work of the OPCW to the wider world. However, despite these accomplishments, it is apparent that the CWCC has yet to become more than a loose grouping of civil society organizations with disparate objectives and priorities.

In recent years, the situation has markedly improved, in large part due to the prioritization of public diplomacy by the current OPCW Director General. This was exemplified in the run-up to and during the Third CWC Review Conference. Firstly through the promotional activities of the Technical Secretariat and the CWCC, far greater numbers of civil society organizations registered and attended this Conference than previously. Furthermore, with the support of the Director General and a number of States Parties (notably Norway); the Chair of the Conference, Ambassador Paturej, brokered amendments to the Rules of Procedure and associated guidelines,[214] facilitating greater potential NGO participation. These measures included an opportunity for NGOs to address plenary sessions, greater access to Conference documents, provision of meeting rooms and technical equipment, a modified/simplified registration process, and the creation of an "NGO Coordinator" to act as a focal point for communication with the Conference Chair and the Secretariat. Consequently, for the first time at a CWC Review Conference, NGOs made statements to a plenary session. The session was opened by Ambassador Paturej, who described the decision for greater NGO participation as a "landmark", whilst the Director General welcomed the "pioneer effort" as a step forward in NGO engagement and said he looked forward to "valuable and constructive inputs".[215] Statements which had been coordinated by the CWC Coalition were made on behalf of 14 NGOs, and included 1 highlighting the issue of wide area RCA means of delivery.[216] Similarly, more side events were held during this Review Conference than for any of its predecessors. Consequently, Guthrie believed that non-governmental presence and presentations were "starting to approach the levels seen in the meetings associated with the Biological Weapons Convention".[217]

The positive contributions made by civil society to the implementation of the Convention, and the Review Conference in particular, were highlighted by a number of States during and following the Conference, including Costa Rica, New Zealand, Norway, the United Kingdom, the United States, and Ireland speaking on behalf of the European Union.[218] Following the significant advances made prior to and during the Third Review Conference by the OPCW in its interactions with key technical and civil society stake-holders, the Third Review Conference: "noted the importance of the contribution to the goals of the Convention that is made by the chemical industry, the scientific community, academia, and civil society organisations engaged with issues relevant to the Convention".[219] And furthermore:

> Encouraged the Secretariat and the States Parties to improve interaction with the chemical industry, the scientific community, academia, and civil society organisations engaged in issues relevant to the Convention, and encouraged the Secretariat and States Parties to develop a more open approach, in conformity with the Rules of Procedure of the policy-making organs with regard to such interaction.[220]

From the mid-1990s onwards, a large number of academics, legal scholars, NGOs, scientific and medical associations and international humanitarian organizations have attempted to alert the international community – particularly the OPCW and to a lesser extent the BTWC States Parties – to the potential dangers of the development, proliferation and misuse of ICA weapons. The research presented to support their interventions has been technical and robust in nature, spanning legal, medical, chemical and life science disciplines and technology-monitoring expertise. These interventions have taken place during annual meetings and Review Conferences of the CWC and BTWC States Parties, as well as through processes such as the IUPAC/OPCW conferences exploring the impact of scientific developments on the CWC, and the ICRC expert meetings on ICA weapons.

In contrast, attempts to raise concerns regarding RCAs and related means of delivery in these fora appear to be more recent and limited in nature; currently restricted to a much smaller number of academics and NGOs from an arms control background. There are intimations that this may change, however, in the wake of the recent widespread inappropriate use of RCAs in many countries, particularly in the Middle East, to suppress popular democratic movements. It appears that a number of health professional bodies and human rights organizations are examining the applicability of the CWC as a potential tool to address this widespread misuse of RCAs, and are exploring the potential of raising their concerns through the OPCW.

12.6. Conclusion

The HAC analysis undertaken in this chapter has highlighted the potential roles that an informed and activist civil society, including members of the scientific and medical communities, can play in the scrutiny of a Government's policy and practice regarding the development and employment of ICA weapons, RCAs and related means of delivery, through open source monitoring and analysis, and, where appropriate, by undertaking investigative field missions to establish the nature of chemical agents utilized in disputed circumstances. However, as of August 2015, such activities have been, and continue to be, limited to a very small and under-resourced group of NGOs.

It appears that some – although still inadequate – attention has been given by both Governments and civil society to the dangers arising from "dual-use" research in the life and chemical sciences, and their potential application to biological and chemical weapons programmes. Such considerations have fostered debate and reflection within the scientific community upon issues of individual researcher responsibility as well as strategies for the regulation of "dual-use" research programmes of potential concern. However, as of August 2015 this discourse appears to have been largely restricted to preventing the diffusion of such technologies to non-State actors of concern and has not adequately addressed the dangers of State development of such weapons. Furthermore, inadequate consideration has been given to contested issues, including how to address the development of the "next generation" of ICA weapons, RCAs and their means of delivery.

Informed civil society – particularly the scientific and medical communities – can play a constructive role in strengthening the existing control regimes by highlighting limitations of these regimes, and developing and promoting policy proposals to address such limitations. Civil society can also highlight the failure of Member States to implement their obligations under the relevant agreements and, where appropriate, can present evidence of such failings to the relevant control regime, and encourage action be taken to address such failings.

It is clear from the proceeding analysis that a comprehensive HAC strategy for the regulation of ICA weapons, RCAs and related means of delivery will require the involvement of informed and activist civil society in monitoring and verification activities; the development and promotion of norms prohibiting participation of scientists and health professionals in weaponization programmes intended for malign application; and far greater active engagement by non-governmental legal, scientific and medical experts, together with human rights organisations and affected communities, in the relevant State policy development processes, both at a national level and in international fora, in particular through the OPCW.

13
Conclusions and Recommendations

13.1. Introduction

As discussed in Chapter 1, holistic arms control (HAC) is an analytical approach consisting of a three-stage process, comprising:

- Stage one – examination of the nature of the weapon or weapons-related technology under review, current and/or potential future scenarios of application (together with attendant human security concerns), and the potential implications of advances in relevant science and technology.
- Stage two – exploration of the full range of potentially applicable control mechanisms, analysing strengths, weaknesses and limitations.
- Stage three – development of a comprehensive strategy to improve existing mechanisms (and/or introduce additional mechanisms) for the effective regulation or prohibition of the weapon or weapons-related technology of concern.

This analytic approach has been applied to the regulation of incapacitating chemical agent (ICA) weapons, riot control agents (RCAs) and their means of delivery. This concluding chapter will summarize the empirical results and attendant analysis, covering stages one and two of the HAC process. Building upon these findings, a strategy for the regulation of such agents and means of delivery in the short-to-medium term (five to ten years) will be proposed, in line with HAC stage three.

13.2. Summary of findings

ICAs, which were analysed in Chapter 2, can be considered as a diverse range of substances whose chemical action on specific core biochemical processes and physiological systems, especially those affecting the higher regulatory activity of the central nervous system, produce a disabling condition (e.g. can cause incapacitation or disorientation, incoherence, hallucination, sedation or loss of consciousness) or, at higher dosages, death. Proponents of ICA weapons have promoted their development and use in certain law enforcement scenarios, particularly hostage situations; and as a possible tool in a variety of military operations, especially where combatants and non-combatants are mixed. Opponents and sceptics have strongly contested the possibility of employing truly "less lethal" ICA weapons and have highlighted the grave dangers to the health of the targeted populations. Further concerns include: the "creeping legitimization" of such weapons with the erosion of the norm against the "weaponization of toxicity"; the risks of their proliferation to both State and non-State actors; their potential use as a lethal force multiplier; their applicability in the facilitation of torture and other human rights violations; and the militarization of the life sciences.

Over the last 50 years, although a number of States have sought to develop ICA weapons, to date only one State – the US – has confirmed weaponizing them for military purposes (now long discontinued). Publicly accessible information indicates China, Israel and the Russian Federation have acquired or developed ICA weapons, and that such weapons are either in the possession, or have been used by the law enforcement or security services of those countries since the coming into force of the CWC in 1997. Although there is evidence of relevant dual-use research in additional countries, the full nature and purpose of such research is often unclear, as are the intended applications to which it will be put. Currently, there is no evidence of concerted attempts by non-State actors, such as terrorist groups, to conduct the R&D of ICAs weapons. Consequently, ICA weapons can be considered as an immature, limited and contested technology; however, one that could be radically affected by advances in science and technology and that may potentially proliferate and be misused by State and non-State actors.

Chapter 3 addressed the properties of RCAs – potent sensory irritants normally of low lethality that produce dose and time-dependent acute site-specific toxicity. Targeted mainly towards the respiratory and mucosal surfaces, their short-term effects are well defined, although the potential for deleterious long-term effects is contested. There is a small discrete range of agents widely stockpiled and regularly utilized by law enforcement officials in the majority of States for dispersal of crowds and/or incapacitating individuals. Credible reports of the misuse of RCAs by law enforcement officials for human rights violations are geographically widespread. RCAs have reportedly been misused to suppress the right to peaceful assembly; in the excessive use of force; and to facilitate ill-treatment and torture. In some instances the misuse of RCAs, particularly when employed in enclosed spaces or in very large amounts, has reportedly resulted in serious injury or death. Although a variety of RCAs were previously employed in armed conflict by a number of States, such use by the majority of States is now obsolete. Research into RCAs and delivery mechanisms continues to be reported in a number of countries. RCAs can, therefore, be considered as a relatively mature and established technology, although one that is subject to continuing potential development.

From this initial review, it is apparent that RCAs and ICA weapons should be considered to be distinct types of technology in terms of their chemical properties, action on target populations, nature of employment, consequent impact on health, and international security and human security risks. These considerations, as well as the existing maturity of research and development and degree of proliferation, have consequently informed the range and nature of application of potential regulatory mechanisms analysed under the second stage of HAC.

Although this analysis has concentrated upon the two classes of chemical agent under review, a second dimension of effective weaponization, namely potential means of agent delivery, has been considered throughout the research process. It was found that the properties of the specific delivery mechanism under consideration – including range, amount and rapidity of agent dispersal – were of critical importance in determining the nature of the weapon's likely use and potential misuse.

It is clear from analysing the results of the HAC stage-two survey, described in chapters 5–12, that a wide range of mechanisms have *potential* utility in the regulation of ICA weapons, RCAs and/or related means of delivery. Whilst all of these mechanisms appear to be *theoretically* applicable to one or more classes of agent or delivery system under consideration, the utility of a number of these mechanisms *in practice* could not be established; in certain cases the relevant regime bodies or States Parties have never publicly considered such application and it is unclear how responsive these regimes would be to taking on such controversial and potentially divisive issues. When such considerations are taken into account, the CWC, BTWC, UN Drugs Control Conventions, human rights law, international humanitarian law, and certain transfer controls appear to be the most clearly and directly relevant and applicable inter-governmental mechanisms available in the present circumstances. A strategy incorporating these regimes and relevant law, together with the potential roles of civil society, will be explored in Section 13.3.

To date, most inter-governmental discourse regarding the regulation of ICA weapons, RCAs and related means of delivery has taken place within the context of the CWC and the attendant control regime. Analysis of this regime clearly shows that the CWC and OPCW certainly have the necessary scope of coverage, appropriate regulatory constraints, structures and mechanisms to facilitate, monitor and enforce implementation with regard to the agents and means of delivery under review, and also benefit from near universal membership. However, the Convention and its attendant regime have a number of important limitations in both design and implementation.

There are weaknesses in the CWC's textual architecture, with ambiguities in a number of Articles detailing State Party obligations. For example, although RCAs are defined under the CWC, the scope and nature of their permissible use in situations of armed conflict and in law enforcement operations are contested by some States, due in part to a lack of definition for "law enforcement" under the Convention. The situation is even more uncertain and contested regarding the employment of ICA weapons in law enforcement.

Furthermore, although States Parties have reportedly employed bilateral consultation mechanisms regarding ICA weapons and RCAs, the potentially powerful multilateral consultation, investigation and fact-finding procedures that could be applied to address such cases of concern under the Convention have never been utilized. The failure by individual States Parties to use such mechanisms is exacerbated by the very circumscribed ability of the OPCW's Technical Secretariat to undertake independent information-gathering and monitoring activities. Similarly, there has been a failure by the OPCW policy-making organs to effectively monitor implementation of the Convention with regard to these agents and to take action where reports of possible breaches of the Convention have become public.

Of the arms control and disarmament treaties, explored in Chapter 6, the BTWC appears to provide the greatest likely potential utility for the regulation of a section of the agents of concern. Article I of the BTWC, together with the extended understandings agreed at successive BTWC Review Conferences, makes it clear that the Convention is comprehensive in its scope and would consequently cover a range of ICAs as well as certain RCAs of biological origin. However, as of August 2015, States Parties have not, collectively addressed these issues, and there are a number of uncertainties regarding the application of the Convention in this area. In addition, there are important limitations on the value of the BTWC (and its control regime) as a tool to regulate ICA weapons and RCAs, arising from its current lack of effective verification and compliance mechanisms, and also the absence of an international organization comparable to the OPCW, which could coordinate such activities and facilitate implementation by States Parties.

Chapter 7 highlighted the significant constraints upon the use of ICA weapons, RCAs and related means of delivery imposed by international humanitarian law (IHL). Indeed, it is clear that the employment of an ICA weapon in armed conflict would either breach a relevant IHL prohibition (such as the prohibitions on SIRUS and deliberate attacks on civilians); or undermine a relevant IHL obligation (such as the duty to protect persons considered *hors de combat*). Similarly, the use of RCAs in armed conflict, in certain proposed scenarios, would appear to contravene the prohibitions on deliberate attacks on civilians and the prohibition on attacks that do not discriminate between civilians and military objectives. However, the potential utility of IHL to the regulation of these agents is curtailed due to limitations in investigation and enforcement procedures, and extremely low levels of State implementation of Article 36 legal reviews of new weapons. Furthermore, IHL is only applicable to situations of armed conflict. Whilst much relevant IHL would extend to non-international armed conflicts, there may well be disagreements as to whether a particular situation is indeed a non-international armed conflict, with the relevant State instead claiming to be involved in law enforcement activities against criminals or terrorist organizations.

It is clear from the review, conducted in Chapter 8, that human rights law is applicable to the employment of ICA weapons, RCAs and related means of delivery, as it regulates the use of force by law enforcement officials and other agents of the State. Human rights law is particularly important to the discussion of the regulation of these agents, as it potentially covers the full "use of force" spectrum, from law enforcement activities through to armed conflict, including counter-terrorist, counter-insurgency and military operations outside armed conflict, where use of these chemicals has been proposed. While several human rights norms may be applicable, the rights to life, to liberty and security, to engage in "peaceful protest", to freedom from torture and other cruel, inhuman or degrading treatment or punishment, and to health, together with attendant obligations on the restraint of force, are the most relevant.

Chapter 10 highlighted the potential utility of regional and pluri-lateral regimes to regulate, and in certain circumstances prohibit, transfer of ICA weapons, RCAs and related means of delivery. However, the agreements highlighted were either found to be politically binding, or, if legally binding, of limited membership. Potentially more effective were legally binding arms embargoes (particularly those introduced by the UN), wielded to address threats to international peace and security arising from armed aggression and, increasingly, gross human rights abuses or breaches of IHL. However, even on the relatively rare occasions when they are utilized, such embargoes currently have limited effectiveness in practice due to ambiguities in their scope of coverage, delayed and patchy implementation, and poor or non-existing monitoring and enforcement measures.

Chapter 9 described how the Single Convention on Narcotic Drugs and the UN Convention on Psychotropic Substances could both prove to be extremely important mechanisms to combat the development, proliferation and use of narcotic and psychotropic substances as ICA weapons intended for either law enforcement or military operations. However, both Conventions were developed and have been applied, as of August 2015, in a crime control context rather than for arms control and disarmament, and it is far from certain whether the States Parties to either instrument will welcome attempts to apply them to such purposes.

In the light of the evident ambiguities and limitations of the existing relevant State-centric control regimes – compounded by inadequate and patchy national implementation, and the failures of States Parties and international organizations to challenge reported treaty violations by certain States – Chapter 12 explored the potential roles that civil society could play in combatting the misuse of ICAs, RCAs and related means of delivery. The application of "societal verification" as a complement to the existing official verification mechanisms, the development of "cultures of responsibility" amongst the medical and life and chemical scientific communities, and the development and advocacy of science-informed policy were potentially valuable activities that scientists, health professionals, academics and other civil society actors could undertake in facilitating the effective regulation of these agents of concern.

13.3. HAC stage three: A proposed strategy for the effective regulation of ICA weapons, RCAs and related means of delivery

The HAC analytical framework seeks to actively explore the utility of employing a multiplicity of mechanisms and to facilitate the active engagement of a variety of relevant actors in the regulation of the weapons or weapons-related technology under review. Consequently, it is an axiomatic principle of HAC that a number of regulatory processes elaborated as part of existing regimes (as well as those independent of such regimes) can be pursued in parallel. However, it is clearly apparent that relevant actors – be they international organizations, States or civil society bodies – have limited human, financial and temporal resources to devote to these issues, and competing demands to address. When analysing the potential responsiveness of particular regimes, factors such as the existing political environment and previous negotiating history concerning such agents, relevant regulatory mechanisms and their application, and the informal and formal regime forum schedules and "agenda space" will have to be considered by the relevant actors.

From the review of the regimes conducted under HAC stage two, it is clear that the CWC (and its attendant control regime) is the most appropriate and probably the most receptive forum, at least in the short term, for the discussion of concerns relating to ICA weapons, RCAs and related means of delivery. However, the results of such discussions are by no means certain, and parallel processes should also be established to explore alternative mechanisms with the BTWC, UN drugs conventions, IHL, human rights instruments, and specific transfer controls potentially yielding positive results in the next five- to ten-year period. Since these are State-centric mechanisms, HAC stage three is drafted primarily in terms of policy proposals for States that are party to the relevant mechanism. However, this section concludes with recommendations for complementary civil society activity in these areas.

13.3.1. The Chemical Weapons Convention

The quinquennial CWC Review Conference, with its mandate to examine long-term issues of concern to the Organization in a strategic manner, and to "take into account any relevant scientific and technological developments",[1] is the most appropriate forum for addressing regulation of ICA weapons, RCAs and related means of delivery. Although the next (4th) Review Conference is not scheduled till 2018, States Parties concerned about these issues should prepare the ground now for fruitful and informed deliberations, by setting out their positions in statements, reports and so on, and facilitating discussion in suitable forums, notably including the Executive Council (EC), the annual Conference of States Parties (CSP) or through the Scientific Advisory Board (SAB). Consideration should be given to the utility of exploring one or more of the following policy options:

(1) Affirm current national practice is to restrict use of toxic chemicals for law enforcement to RCAs; where such restriction does not exist, States should introduce national moratoria halting the initiation or continuation of the development, acquisition, stockpiling, transfer and use of ICA weapons for law enforcement purposes. In addition, if requisite agreement were forthcoming, a group of like-minded States could introduce a moratorium on such activities at the pluri-lateral level. Such moratoria should remain in place until CWC States Parties collectively determine whether or not the use of ICA weapons in law enforcement is permitted under the Convention. States should also clearly reaffirm the existing prohibition on the use of the toxic properties of all chemicals (including those promoted as RCAs and ICA weapons) as a method of warfare.

(2) Initiate mechanisms to facilitate discussion and make recommendations to OPCW policy organs on currently contested or ambiguous issues:

An Open Ended Working Group or some other formal mechanism could be established to make recommendations on a range of currently contested or ambiguous issues, for consideration by a future CSP or Review Conference. Such formal processes would be open to all States Parties who wished to participate and would reach their conclusions by consensus.

Alternatively, States Parties could initiate a process of informal meetings of experts similar to the model developed by the BTWC States Parties in 2002 to "discuss and promote common understandings and promote effective action" on BTWC implementation measures.[2] As part of this informal process, expertise could be drawn from a range of relevant State sectors, including national implementation officials, scientific advisors, law enforcement officials, and experts in IHL and IHRL. These informal expert meetings could run in parallel or prior to the formal mechanism and could present recommendations to the formal mechanism or directly to an appropriate OPCW body. In addition to any OPCW process, it would be highly beneficial if informal inter-governmental consultation mechanisms on these issues were established.[3] Among the issues that these formal and/or informal mechanisms could explore are:

(i) terms undefined or inadequately defined under the CWC: a suitable mechanism could be established to:

- define the terms "law enforcement" and "method of warfare" as used in the CWC, explore the range of activities contained within each term and determine where activities such as counter-insurgency operations should lie;
- identify which chemicals should be considered as toxic chemicals in the sense of having a "chemical action on life processes that can cause temporary incapacitation in human beings or other animals", in particular the position of malodorants should be addressed.

(ii) status of ICA weapons under the CWC: In its National Working Paper presented to the 2nd CWC Review Conference, Switzerland called for: "... a mandate for a discussion of, *inter alia*, an agreed definition of incapacitating agents, the status of incapacitating agents under the Convention, and possible transparency measures...".[4] If such a proposal were to be introduced and agreed at the 4th Review Conference, it is envisaged that a suitable mechanism would potentially come to a determination *either* that development, stockpiling, transfer and use of ICA weapons for law enforcement are *prohibited* under the CWC *or* that such actions are *permitted*, but should be stringently regulated under the Convention. If the latter position is taken, then the mechanism could also:

- clarify under what limited circumstances and with what constraints such use would be permissible;
- propose options for reporting and transparency measures applicable to such agents and their means of delivery;
- explore the implications for the verification regime.

(iii) regulation of RCAs under the CWC: A suitable mechanism could be established to:

- identify the range of chemical agents that are covered by the term RCA, as defined under the CWC, and give particular consideration to whether malodorants should be classed as RCAs;
- clarify the specific restricted circumstances under which use of RCAs by military personnel may be permissible, and identify those circumstances under which use is prohibited under the CWC;
- clarify the CWC limitations on the use of RCAs for law enforcement, specifically taking into account the CWC's constraints upon "types and quantities";
- explore the limitations on the development, production, stockpiling, transfer and use of RCAs arising from existing obligations under relevant international law including international human rights law and IHL.

(iv) regulation of law enforcement means of delivery: A suitable mechanism could develop criteria for determining which means of delivering and dispersing toxic chemicals are inappropriate for law enforcement purposes and would consequently breach Article II.1.[5] Potential reporting, information-sharing and verification mechanisms applicable to such means of delivery could also be explored. In addition, a guidance document could be developed detailing those means of delivery considered inappropriate for law enforcement purposes and consequently prohibited. Such prohibited means of delivery should, as a minimum, include: artillery shells, aerial bombs, large calibre mortar shells and cluster munitions. This guidance document could be reviewed regularly in a suitable forum, such as a CSP or Review Conference, to take into account developments in relevant science and technology.

(v) reporting and transparency mechanisms for toxic chemicals utilised in law enforcement: A suitable mechanism could explore and develop recommendations for extending the existing RCA reporting and transparency obligations,[6] in order to cover all toxic chemicals held by States Parties for law enforcement purposes.[7] The working group could also consider whether existing information requirements are adequate or should be expanded to include, for example:

- name/CAS number of each type of toxic chemical and quantities held;
- nature and quantities of the associated munitions, means of delivery or dispersal;
- authorities holding stockpiles and permitted to use toxic chemicals and associated munitions, means of delivery or dispersal;
- nature of intended use e.g. incapacitating violent individuals, riot control and crowd dispersal, hostage situations;
- decisions by States Parties not to introduce certain toxic chemicals (e.g. ICAs) for law enforcement purposes and their rationale.

(vi) improve OPCW monitoring of science and technology of relevance to ICA weapon and RCA development: In 2011, the report of a high-level expert panel convened by the OPCW Director General to explore the future priorities of the Organization recommended that the OPCW should "improve and widen the scope of monitoring and evaluating developments in chemical science and technology..."[8] Subsequently, these issues were also included in the 2015 Technical Secretariat discussion paper – The OPCW in 2025: Ensuring A World Free of Chemical Weapons.[9] Amongst the measures the Organization may wish to consider are those to:

- Ensure more frequent and considered review by CWC States Parties of relevant advances in science and technology and the implications for the Convention. In addition to the broad-scope review currently undertaken every five years in preparation for and during the Review Conference,

more limited reviews of specific issues or technologies of potential concern could be prepared by the TS/SAB for consideration by States Parties at the annual CSPs.
- Further enhance the Technical Secretariat's capability to monitor advances in science and technology of potential concern to the Organization, specifically including those that could be utilized for development of ICA weapons, RCAs and their means of delivery; and establish suitable mechanisms, allowing it to bring relevant concerns to the attention of the States Parties and appropriate OPCW organs.

In addition to the foregoing proposed mechanisms, all States Parties should utilize existing CWC consultation, investigation and fact-finding mechanisms where activities of potential concern are reported, such as the alleged use of ICA weapons or RCAs by law enforcement, security or military forces, for human rights violations or breaches of IHL. In addition to obtaining information regarding the specific alleged incidents that have raised concerns, clarification could also be sought concerning the nature and quantities of ICA weapons or RCAs and related means of delivery developed and stockpiled, and the entities holding such agents; the range of intended uses to which they can be put; and the political and legal controls on development, stockpiling, deployment and use. If bilateral consultations with the relevant States Parties are not fruitful, concerned States Parties could consider a formal request under Article IX of the CWC.

There appears to be a growing recognition that the OPCW needs to undergo an organizational-wide process of reflection and adaptation to meet challenges in both the external environment (e.g. advances in science and technology, globalization of the chemical industry, changes in the nature of armed conflict, etc.) as well as those resulting from its successful advance towards the eventual destruction of all existing declared chemical weapons stockpiles. However, it is unclear whether the necessary political will currently exists in State capitals for the Organization to effectively address the "difficult" questions regarding regulation of ICA weapons, RCAs and related means of delivery. It is doubtful whether progress will be made on these issues in the near future unless at least one State (and ideally a group of like-minded States) champions such a cause. Whilst a growing number of States (including Australia, Germany, Norway, Switzerland, the UK and the US) appear willing to raise the question of ICA weapons in such forums, the position with regard to RCAs and related means of delivery is less certain. In the light of such uncertainty, alternative routes for regulation outside of the CWC/OCPW are explored in Sections 13.3.2–13.3.7.

13.3.2. Biological and Toxin Weapon Convention

BTWC States Parties could, at a Meeting of States Parties (MSP) or at the forthcoming 8th Review Conference (to be held in December 2016), affirm that ICA weapons and RCAs of biological origin and their synthetic analogues are covered under the scope of the Convention, and that the use of such agents and associated means of delivery for "hostile purposes or in armed conflict" is prohibited under the Convention. In addition, States Parties could give consideration to initiating a suitable formal or informal process to collectively address how ICA weapons and RCAs of biological origin intended for counter-terrorist, counter-insurgency or military operations short of armed conflict, should be regulated or prohibited by the BTWC.

Additionally, the States Parties and the relevant organizations of the BTWC and the CWC could improve their coordination to address the implications to both treaties of the convergence of the chemical and life sciences and related technologies, including with respect to the development of ICA weapons and RCAs. Furthermore, as part of a post-Review Conference intersessional process, BTWC States Parties could include a focus on monitoring and assessing the risks of misuse of advances in relevant converging life and chemical science and technologies, including those applicable to ICA weapon or RCA development.

13.3.3. International humanitarian law

States could explore the effective application of existing constraints imposed by IHL instruments (principally the Four Geneva Conventions and Additional Protocols) and customary IHL upon the development, acquisition or use of ICA weapons, RCAs and related means of delivery in armed conflict; and the consequent implications for State actions in this area. Of particular importance would be measures to promote and facilitate the effective implementation of States' obligations under Article 36 of Additional Protocol I to conduct reviews of any new weapon (including ICA weapons, RCAs and related means of delivery) developed or acquired, so as to determine its compatibility with the principles and rules of IHL.

One possible forum for such deliberations could be the quadrennial International Conferences of the Red Cross and Red Crescent Movement, which bring together the States party to the Geneva Conventions, the ICRC, the International Federation of Red Cross and Red Crescent Societies and all recognized National Red Cross and Red Crescent Societies to examine and decide upon humanitarian matters of common interest; the next (32nd) International Conference will be held in 2017.

13.3.4. Human rights law

States could explore the constraints on the employment of ICA weapons and RCAs arising from international and regional human rights instruments and customary international human rights law. States could also bring cases of reported ICA weapons and RCA misuse to the attention of the appropriate human rights mechanisms, including: UN Special Procedures and the UN Human Rights Council; relevant international and regional treaty bodies (e.g. UN Human Rights Committee under the International Covenant on Civil and Political Rights, UN Committee against Torture under the

UN Convention Against Torture); regional judicial mechanisms capable of delivering binding legal judgements regarding violations of regional treaties (e.g. European Court of Human Rights, Inter-American Court, African Commission on Human and Peoples Rights). Since a number of such regional judicial mechanisms are potentially open to individual petition, victims and their families can also directly seek redress in cases of agent misuse, and civil society organizations can attempt to employ such mechanisms to develop human rights case law on these issues.

Furthermore, States could consider requesting that a suitable body, such as the UN Human Rights Council, the Office of the UN High Commissioner for Human Rights, or the UN Crime Congress, develops guidance/procedures for evaluating the human rights compatibility or incompatibility of all proposed "less lethal" weapons (which certainly cover RCAs and which some States may consider could potentially include certain ICA weapons). If appropriate, the relevant body could also recommend constraints on the use of any "less lethal" weapons deemed compatible with human rights standards,[10] and develop guidelines for monitoring and ensuring subsequent use is in accordance with human rights law.

Alternatively, States, acting through the relevant UN procedures, could seek to initiate an investigation by relevant UN Special Rapporteur(s) (such as those on Torture, or Extra-Judicial Executions, or Counter-Terrorism and Human Rights) on the human rights implications of the development and use of ICA weapons and RCAs for law enforcement. Such a study could be undertaken in the context of a broader study on the use of force in law enforcement operations.

13.3.5. United Nations Drug Control Conventions

Both the Single Convention on Narcotic Drugs and the UN Convention on Psychotropic Substances restrict the legitimate use of a range of ICAs to "medical and scientific purposes". States Parties to these Conventions could seek to formally establish the implications of these restrictions upon the development, stockpiling, transfer and use of ICA weapons intended for law enforcement or military applications. If appropriate, and required, State Parties could bring forward clarificatory amendments (or agree common understandings) through the appropriate Convention mechanisms to explicitly prohibit or constrain such activities.

The reporting (and potentially the consultation/investigatory) mechanisms of both Conventions could be considered by States Parties as potential routes for obtaining information relevant to the regulation of ICAs. Where a State Party has a concern about the development, stockpiling, transfer or use of a narcotic drug or psychotropic substance potentially intended for employment as an ICA weapon in either military operations or law enforcement, they could consider bringing their concerns before the relevant Drug Convention bodies, for example, International Narcotics Control Board, or the Commission on Narcotic Drugs of the Economic and Social Council of the UN.

Although the international governmental and non-governmental arms control communities have previously paid little attention to the potential applicability of the Drug Conventions to the regulation of ICA weapons intended for law enforcement or military purposes, this situation may be changing.[11] It is certainly an area meriting further research and engagement with the State Parties of both Conventions.

13.3.6. Transfer controls

Individual States could, if they have not done so, establish national transfer controls that restrict the import, export, transit, trans-shipment and brokering of ICA weapons, RCAs and/or related means of delivery. Whilst certain regions have developed instruments and attendant transfer control regimes (e.g. EU Council Common Position 2008/944 and EC Regulation 1236/2005) to explicitly regulate the transfer of certain ICAs, RCAs and related means of delivery, and allow for additional items to be added to the relevant control lists, the majority of regions have not. States that are members of the aforementioned regimes could strive to ensure existing control lists are regularly updated so as to encompass the full range of relevant agents/means of delivery of concern, and that restrictions on the transfer of such items are strengthened and effectively implemented. States outside these control regimes should give consideration to aligning their national control regimes to them and/or seeking to introduce similar measures in their relevant regional organizations.

In the long term, consideration could be given to the development of an international instrument restricting and/or prohibiting the transfer of security equipment (potentially including ICA weapons, RCAs and related means of delivery) that could be utilized for torture, ill-treatment or the death penalty. Alternatively, the possibility of introducing measures to widen the scope of coverage of the Arms Trade Treaty to include relevant security equipment, could be explored. States could also seek to establish appropriate mechanisms to review introduction, monitoring and implementation of UN and regional arms embargoes and, where necessary, ensure that the scope of coverage of such embargoes clearly includes ICA weapons, RCAs and related means of delivery.

13.3.7. Civil society

Concerned civil society actors – particularly those in the life and chemical science communities – can continue fostering a "culture of responsibility" in science by developing, promulgating and applying codes of conduct, pledges and conventions regulating "dual-use" research, but focusing now upon initiatives explicitly combating the misapplication of such research by State weapons programmes. Complimentary activities include promoting the non-participation in and "whistle-blowing" on RCA and ICA weapons programmes of concern; and education and awareness-raising amongst scientific communities, State officials and the general public of the dangers of such development and use.

Individual life and chemical scientists and professional associations, as well as NGOs informed by such expertise, can play important constructive roles in highlighting existing limitations in the BTWC, CWC and attendant control regimes with regard to ICA weapons, RCAs and related means of delivery, and developing and promoting possible science-informed policy responses; undertaking societal monitoring and verification of existing implementation by States Parties of relevant instruments and highlighting R&D activities of concern; predicting research trajectories in relevant scientific disciplines and warning of potential future threats.

In certain CWC States Parties, there are limited mechanisms facilitating Government engagement with a narrow range of academics, NGOs and scientific bodies perceived to have relevant expertise in CBW arms control and disarmament. Where these exist they provide important opportunities for at least some civil society actors to bring concrete proposals to address ICA weapons, RCAs and their means of delivery to the attention of relevant State officials.

In order to foster more informed discussions and facilitate appropriate policy development, in relevant national and OPCW forums, efforts can be made through coordinating bodies such as the Chemical Weapons Convention Coalition (CCWC) and by the OPCW itself, to interact more fully with a far wider range of civil society actors, including human rights, IHL and humanitarian aid organizations, as well as those local communities, pro-democracy and civil liberties movements that have suffered, are at risk from, or are concerned about the misuse of RCAs or ICA weapons. In addition, it would be beneficial if these civil society organizations and communities now sought to engage with the OPCW on these issues, for example by attending CSPs and Review Conferences.

One community of particular relevance and importance are health professionals. There are a range of declarations and regulations adopted by the World Medical Association (WMA) that guide health professionals in situations of conflict and unrest; forbid physician involvement in development of CBWs, and prohibit their participation in torture, ill-treatment and other forms of human rights abuse. However, there are no widely accepted guidelines specifically determining the permissibility or non-permissibility of medical involvement in the development of ICA weapons and RCAs intended for law enforcement or certain military operations. Given the potential importance of medical participation to the development, testing and utilization of such weapons, the introduction of clear guidance by national medical associations, and subsequently by the WMA, constraining the involvement of health professionals in such activities is needed. In addition, health professionals working together with the broader medical–scientific community can undertake and bring to the attention of relevant control regimes and the public the results of peer-reviewed studies on the short-term and long-term health consequences of the use and misuse of these agents.

Notes

1 Introduction

1. Üzümcü, A., Director-General, Organisation for the Prohibition of Chemical Weapons (OPCW), 2013, Working Together for a World Free of Chemical Weapons, and Beyond, Nobel Peace Prize Lecture, 10 December 2013.
2. See for example: Altmann, J. Preventive Arms Control: Concept and Design, in Altmann, J. (ed.), *Military Nanotechnology: Potential Applications and Preventive Arms Control*, Routledge, London, 2006.
3. See for example: McLeish, C. and Rappert, B. (eds), *A Web of Prevention: Biological Weapons, Life Sciences, and the Governance of Research*, Earthscan, London, 2007.

2 Incapacitating Chemical Agent Weapons

1. Dando, M. Scientific Outlook for the Development of Incapacitants, in Pearson, A., Chevrier, M. and Wheelis, M. (eds), *Incapacitating Biochemical Weapons*, Lanham, MD: Lexington Books, 2007, p. 125.
2. Aas, P. The Threat of Mid-Spectrum Chemical Warfare Agents, *Prehospital and Disaster Medicine*, volume 18, number 4, 2003, pp. 306–312.
3. Dando, M. (2007) *op.cit.*, pp. 125–126.
4. Davison, N. "Off the Rocker" and "On the Floor": The Continued Development of Biochemical Incapacitating Weapons, *Bradford Science and Technology Report No. 8*, Bradford Disarmament Research Centre, August 2007, pp. 2–4; Davison, N. *"Non-Lethal" Weapons*, Basingstoke, UK: Palgrave Macmillan, 2009.
5. Hemsley, J. *The Soviet Biochemical Threat to NATO*, RUSI Defence Studies, Palgrave Macmillan, 1987, p. 10.
6. Pearson, G. Relevant Scientific and Technological Developments for the First CWC Review Conference: The BTWC Review Conference Experience, *CWC Review Conference Paper No. 1*. Department of Peace Studies, University of Bradford. August 2002.
7. For further discussion of this term, see Aas, P. (2003) *op.cit.*
8. For example, see US Department of Defense, *Department of Defense Dictionary of Military and Associated Terms*, Joint Publication 1–02, 8th November 2010, as amended through 15th February 2012, p. 155.
9. For example, *NATO Glossary of Terms and Definitions (English and French)*, NATO document AAP-6(2012), 2012, p. 2-I-2.
10. Table modified from Pearson, G. (2002) *op.cit.*, p. 5.
11. Spiez Laboratory, *Technical Workshop on Incapacitating Chemical Agents*, Spiez, Switzerland, 8–9 September 2011, 2012, p. 10; see also International Committee of the Red Cross, *Toxic Chemicals as Weapons for Law Enforcement, A Threat to Life and International Law?*, ICRC, Geneva, September 2012, p. 2.; The Royal Society, *Brain Waves Module 3: Neuroscience, Conflict and Security*, 2012, pp. 44–45.

12. Royal Society, Brain Waves Module 3: *Neuroscience, Conflict and Security*, RS Policy document 06/11, February 2012, pp. 44–45; see also: Royal Society, The Chemical Weapons Convention and convergent trends in science and technology, RS seminar in OPCW, 18 February 2013, p. 2.
13. Klotz, L., Furmanski, M. and Wheelis, M. Beware the Siren's Song: Why "Non-Lethal" Incapacitating Chemical Agents are Lethal. Federation of American Scientists, 2003.
14. See for example: OPCW, Conference of the States Parties, *Report of the Scientific Advisory Board on Developments in Science and Technology for the Third Special Session of the Conference of the States Parties to Review the Operation of the Chemical Weapons Convention*, Third Review Conference RC-3/DG.1, 8th–19th April 2013, 29th October 2012; Royal Society (February 2012) *op.cit.*, p. 44. See also Royal Society (18th February 2013) *op.cit.*, p. 2; Klotz, L., Furmanski, M. and Wheelis, M. (2003) *op.cit.*
15. Lakoski, J., Murray, W. and Kenny, J. *The Advantages and Limitations of Calmatives for Use as a Non-lethal Technique*, College of Medicine Applied Research Laboratory, Pennsylvania State University, 3rd October 2000.
16. Table modified from: Lakoski, J., Murray, W. and Kenny, J. (2000) *op.cit.*, pp. 15–16 and subsequent discussion in study: pp. 16–47.
17. See Aas, P. (2003) *op.cit.*, p. 309.
18. See, for example, Fenton, G., Current and Prospective Military and Law Enforcement Use of Chemical Agents for Incapacitation, in Pearson, A., Chevrier, M. and Wheelis, M. (eds), (2007) *op.cit.*, pp. 103–123; Whitbred, G. *Offensive Use of Chemical Technologies by US Special Operations Forces in the Global War on Terrorism*, Maxwell Paper Number 37, Maxwell Air Force Base, Alabama: Air University Press, 2006. It should be noted that other authors have questioned the utility of ICA weapons in certain proposed scenarios such as premeditated hostage situations, due to the availability of counter-measures. See Wheelis, M., in Pearson, A., Chevrier, M. and Wheelis, M. (eds), (2007) *op.cit.*, p. 6.
19. Ekeus panel, *Report of the Advisory Panel on Future Priorities of the Organisation for the Prohibition of Chemical Weapons*, OPCW Director General, S/951/2011, 25th July 2011, paragraph 13.
20. Perry Robinson, J. *Non-lethal Warfare and the Chemical Weapons Convention, Further Harvard Sussex Program submission to the OPCW Open-Ended Working Group on Preparations for the Second CWC Review Conference*, October 2007, available at http://www.sussex.ac.uk/Units/spru/hsp/Papers/421rev3.pdf (accessed 31st July 2009), p. 32.
21. Perry Robinson, J. Difficulties facing the Chemical Weapons Convention, *International Affairs*, volume 84, number 2, March 2008, pp. 223–239, at p. 238.
22. Perry Robinson, J. (March 2008) *op.cit.*, p. 238; Perry Robinson, J. (October 2007) *op.cit.*, p. 32; Perry Robinson, J., correspondence with the author, 13th April 2008.
23. Mogl, S. speaker's summary, session 5, International Committee of the Red Cross (ICRC), Expert meeting, Incapacitating chemical agents, implications for international law, Montreux, Switzerland, 24th–26th March 2010, p. 62.
24. Pearson, A. Incapacitating Biochemical Weapons: Science, Technology, and Policy for the 21st Century, *Nonproliferation Review*, volume 13, number 2, July 2006, p. 172.
25. ICRC (September 2012) *op.cit.*, p. 4.
26. *Ibid.*
27. ICRC (September 2012) *op.cit.*, p. 4.

28. *Ibid.*
29. See for example: Wheelis, M. and Dando, M. Neurobiology: A Case Study of the Imminent Militarization of Biology, *International Review of the Red Cross*, volume 87, number 859, September 2005, p. 564; Pearson, A (2006) *op.cit.*, p. 169.
30. See for example: Pearson, A. (2006) *op.cit.*, p. 187, footnote 137; Perry Robinson, J. (March 2008) *op.cit.* Also, see Chapter 3 of this publication for a discussion of RCA use by private military and security companies.
31. Koplow, D. The Russians and the Chechens in Moscow in 2002, in *Non-lethal Weapons: The Law and Policy of Revolutionary Technologies for the Military and Law Enforcement*, Cambridge University Press, 10th April 2006, pp. 100–113.
32. Crowley, M. The Use of Incapacitants in Law Enforcement, in Casey-Maslen, S. (ed.), *Weapons Under International Human Rights Law*, Cambridge University Press, 2014, pp. 369–374.
33. See for example: Amnesty International, *Death Sentences and Executions 2013*, ACT 50/001/2014, March 2014.
34. Wheelis, M. and Dando M. (2005) *op.cit.*, pp. 553–571.
35. Perry Robinson, J. (October 2007) *op.cit.*, p. 32.
36. Dando, M. Biologists Napping while Work Militarized, *Nature*, volume 460, number 7258, 20th August 2009, p. 951. Dando's concerns were echoed in an accompanying *Nature* editorial entitled: "A Question of Control: Scientists Must Address the Ethics of using Neuroactive Compounds to Quash Domestic Crises." *Nature* (20th August 2009) *op.cit.*, p. 933.
37. ICRC (September 2012) *op.cit.*, pp. 4–5.
38. British Medical Association Board of Science and BMA Science & Education department, *The Use of Drugs as Weapons: The Concerns and Responsibilities of Healthcare Professionals*, London: BMA, May 2007, p. 1.
39. Perry Robinson, J. (October 2007) *op.cit.*, p. 31.
40. For descriptions of the incident see, e.g., Amnesty International, Amnesty International 2003 Annual Report, London, 2003, entry for the Russian Federation, p. 208; Amnesty International, Rough Justice: The Law and Human Rights in the Russian Federation, AI Index EUR 46/054/2003, October 2003; Koplow, D. (2006) *op.cit.*; Pearson, A., Chevrier, M. and Wheelis, M. (eds), (2007) *op.cit.*; see also BBC news coverage, including: How Special Forces Ended Siege, 29th October 2002, and BBC 2, Horizon: The Moscow Theatre Siege (broadcast 15th January 2004), transcript on: http://www.bbc.co.uk/science/horizon/2004/moscowtheatretrans.shtml (accessed 1st April 2015).
41. The Spetsnaz "Alpha Team" that conducted the assault was a hybrid commando unit of the Federal Security Service (FSB), according to BBC News, "Spetsnaz: Russia's Elite Force", 28 October 2002. This 1,500–2,000-strong anti-terrorist unit had seen extensive action in Afghanistan and Chechnya. As cited in Koplow, D. (2006) *op.cit.*
42. Dunlop, J. B. *The 2002 Dubrokvka and 2004 Beslan Hostage Crises, a Critique of Russian Counter-Terrorism, Soviet and Post-Soviet Politics and Society*, Verlad, Stuttgart, 2006, pp. 145–146.
43. Wheelis, M. Human Impact of Incapacitating Chemical Agents in: ICRC, Expert Meeting: Incapacitating Chemical Agents, Implications for International Law, Montreux, Switzerland, 24–26 March 2010, October 2010; Levin, D. and Selivanov, V. Medical and Biological Issues of NLW Development and Application, Proceedings of the Fifth European Symposium on Non-Lethal Weapons,

11th–13th May 2009, Ettlingen, Germany, European Working Group on Non-Lethal Weapons, V23, p. 7; Four Years Later, Moscow Hostages Suffering, CBS Evening News, 21st October 2006.
44. See for example: Human Rights Watch, press release: Independent Commission of Inquiry Must Investigate Raid on Moscow Theater: Inadequate Protection for Consequences of Gas Violates Obligation to Protect Life, 30th October 2002, Human Rights Watch.
45. ITAR-TASS, from Moscow in English, 2112 hrs GMT 30th October 2002, as in FBIS-SOV-2002-1030, "Russian Experts Discuss Use of Fentanyl in Hostage Crisis", as cited by Perry Robinson, J Disabling Chemical Weapons, A Documented Chronology of Events, 1945–2011, 20th November 2012, (limited distribution, copy circulated to the author), Reference 021026.
46. Alison, S. [from Moscow for Reuters], 1257 hrs ET 30th October 2002, "Russian Confirms Siege Gas Based on Opiate Fentanyl", as cited in Perry Robinson, J. (20th November 2012) *op.cit.*, Reference 021026.
47. Amnesty International (October 2003) *op.cit.*, p. 53.
48. Riches, J., Read, R., Black, R., Cooper, N. and Timperley, C. Analysis of Clothing and Urine from Moscow Theatre Siege Casualties Reveals Carfentanil and Remifentanil Use, *Journal of Analytical Toxicology*, volume 36, 2012, pp. 647–656.
49. Wheelis, M. Feasibility of "Incapacitating Chemical Agents", International Committee of the Red Cross, Expert Meeting: Incapacitating Chemical Agents, Implications for International Law, Montreux, Switzerland, 24th–26th March 2010, p. 22.
50. Interview with Professor Mark Wheelis, by email, correspondence dated 13th July 2010.
51. Interview with Professor Mark Wheelis, by email, correspondence dated 13th July 2010.
52. Whilst certain national, pluri-lateral and civil society bodies have defined non-lethal or less lethal weapons, there is currently no internationally agreed definition of such weapons. For some examples of existing definitions, see: UK College of Policing, Use of force, firearms and less lethal weapons, 23 October 2013, available at: https://www.app.college.police.uk/app-content/armed-policing/use-of-force-firearms-and-less-lethal-weapons/ (accessed 2 October 2015); NATO, Research and Technology Organisation (RTO) Technical Report TR-SAS-060 Non-Lethal Weapons Effectiveness Assessment Development and Verification Study, October 2009, available at: http://ftp.rta.nato.int/public//Pub FullText/RTO/TR/RTO-TR-SAS-060///$$TR-SAS-060-ALL.pdf (accessed 2 October 2015), p.1–1; Amnesty International, Use of Force, guidelines for implementation of the UN Basic Principles on the Use of Force and Firearms by Law Enforcement Officials, August 2015, available at https://anistia.org.br/wp-content/uploads/2015/09/Amnesty_USE-OF-FORCE_Final_web.pdf (accessed 2 October 2015) pp. 23, 132–3.

In recognition of the continuing contested discourse over the nature, scope and application of the terms non-lethal or less lethal weapon, the term "less lethal" will be placed in quotation marks and used by the author during this publication – unless quoted directly from specific individuals or organizations cited in the text.
53. Klotz, L., Furmanski, M. and Wheelis, M. Beware the Siren's Song: Why "Non-Lethal" Incapacitating Chemical Agents are Lethal. Federation of American Scientists, 2003.
54. Klotz, L., Furmanski, M. and Wheelis, M. (2003) *op.cit.*, p. 7.

55. Pearson, A. Late and Post-Cold War Research and Development of Incapacitating Biochemical Weapons, in Pearson, A., Chevrier, M. and Wheelis, M. (2007) *op.cit.*, p. 70. Furthermore, Pearson has noted that "... unavoidable differences in exposure time and agent distribution after an agent is disseminated in the field make the uniform delivery of precisely controlled doses of incapacitating agents nearly impossible. This only encourages users to deliver more agent than needed to incapacitate most individuals in order to compensate for those individuals who inevitably would not receive enough. This problem is complicated even more by the need for rapid incapacitation in most scenarios, as this requires the delivery of higher doses..." Pearson, A. (2007) *op.cit.*, p. 70.
56. British Medical Association (2007) *op.cit.*, p. 1.
57. Royal Society (2012) *op.cit.*, p. iv. See also Spiez Laboratory (2012) *op.cit.*
58. Pearson, A. Could Incapacitating Weapons Become "Everyday" Weapons? 18th March 2008, Round Table On: The Expanding Range of Biowarfare Threats, *Bulletin of the Atomic Scientists*.
59. *Ibid.*
60. For further information concerning historical State ICA weapons R&D activities see Crowley, M. and Dando, M. *Down the Slippery Slope? A Study of Contemporary Dual-Use Research Potentially Applicable to Incapacitating Agent Weapons*, Biochemical Security 2030 project/Bradford Non-lethal Weapons Research Project, October 2014; Perry Robinson, J., Incapacitating Chemical Agents in Context: An Historical Overview of States' Policy, pp. 89–96 in: ICRC 2012 expert meeting report (January 2013) *op.cit.*; Royal Society (2012) *op.cit.*, pp. 10–13; Crowley, M. (2009) *op.cit.*; Dando, M. and Furmanski, M. Midspectrum Incapacitant Programs, in Wheelis, M., Rózsa, L. and Dando, M. (eds), *Deadly Cultures: Biological Weapons since 1945*, 2006.
61. Ketchum, J. S. and Sidell, F. R. Incapacitating Agents, pp. 287–305, in Sidell, F. R., Takafuji, E. T. and Franz, D. R. (eds), *Military Aspects of Chemical and Biological Warfare*. Office of the Surgeon General, US Army, Washington DC, 1997, p. 291.
62. Ketchum, J. S. and Sidell, F. R. (1997) *op.cit.*, p. 294.
63. Ketchum, J. S. *Chemical Warfare: Secrets Almost Forgotten*. Private publication, United States. ISBN: 1-4243-0080-0., 2006.
64. Furmanksi, M. Historical Military Interest in Low-lethality Biochemical Agents, in Pearson, A., Chevrier, M. and Wheelis, M. (eds) (2007) *op.cit.*, p. 54.
65. HQ U.S.M.A.V., Command History 1964, volume 35, p. 133, as cited in: Furmanksi, M. (2007) *op.cit.*, p. 54.
66. Perry Robinson has documented a number of alleged but unconfirmed reports of BZ employment by US military forces in the Vietnam War. See Perry Robinson, J. (20th November 2012) *op.cit.*, entry 660314.
67. For a discussion of the involvement of US pharmaceutical and chemical companies, and research institutes in the US military ICA weapons development programme at this time, see Perry Robinson, J. (20th November 2012) *op.cit.*, entries 611100, 660600, 670600, 680600, 690800.
68. Further discussion see Perry Robinson, J. (20th November 2012) *op.cit.*
69. Perry Robinson, J. (20th November 2012) *op.cit.*, chronology reference: 761000, p. 85.
70. Dando, M. and Furmanski, M. (2006) *op.cit.*, p. 250.
71. Perry Robinson, J. Bringing the CBW Conventions Closer Together, *CBW Conventions Bulletin*, issue 80, September 2008, p. 3.
72. Ulrich, R., Wilhelmsen, C. and Krakauer, T. Staphyloccocal Enterotoxin B and Related Toxins, in: *Medical Aspects of Biological Warfare*, Textbooks of Military

Medicine, 2007, Office of the US Surgeon General, Department of the Army, pp. 311–322.
73. Ulrich, R., Wilhelmsen, C. and Krakauer, T. (2007) *op.cit.*, p. 312.
74. Perry Robinson, J. (September 2008) *op.cit.*, p. 3.
75. Guillemin, J. *Biological Weapons: From the Invention of State-Sponsored Programs to Contemporary Bioterrorism*, Columbia University Press, 2005, pp. 122–127.
76. Royal Society *Brain Waves Module 3* (February 2012) *op.cit.*, pp. 46–50.
77. *Ibid.*, p. 46.
78. Balali-Mood, M., Steyn, P., Sydnes, L., Trapp, R. International Union of Pure and Applied Chemistry (IUPAC), *Impact of Scientific Developments on the Chemical Weapons Convention (IUPAC Technical Report)*, January 2008, p. 185.
79. National Research Council, *Emerging Cognitive Neuroscience and Related Technologies*, 2008, available at http://www.nap.edu/openbook.php?record_id=12177 (accessed 30th July 2009).
80. National Research Council (2008) *op.cit.*, p. 136.
81. National Research Council (2008) *op.cit.*, p. 138.
82. International Committee of the Red Cross, Expert Meeting: Incapacitating Chemical Agents, Implications for International Law, Montreux, Switzerland, 24th–26th March 2010, p. 3.
83. Interview with senior US State Department official, 27th May 2010. [Interviewee was speaking in his private capacity].
84. Interview with senior US State Department official, 27th May 2010. [Interviewee was speaking in his private capacity].
85. Interview with senior US State Department official, 27th May 2010. [Interviewee was speaking in his private capacity].
86. Dual use is a concept that can be applied to the tangible and intangible features of a technology that enable it to be utilized for both hostile and peaceful ends with no, or only minor, modifications. The hostile use of a specific technology does not arise automatically from the inherent properties of that technology, but requires the active intervention of relevant actors. [For further discussion see Molas-Gallart, J. and Perry Robinson, J. "Assessment of Dual-use Technologies in the Context of European Security and Defence", Report for the Scientific and Technological Options Assessment (STOA), European Parliament, 1997; McLeash, C. "Reflecting on the Dual-Use Problem," in Rappert, B. and McLeish, C. (eds), *A Web of Prevention: Biological Weapons, Life Sciences, and the Governance of Research*, 2007, Routledge, UK.]
87. For further information about the survey methodology and the survey findings incorporating detailed country case studies see Crowley, M. and Dando, M. "Down the Slippery Slope? A Study of Contemporary Dual-Use Research Potentially Applicable to Incapacitating Agent Weapons," Biochemical Security 2030 Project/Bradford Non-lethal Weapons Research Project, October 2014.
88. Security, Anti-Riot Weapons and Ammunition Brochure, China North Industries Corporation (NORINCO), undated, brochure distributed at MILIPOL security exhibition, Paris, 1995 (copy on file with the Omega Research Foundation), p. 11.
89. *Jane's Police and Security Equipment 1995–1996*, ed. Hogg, I., Janes Information Group Limited, Coulsden, Surrey, 1996, p. 306.
90. NORINCO brochure (undated) *op.cit.* [distributed at MILIPOL 1995], p. 11.
91. State 9619 Plant company brochure, undated, distributed at Asia Pacific China Police Expo 2004, [23rd–26th June 2004, Beijing Exhibition Centre, Beijing,

China], and at Asia Pacific China Police Expo 2006 [24th–27th May 2006, Beijing Exhibition Centre, Beijing, China] (copies of both brochures on file with the Omega Research Foundation).
92. State 9616 Plant company brochure (undated) *op.cit.*, p. 9.
93. State 9616 Plant company brochure (undated) *op.cit.*, p. 9.
94. State 9616 Plant company brochure (undated) *op.cit.*, p. 9.
95. Arthur, G. New Equipment in Hong Kong, *Defence Review Asia*, 19th March 2012, p. 35.
96. Email correspondence to Dr M. Crowley, BNWLRP, from Mr G. Arthur, 16th August 2014. For a photograph showing the display of the BBQ-901 tranquiliser gun at a PLA event in May 2011, see Crowley, M. and Dando, M. (October 2014) *op.cit.* p.18.
97. Guo Ji-Wei and Xue-Sen Yang, Ultramicro, Nonlethal and Reversible: Looking Ahead to Military Biotechnology, *Military Review*, July–August 2005 p. 75.
98. Guo Ji-Wei and Xue-sen Yang (2005) *op.cit.*, p. 75.
99. Qi, L., Cheng, Z., Zuo, G. Li, S. and Fan, Q., Oxidative Degradation of Fentanyl in Aqueous Solutions of Peroxides and Hypochlorites, *Defence Science Journal*, volume 61, number 1, January 2011, pp. 30–35.
100. Qi, L. et al. (2011) *op.cit.*, p. 30.
101. See Czech Republic Council for Research, Development and Innovation, available at http://www.isvav.cz/h12/resultDetail.do?rowId=RIV%2F60162694%3AG 16__%2F02%3A00000625%21RIV%2F2003%2FMO0%2FG16003%2FN (accessed 20th August 2014).
102. Speakers' biographies, Jane's Less-Lethal Weapons 2005 Conference, 26th–27th October 2005, Royal Armouries Museum, Leeds, UK (Copy held by the authors).
103. *Ibid.*
104. Hess, L., Schreiberova, J. and Fusek, J. Pharmacological Non-Lethal Weapons, Proceedings of the 3rd European Symposium on Non-Lethal Weapons, 10th–12th May 2005, Ettlingen, Germany, European Working Group on Non-Lethal Weapons, Pfinztal: Fraunhofer ICT, V23.
105. Hess, L. et al. (10th–12th May 2005) *op.cit.*, pp. 4–8.
106. Hess, L. et al. (10th–12th May 2005) *op.cit.*, pp. 8–9.
107. Hess, L. et al. (10th–12th May 2005) *op.cit.*, p. 12.
108. Hess, L. et al. (10th–12th May 2005) *op.cit.*, pp. 11–12.
109. Hess, L. et al. (10th–12th May 2005) *op.cit.*, pp. 10–14.
110. Hess, L. et al. (10th–12th May 2005) *op.cit.*, p. 14.
111. Hess, L., Schreiberova, J. and Fusek, J. Ultrapotent Opioids as Non-Lethal Weapons paper given at: Meeting of NATO RTO TG-004, 23rd–26th May 2005, University of Defence, Faculty of Military Health Sciences, Hradec Kralove, Czech Republic.
112. Hess, L. et al. (23rd–26th May 2005) *op.cit.*, p. 1.
113. Hess, L. et al. (23rd–26th May 2005) *op.cit.*, p. 1.
114. Hess, L. et al. (23rd–26th May 2005) *op.cit.*, p. 4.
115. Hess, L., Schreiberova, J. and Fusek, J., Pharmacological Non-Lethal Weapons, Jane's Less-Lethal Weapons 2005 Conference, 26th–27th October 2005, Royal Armouries Museum, Leeds, UK.
116. Hess, L., Schreiberova, J. and Fusek, J. (October 2005) *op.cit.*
117. Hess, L., Schreiberova, J. and Fusek, J. Pharmacological Non-Lethal Weapons, [PowerPoint presentation] (October 2006) *op.cit.*, slide 32.

118. Hess, L., Schreiberová, J., Málek, J., Votava, M., Fusek and J. Drug-Induced Loss of Aggressiveness in the Macaque Rhesus, Proceedings of 4th European Symposium on Non-Lethal Weapons, 21st–23rd May 2007, Ettlingen, Germany, European Working Group on Non-Lethal Weapons, Pfinztal: Fraunhofer ICT, V15, p. 6.
119. Hess, L. et al. (2007) *op.cit.*, p. 7.
120. Hess, L. et al. (2007) *op.cit.*, p. 7.
121. Hess, L., Votava, M., Schreiberová, J., Málek, J. and Horáček, M. Experience with a Naphthylmedetomidine – Ketamine – Hyaluronidase Combination in Inducing Immobilization in Anthropoid Apes, *Journal of Medical Primatology*, volume 39, number 3, June 2010, pp. 151–159.
122. Votava, M., Hess, L., Schreiberová, J., Málek, J. and Štein, K. Short Term Pharmacological Immobilization in Macaque Monkeys, *Veterinary Anaesthesia and Analgesia*, volume 38, issue 5, September 2011, pp. 490–493.
123. Hess, L., Votava, M., Slíva, J., Málek, J., Kurzová, A. and Štein, K. Ephedrine Accelerates Psychomotor Recovery from Anesthesia in Macaque Monkeys, *Journal of Medical Primatology*, volume 41, issue 4, August 2012, pp. 251–255.
124. Czech Republic, National Authority of the CWC, Reply to the University of Bradford, Re: Request for information concerning research potentially related to incapacitating chemical agents, 14th July 2014, p. 6.
125. Czech Republic (14th July 2014) *op.cit.*, p. 4.
126. Czech Republic (14th July 2014) *op.cit.*, p. 5.
127. Czech Republic (14th July 2014) *op.cit.*, p. 6.
128. Czech Republic (14th July 2014) *op.cit.*, p. 5.
129. Gupta, P. K., Ganesan, K., Pande, A. and Malhotra, R.C. A Convenient One-Pot Synthesis of Fentanyl, *Journal of Chemical Research*, July 2005, pp. 452–453.
130. Gupta P. K. et al. (July 2005) *op.cit.*, p. 452.
131. Gupta P. K. et al. (July 2005) *op.cit.*, p. 452.
132. Gupta, P. K., Ganesan, K., Gutch, P. K., Manral, L. and Dubey, D. K. Vapor Pressure and Enthalpy of Vaporization of Fentanyl, *Journal of Chemical & Engineering Data*, volume 53, number 3, 2008, pp. 841–845.
133. Gupta, P. K. et al. (2008) *op.cit.*, p. 844.
134. Manral, L., Gupta, P. K., Suryanarayana, M. V. S., Ganesan, K. and Malhotra, R. C. Thermal Behaviour of Fentanyl and Its Analogues during Flash Pyrolysis, *Journal of Thermal Analysis and Calorimetry*, May 2009, volume 96, issue 2, pp. 531–534.
135. Manral, L. et al. (May 2009) *op.cit.*, p. 531.
136. Manral, L, Muniappan, N., Gupta, P. K., Ganesan, K., Malhotra, R. C. and Vijayaraghavan, R. Effect of Exposure to Fentanyl Aerosol in Mice on Breathing Pattern and Respiratory Variables, *Drug and Chemical Toxicology*, volume 32, issue 2, 2009, pp. 108–113.
137. Manral, L. et al. (2009) *op.cit.*, p. 109.
138. Manral, L. et al. (2009) *op.cit.*, p. 112.
139. Concentration of a chemical that depressed 50% of respiratory frequency in the test animals.
140. Concentration of a chemical that kills 50% of the test animals during the observation period.
141. Manral, L. et al. (2009) *op.cit.*, p. 112.
142. Yadav, P., Chauhan, J. S. Ganesan, K., Gupta, P. K., Chauhan, D. and Gokulan, P. D. Synthetic Methodology and Structure Activity Relationship Study of

N-[1-(2-Phenylethyl)-Piperidin-4-yl]-Propionamides, *Pelagia Research Library Der Pharmacia Sinica*, volume 1, issue 3, 2010, pp. 126–139.
143. Gupta, P. K., Yadav, S. K., Bhutia, Y. D., Singh, P., Rao, P. Gujar, N. L., Ganesan, K. and Bhattacharya, R. Synthesis and Comparative Bioefficacy of N-(1-Phenethyl-4-Piperidinyl) Propionanilide (Fentanyl) and Its 1-Substituted Analogs in Swiss Albino Mice, *Medicinal Chemistry Research*, volume 22, issue 8, August 2013, pp. 3888–3896.
144. Gupta, P. K. et al. (2013) *op.cit.*, p. 3889.
145. Website of the Defence Research & Development Organisation, Ministry of Defence, Government of India, available at http://www.drdo.gov.in/drdo/English/index.jsp?pg=homebody.jsp (accessed 1st April 2015).
146. *Ibid.*
147. Jain, A. K., Gupta, P. K., Ganesan, K., Pande, A. and Malhotra, R. C. Rapid Solvent-Free Synthesis of Aromatic Hydrazides Under Microwave Irradiation, *Defence Science Journal*, volume 57, number 2, March 2007, pp. 267–270.
148. *DRDO Newsletter*, volume 30, number 4, April 2010, Defence Research & Development Organisation, available at http://drdo.gov.in/drdo/pub/newsletter/2010/apr_10.pdf (accessed 1st April 2015).
149. *DRDO Newsletter* (April 2010) *op.cit.*, p. 12.
150. Correspondence to Dr M. Crowley, BNLWRP, from R. K. Singh, Deputy Chief of Mission, Embassy of India, The Hague, forwarding the response of the Indian CWC National Authority, 22nd July 2014.
151. Indian CWC National Authority (22nd July 2014) *op.cit.*
152. Indian CWC National Authority (22nd July 2014) *op.cit.*
153. See OPCW, Chemical Weapons Convention (1993) *op.cit.*, Article X(4) for relevant State reporting requirements.
154. See for example: Nezamoleslam, T., Javahery, B., Shakiba, N. and Fakhraian, H. Structure-Activity Relationship, Atomic Electron Density and Conformational Investigation of Fentanyl Analogues, *Journal of Passive Defence Science &Technology*, volume 1, 2010, pp. 23–32; Kamranpey, H. Aerosolisation of Medetomidine Hydrochloride as an Incapacitating Agent, *Journal of Passive Defence Science & Technology*, volume 3, 2011, pp. 51–6; Abazari, M. S. Investigating the Phase Behavior of Medetomidine Hydrochloride, Ketamine Hydrochloride and Sevoflurane in the Presence of Ethanol and Propellant, *Journal of Passive Defence Science & Technology* 1, 2013, pp. 65–70. For further discussion see Crowley, M. and Dando, M. (2014) *op.cit.*, pp. 34–38.
155. Nuclear Threat Initiative, Imam Hussein University (IHU), http://www.nti.org/facilities/251/ (accessed 25th June 2015); Imam Hussein University, available at https://en.wikipedia.org/wiki/Imam_ Hossein_University (accessed 25th June 2015). For more information see Imam Hossein Comprehensive University website, available at http://www.ihu.ac.ir/?q=fa/node/1 (accessed 25th June 2015).
156. Nuclear Threat Initiative, Imam Hussein University (IHU), available at http://www.nti.org/facilities/251/ (accessed 25th June 2015). See also Commander: Enemies Trying to Downplay Persian Gulf's Importance, FARS News Agency, 2012, available at http://www.highbeam.com (accessed 20th August 2014); Zarifmanesh: Universities are front line in fight against "Global Arrogance", Sepah News, 30th January 2013, as cited in: Lucas, S. and Paraszczuk, P. The Resistance Economy, in: L'économie réelle de l'Iran: Au-delà des chiffres (ed.), Makinsky, M., Editions L'Harmattan, Paris, 2014, also cited in: Iran Military

News, available at http://iranmilitarynews.org/tag/ brigadiergeneral/ (accessed 9 th May 2014).
157. Correspondence to Dr M. Crowley, BNLWRP, from Dr H. Farajvand, Secretary of the National Authority for the CWC, Ministry of Foreign Affairs of the Islamic Republic of Iran, 15 th July 2014
158. *Ibid.*
159. Knip, K. Biologie in Ness-Ziona, *NRC Handelsband*, 27th February 1999, available at http://retro.nrc.nl/W2/Lab/Ziona/inhoud.html (accessed 19th June 2014). Knip's research report is in Dutch. A brief overview of his findings is contained in Cohen, A. Israel and Chemical/Biological Weapons: History, Deterrence, and Arms Control, *The Nonproliferation Review*, Fall–Winter 2001, pp. 38–39.
160. For details of the original papers see Crowley, M. and Dando, M. (2014) *op.cit.*, p. 39.
161. Normack, M., Lindblad, A., Norqvist, A., Sandstrom B. and Waldenstrom, L. Israel and WMD: Incentives and Capabilities, Swedish Defence Research Agency (FOI), December 2005, p. 41.
162. Normack, M. et al. (December 2005) *op.cit.*, p. 41.
163. See Harvard Sussex Programme, News Chronology, CBW Conventions Bulletin, 38, December 1997, p. 29; Physician Member of Hit Team, Paper Says, *Canadian Medical Association Journal*, volume 157, number 11, December 1997, p. 1504; Beyer, L. Don't Try This at Home – Or in Aman, *Time*, volume 150, number 17, 27th October 1997, p. 27; Ginsburg, M. "Should There Be a Need": The Inside Story of Israel's Chemical and Biological Arsenal, *Times of Israel*, 17th September 2013; Cowell, A. The Daring Attack That Blew Up in Israel's Face, *New York Times*, 15th October 1997.
164. Israeli Intelligence Agencies Come Under Fire, *Jane's Intelligence Review*, 1st January 1998.
165. *Jane's Intelligence Review* (1st January 1998) *op.cit.*
166. Cowell, A. *New York Times* (15th October 1997) *op.cit.*
167. Cowell, A. *New York Times* (15th October 1997) *op.cit.*
168. Harvard Sussex Programme CBW Conventions Bulletin (December 1997) *op.cit.*, p. 29.
169. Amnesty International, Attempt to Kill Hamas Leader Follows a Pattern of Extrajudicial Killings, News Service 168/97, 8th October 1997, AI Index: MDE 15/89/97; Harvard Sussex Programme CBW Conventions Bulletin (December 1997) *op.cit.*, p. 29.
170. *Jane's Intelligence Review* (1st January 1998) *op.cit.*
171. Beyer, L. Hamad, J. and Klein, A. What Went Wrong? The Botched Hit on a Hamas Leader in Jordan Is the Latest Big Problem for Israel's Benjamin Netanyahu, *Time Magazine*, 27th October 1997, p. 52: As cited in: Normack, M. et al. (December 2005) *op.cit.*
172. For further discussion see: Crowley, M. and Dando, M. (2014) *op.cit.*, p. 41.
173. Klochikhin, V., Pirumov, V., Putilov, A. and Selivanov, V. The Complex Forecast of Perspectives of NLW for European Application. Proceedings of the 2nd European Symposium on Non-Lethal Weapons, Ettlingen, Germany, 13–14th May 2003, V16, Pfinztal: Fraunhofer ICT, p. 3.
174. *Ibid.*
175. Klochikhin, V., Lushnikov, A., Zagayonov, V., Putilov, A., Selivanov, V. and Zatekvakhin, M. Principles of Modelling of the Scenario of Calmative

Application in a Building with Deterred Hostages, Proceedings of the 3rd European Symposium on Non-Lethal Weapons, Ettlingen, Germany, 10–12th May 2005, V17, Pfinztal: Fraunhofer ICT, p. 3.
176. Klochikhin, V. et al. (2005) *op.cit.*, p. 3.
177. Klochikhin, V. et al. (2005) *op.cit.*, pp. 3–4.
178. Klochikhin, V. and Selivanov, V. Report on the 1st Phase of the Project "Gas Flow", Presentation in MBDA, 24th–27th November 2009, London. [Copy of presentation held by authors]. Further details of this London meeting are not available.
179. Klochikhin, V. and Selivanov, V. (2009) *op.cit.*, slide 77.
180. For details of relevant Russian scientific papers see Riches, J. et al. (2012) *op.cit.*
181. For details of the original papers see Crowley, M. and Dando, M. (2014) *op.cit.*, p. 49.
182. For further information see Walker, J., "Inappropriately Hilarious": An Historical Overview of the Interest In and Use of Incapacitating Chemical Agents, March 2010; Maclean, A. *Historical Survey of the Porton Down Volunteer Programme*, Ministry of Defence, June 2006; Dando, M. and Furmanski, M. Midspectrum Incapacitant Programs, in Wheelis, M., Rózsa, L. and Dando, M. (eds), *Deadly Cultures: Biological Weapons Since 1945*. Cambridge: Harvard University Press, 2006.
183. Three UK research papers on fentanyl and related analgesic chemicals produced by the Chemical Defence Establishment were cited in: Agent Research Studies: 1966–1990, US Army Armament Munitions Command, Chemical Research, Development & Engineering Center, Aberdeen Proving Ground MD, report CRDEC-TR-345, April 1992, declassified with redactions from CONFIDENTIAL.
184. Donnelly, T. Less Lethal Technologies: Initial Prioritization and Evaluation, UK Home Office, Policing and Crime Reduction Group, Police Scientific Development Branch, publication no. 12/01. For further discussion of this assessment process, see Crowley, M. and Dando, M. (2014) *op.cit.*, pp. 56–58.
185. Northern Ireland Office, Patten Report Recommendations 69 and 70 Relating to Public Order Equipment. A Research Programme into Alternative Policing Approaches Towards the Management of Conflict. Fourth Report prepared by the Steering Group led by the Northern Ireland Office, in consultation with the Association of Chief Police Officers. Belfast: Northern Ireland Office, January 2004. [For further analysis see: Crowley, M. and Dando, M. (2014) *op.cit.*, pp. 54–60].
186. OPCW, Conference of States Parties, United Kingdom: Statement by Mr Alistair Burt, Parliamentary Under Secretary of State for Foreign and Commonwealth Affairs, Third Review Conference, RC-3/NAT.22, 8th–19th April 2013, 9th April 2013.
187. Biomedical Sciences Department, CBD Sector [of Defence Evaluation and Research Agency] Porton Down, An Overview of Research Carried out on Glycollates and Related Compounds at CBD Porton Down, DERA/CBD/CR990418, September 1999; Riches, J. R., Read, R. W., Black, R. M., Cooper, N. J., Timperley, and C. M. Analysis of clothing and urine from Moscow theatre siege casualties reveals carfentanil and remifentanil use, *Journal of Analytical Toxicology*, volume 36, number 9, November 2012.
188. A copy of all the annual Article X declarations submitted by the UK government to the OPCW Technical Secretariat covering calendar years from 1997 to 2013

was provided to the author by the UK Government. The UK government released a copy of its first (1997) annual Article X declaration to the OPCW in a response to a Parliamentary Question, and deposited a hard copy in the House of Commons library [See UK Ministry of Defence, Monday 27th April 1998 response of Secretary of State for Defence, Dr J. Reid to Parliamentary Question by Mr R. Sedgemoor, Hansard, 26671]. Hard copies of subsequent UK annual Article X declarations have also been deposited in the House of Commons library.

189. United States Army, Topic CBD00-108, Chemical Immobilizing Agents for Non-Lethal Applications, Small Business Innovation Research Solicitation, CBD 00.1, December 1999.
190. United States Army, SBIRS CBD00-108 (December 1999) *op.cit.*
191. United States Army, SBIRS CBD00-108 (December 1999) *op.cit.*
192. United States Army, SBIRS CBD00-108 (December 1999) *op.cit.*
193. United States Army, SBIRS CBD00-108 (December 1999) *op.cit.*
194. United States Army, SBIRS CBD00-108 (December 1999) *op.cit.*
195. United States Army, SBIRS CBD00-108 (December 1999) *op.cit.*
196. United States Army, CBD, 26 Phase I Selections from the 00.1 Solicitation.
197. Ruppe, D. United States: US Military Studying Nonlethal Chemicals, *Global Security Newswire*, 4th November 2002.
198. Lakoski, J., Bosseau Murray, W. and Kenny, J. (2000) *op.cit.*, p. 2.
199. Lakoski, J., Bosseau Murray, W. and Kenny, J. (2000) *op.cit.*, pp. 15–45.
200. National Institute of Justice, Grant No. 2001-RD-CX-K002. Details from NIJ Research Portfolio available December 2006 at: http://nij.ncjrs.org/portfolio/ as cited in Davison, N. (2007) *op.cit.*, p. 24.
201. In February 2003, a presentation by the Senior Program Manager for the NIJ Less-Than-Lethal Technology Program indicated that the project had been reviewed by a liability panel and that work was progressing at PSU. Cecconi, J. (2003) Less-Than-Lethal Program. Presentation to the 2003 National Institute of Justice Annual Technology Conference, as cited in Davison, N. (2007) *op.cit.*, p. 24.
202. National Research Council, An Assessment of Non-Lethal Weapons Science and Technology, Committee for an Assessment of Non-Lethal Weapons Science and Technology, National Research Council, Division on Engineering and Physical Sciences, Naval Studies Board Washington, DC: National Academies Press, 2003.
203. National Research Council (2003) *op.cit.*, p. 107.
204. Advances in Bioscience for Airmen Performance BAA-09-02-RH, Air Force Office of Scientific Research (AFOSR), Department of Defense, original grants notice posted 1st October 2009. This, together with subsequent revisions, is available from https://www.fbo.gov/?s=opportunity&mode=form&tab=core&id=0b237485b3d66e02ad7e4b94588069e0&_cview=0 (accessed 1st April 2015).
205. The three other "technical mission areas" were: Applied Biotechnology – Goal is to develop and exploit advances in biotechnology and associated nanotechnologies to enhance performance and situational awareness of the force; Vulnerability Analysis – Goal is to rapidly identify human threat conditions, and sustain/expand airmen performance in stressful environments. It includes research in physical and physiological biosignatures, neuroscience, anthropometry, biomechanics, human modelling, database networking and data mining; Counterproliferation – Goal is to improve the Air Force's ability to locate, identify, track, target and destroy biological warfare agents (BWA) and other weapons of mass destruction (WMD), as well as anticipate and mitigate

WMD effects on Air Force operations. Air Force Office of Scientific Research (1st October 2009) *op.cit.*
206. Air Force Office of Scientific Research (1st October 2009) *op.cit.*
207. Air Force Office of Scientific Research (1st October 2009) *op.cit.*
208. A contract was awarded on 23rd August 2012 for advanced ammonium nitrate detection prototype development (FA8650-12-C-6270) and a second on 19th June 2013 to conduct research for technology integration (FA8650-13-C-6398).
209. Correspondence to Dr M. Crowley, BNLWRP, from Ms. Cynthia O. Smith, Department of Defense Spokeswoman, the Defense Press Office, Office of the Assistant Secretary of Defense (Public Affairs), United States of America, 11th September 2014.
210. See National Institute of Justice, Community Acceptance Panel – Riot Control Agents, 30th April 2007, available at http://www.nij.gov/topics/technology/less-lethal/pages/riot-control-agents.aspx (accessed 1st April 2015); Weiss, D. Calming Down: Could Sedative Drugs Be a Less-Lethal Option?, *NIJ Journal* number 261, 2007, pp. 42–46; Davison, N. Marketing New Chemical Weapons, *Bulletin of the Atomic Scientists*, 29th June 2009.
211. It is notable that the panel included the Director of the Joint Non-Lethal Weapons Directorate (JNLWD), the Riot Control Agents Program Manager from the US Army RDECOM-ARDEC, and the Associate Director of the Institute for Non-Lethal Defense Technologies, PSU, who had been one of the authors of the 2000 Pennsylvania State Report exploring the utility of a range of potential ICAs. Details of the panel's 26 members can be found at http://www.nij.gov/topics/technology/less-lethal/pages/riot-control-agents.aspx (accessed 1st April 2015).
212. It is interesting to note that the NIJ report of this meeting appeared to class "calmatives" as RCAs. It is unclear whether this was indicative of an official NIJ position on this issue.
213. National Institute of Justice (30th April 2007) *op.cit.*
214. Weiss, D. (2007) *op.cit.*, pp. 42–43.
215. National Institute of Justice, NIJ Awards in FY2007, Operationalizing Calmatives – Legal Issues, Concepts and Technologies, available at http://www.nij.gov/funding/awards/Pages/2007.aspx#less-lethaltechnologies (accessed 1st April 2015).
216. Weiss, D. (2007) *op.cit.*, p. 43.
217. See for example, LECTEC Annual Report 2010, September 2010, available at https://www.justnet.org/pdf/LECTAC-2010-Report.pdf (accessed 1st April 2015), p. iii.
218. LECTEC Annual Report 2010 (September 2010) *op.cit.*, p. iii.
219. LECTEC Annual Report 2010 (September 2010) *op.cit.*, p. 34.
220. LECTEC Annual Report 2010 (September 2010) *op.cit.*, p. 34.
221. LECTEC Annual Report 2010 (September 2010) *op.cit.*, p. 34.
222. According to the JUSTNET – the website of the National Law Enforcement and Corrections Technology Centre: "NIJ is currently working to re-define its research, development, test and evaluation (RDT&E) process, and the role that various working and advisory groups have in that process, to ensure that the process is cost effective, sustainable, and continues to meet the needs of public safety practitioners. As such, many of the Technology Working Groups and LECTAC are currently under review and are not actively meeting at this time. [Dated 2010]." See https://www.justnet.org/our_centers/fact_sheets/lectac.html (accessed 11th August 2014).

223. See for example: OPCW, Executive Council, Seventy-Second Session, United States of America: Statement by Ambassador Robert P. Mikulak, United States Delegation to the OPCW, at the Seventy-Second Session of the Executive Council, EC-72/NAT.8, 6th and 7th May 2013, 6th May 2013; OPCW, United States of America: Statement by H.E. Ambassador Robert P. Mikulak, Permanent Representative of the United States of America to the OPCW at the Seventy-Eighth Session of the Executive Council, EC-78/NAT.10, 17th March 2015.
224. OPCW, United States, Statement by Under Secretary Rose E Gottemoeller to the 19th Session of the Conference of the States Parties, Organisation for the Prohibition of Chemical Weapons, The Hague, Netherlands, 3rd December 2014.
225. See The Royal Society, *Brain Waves, Module 1: Neuroscience, Society and Policy*, January 2011; The Royal Society, *Brain Waves, Module 3* (2012) *op.cit.*
226. See for example, Andreasen, N. *Brave New Brain: Conquering Mental Illness in the Era of the Genome*, Oxford University Press, US, 2004.
227. Wheelis, M. and Dando, M. (2005) *op.cit.*, p. 10.
228. National Research Council, *Globalization, Biosecurity, and the Future of the Life Sciences*, 2006 Executive Summary, Recommendation 2, pp. 8–9, available at http://www.nap.edu/catalog.php?record_id=11567 (accessed 30th July 2009).
229. McCreight, R. *Protecting Our National Neuroscience Infrastructure: Implications for Homeland Security*, National Security and the Future of Strategic Weapons, George Washington University, Institute of Crisis, Disaster and Risk Management, January 2007, available at https://www.chds.us/?acct:file&mode=dl&h&w&drm=resources%2Fuapi%2Fsummit07&f=ProtectingNeuroscienceInfrastructure.ppt&altf=ProtectingNeuroscienceInfrastructure.ppt (accessed 8th July 2015), slide 9.
230. McCreight, R. (2007) *op.cit.*, Slide 14.
231. Trapp, R. "Incapacitating Chemical Agents": Some Thoughts on Possible Strategies and Recommendations in: ICRC (2010) *op.cit.*, p. 65.
232. National Research Council (2008) *op.cit.*, p. 135.
233. National Research Council (2008) *op.cit.*, p. 137.
234. There have been isolated reports of small-scale use of *"sleeping gas"* by criminals in France, Italy and Spain. See for example: Hooper, J. 'Sleeping gas' thieves target super-rich at Italian billionaires' resort, the Guardian, 30th August 2011; BBC world news, Italian thieves use sleeping gas on Costa Smeralda, 31st August 2011, available at http://www.bbc.co.uk/news/world-europe-14734741 (accessed 25th June 2015).

3 Riot Control Agents

1. For further detailed discussion see Olajos, J. and Salem, H. Riot Control Agents: Pharmacology, Toxicology, Biochemistry and Chemistry, *Journal of Applied Toxicology*, volume 21, 2001, pp. 355–391; Salem, H., Gutting, B., Kluchinsky, T., Boardman, C., Tuorinsky, S. and Hout, J. Riot Control Agents, in *Medical Aspects of Chemical Warfare*, Borden Institute, Office of The Surgeon General, AMEDD Center & School, US Army, 2008, pp. 441–482.
2. Olajos, J. and Salem, H. (2001) *op.cit.*, p. 356; Salem, H. et al. (2008) *op.cit.*, p. 442; Sutherland, R. *Chemical and Biochemical Non-Lethal Weapons, Political and Technical Aspects*, SIPRI Policy Paper 26, Stockholm, Sweden: SIPRI, 2008, p. 12.
3. Sutherland, R. (2008) *op.cit.*, p. 12.

4. Table modified from Olajos, J. and Salem, H. (2001) *op.cit.*, p. 379. The inhalation toxicity is expressed by the notation Ct. It is defined as the product of the concentration in mg m^{-3} multiplied by the exposure time (t) in minutes. The terms LCt50 and ICt50 describe the airborne dosages that are lethal (L) or incapacitating (I) to 50% of the exposed population.
5. Salem, H., Gutting, B. and Kluchinsky, T. et al. (2008) *op.cit.*, p. 443.
6. Fumanski, M. Historical Military Interest in Low-Lethality Biochemical Agents, in Pearson, A., Chevrier, M. and Wheelis, M. (eds), *Incapacitating Biochemical Weapons*. Lanham, MD: Lexington Books, 2007, p. 38.
7. Edgewood Arsenal, History of Research and Development of the Chemical Warfare Service through 1945, Edgewood Arsenal Medical Research Laboratory, 1966, EASP, pp. 400–427, as cited in: Salem, H., Gutting, B. and Kluchinsky, T. et al. (2008) *op.cit.*, p. 479.
8. *Ibid.*, p. 460.
9. OPCW, Report of the OPCW on the Implementation of the Convention on the Prohibition of the Development, Production, Stockpiling and Use of Chemicals and Their Destruction in 2013, 19th Session of Conference of States Parties, C-19/4, 3rd December 2014, Annex 7, p. 57.
10. Sutherland, R. (2008) *op.cit.*, p. 14.
11. See for example: Thorburn, K. M. Injuries After Use of the Lachrimatory Agent Chloroacetophenone in a Confined Space, *Archive of Environmental Health*, volume 37, 1982, pp. 182–186; Stein, A. and Kirwan, W. Chloracetophenone (Tear Gas) Poisoning: A Clinico-pathologic Report, *Journal of Forensic Science*, volume 9, 1964, pp. 374–382. As cited in: Salem, H., Gutting, B. and Kluchinsky, T. et al. (2008) *op.cit.*, p. 480.
12. Hu, H., Fine, J., Epstein, P., Kelsey, K., Reynolds, P. and Walker, B. Tear Gas: Harassing Agent or Toxic Chemical? *Journal of the American Medical Association*, volume 262, 1989, pp. 660–663.
13. Salem, H., Gutting, B. and Kluchinsky, T. et al. (2008) *op.cit.*, p. 444.
14. Sutherland, R. (2008) *op.cit.*, p. 13.
15. For further discussion see, for example: Omega Research Foundation, *Crowd Control Technologies: An Assessment of Crowd Control Technology Options for the European Union*, Luxembourg: European Parliament, Directorate General for Research, Directorate A, The STOA Programme, June 2000.
16. OPCW (3rd December 2014) *op.cit.*, p. 57.
17. Hu, H., Fine, J. and Epstein, P. et al. (1989) *op.cit.*, p. 661.
18. Kluchinsky, T. A. Jr., Savage, P. B., Fitz, R., Smith, P. A. Liberation of Hydrogen Cyanide and Hydrogen Chloride During High Temperature Dispersion of CS Riot Control Agent, *AIHA J (Fairfax, Va)*, volume 63, 2002, pp. 493–496; Kluchinsky, T. A. Jr., Savage, P. B., Sheely, M. V., Thomas, R. J., Smith, P. A. Identification of CS-Derived Compounds Formed During Heat-Dispersion of CS Riot Control Agent, *Journal of Microcolumn Separations*, volume 13, number 5, September 2001, pp. 186–190, as cited in Salem, H., Gutting, B. and Kluchinsky, T. et al. (2008) *op.cit.*, p. 474.
19. See for example: Flame, Riot Control Agents, and Herbicide Operations, U.S. Marine Corps, Field Manual, U.S. Department of the Army, Washington, DC, FM 3-11.11/MCRP 3-3.7.2 C1 (10th March 2003), available at https://fas.org/irp/doddir/army/fm3-11-11-excerpt.pdf (accessed 2nd July 2015). See also Nonlethal Weapons: Terms and References, Bunker, R. (ed.), INSS Occasional Paper 15, U.S. Air Force Institute for National Security Studies USAF Academy, Colorado, July

1997, available at http://www.usafa.edu/df/inss/OCP/ocp15.pdf (accessed 2nd July 2015).
20. CSX is comprised of one gram of powdered CS dissolved in 99 grams of trioctylphosphite (TOF). U.S. Marine Corps Field Manual (10th March 2003) *op.cit.*, p. 6–2.
21. CS1 is a free-flowing agent powder comprised of 95% crystalline CS blended with 5% silica aerogel. U.S. Marine Corps Field Manual (10th March 2003) *op.cit.*, pp. 6–2.
22. U.S. Marine Corps Field Manual (10th March 2003) *op.cit.*, pp. 6–2.
23. US Air Force, INSS (July 1997) *op.cit.*, p. 26; see also U.S. Marine Corps Field Manual (10th March 2003) *op.cit.*, pp. 6–2.
24. Munavalli, S., Rohrbaugh, D. and Durst, H. *Chemistry, Biochemistry, Pharmacology, and Toxicology of CS and Synthesis of Its Novel Analogs, Final Report*, May 2003–September 2004, Published October 2007, Science Applications International Corp (SAIC)/Edgewood Chemical Biological Centre, ECBC-TR-540.
25. Munavalli, S., Rohrbaugh, D. and Durst, H. (2007) *op.cit.*, p. 19.
26. Salem, H., Gutting, B. and Kluchinsky, T. et al. (2008) *op.cit.*, p. 466.
27. Sutherland, R. (2008) *op.cit.*, p. 15.
28. Salem, H., Gutting, B. and Kluchinsky, T. et al. (2008) *op.cit.*, p. 466.
29. OPCW (3rd December 2014) *op.cit.*, p. 57.
30. UK House of Commons Written Answers for 12th March 1998, pt. 17.
31. For further discussion see Walker, J. A Tale of Two Riot Control Agents: U.K. Attitudes to CS and CR in Warfare and Law Enforcement 1969–1975, *CBW Conventions Bulletin*, number 86, February 2010, p. 10; Morrison, C. and Bright, M. Secret Gas Was Issued for IRA Prison Riots, *The Observer*, 23rd January 2005.
32. Walker, J. *Britain and Disarmament: The UK and Nuclear, Biological and Chemical Weapons Arms Control and Programmes 1956–1975*, Farnham: Ashgate Publishing Limited, 2012, p. 44.
33. Salem, H., Gutting, B. and Kluchinsky, T. et al. (2008) *op.cit.*, pp. 452–454.; Amnesty International, *Pain Merchants: Security Equipment and Its Use in Torture and Other Ill-Treatment*, ACT 40/008/2003, London: Amnesty International, 2nd December 2003, pp. 63–67.
34. OPCW (3rd December 2014) *op.cit.*, p. 57.
35. Salem, H. et al. (2008) *op.cit.*, p. 453.
36. *Ibid.*
37. See for example: Amnesty International, *USA: Updated Briefing to the Human Rights Committee on the Implementation of the International Covenant on Civil and Political Rights*, AMR 51/111/2006, 12th July 2006, p. 4.
38. Sutherland, R. (2008) *op.cit.*, pp. 21–22.
39. Amnesty International (2nd December 2003) *op.cit.*, p. 63; Written Response to Parliamentary Question by Lord Henley (Parliamentary Under Secretary of State, Environment, Food and Rural Affairs), *Hansard*, HL Deb, 15th March 2012, c113W.
40. Salem, H. et al. (2008) *op.cit.*, p. 464.
41. *Ibid.*
42. *Ibid.*
43. Sutherland, R. (2008) *op.cit.*, pp. 14–15.
44. Salem, H. et al. (2008) *op.cit.*, p. 464.
45. World Health Organization *Health Aspects of Chemical and Biological Weapons*, 1st Edition, Geneva, Switzerland: WHO, 1970, pp. 24, 55, as cited in: Salem, H., Gutting, B. and Kluchinsky, T. et al. (2008) *op.cit.*, p. 464.

46. Sutherland, R. (2008) *op.cit.*, pp. 14–15.
47. See 001106 & 000315–16, Harvard Sussex Events Database, Retrieved 7th July 2009; OPCW, Report of the Third Session of the Scientific Advisory Board, SAB III/I, 27th April 2000.
48. See for example, Salem. H, Gutting, B. and Kluchinsky, T. et al. (2008) *op.cit.*, p. 464.
49. OPCW, Report of the OPCW on the Implementation of the Convention on the Prohibition of the Development, Production, Stockpiling and Use of Chemical Weapons and on Their Destruction in 2007, C-13/4, 3rd December 2008, Annex 3, p. 30.
50. See for example: OPCW (3rd December 2013) *op.cit.*, Annex 7, p. 57.
51. Sutherland, R. (2008) *op.cit.*, p. 15.
52. Salem, H. et al. (2008) *op.cit.*, p. 455.
53. *Ibid.*, p. 456.
54. *Ibid.*, p. 457.
55. Article II.7 of the CWC defines RCAs as, *inter alia*, "Any chemical not listed in a Schedule...". Therefore only chemicals not listed under the CWC Schedules can be utilized as RCAs.
56. OPCW, Report of the OPCW on the Implementation of the Convention on the Prohibition of the Development, Production, Stockpiling and Use of Chemical Weapons and on Their Destruction in the Year 2000, C-VI/5, 17th May 2001, p. 15.
57. OPCW, Report of the OPCW on the Implementation of the Convention on the Prohibition of the Development, Production, Stockpiling and Use of Chemical Weapons and on Their Destruction in the Year 2001, C-7/3, 10th October 2002, p. 50.
58. OPCW, Report of the OPCW on the Implementation of the Convention on the Prohibition of the Development, Production, Stockpiling and Use of Chemical Weapons and on Their Destruction in 2002, C-8/5, 22nd October 2003, p. 49.
59. See for example: OPCW (3rd December 2014) *op.cit.*, Annex 7, p. 57.
60. Physicians for Human Rights, *Bloody Sunday: Trauma in Tblisi, the Events of April 9, 1989 and Their Aftermath*, February 1990; See also 9th April and 24th May 1989 entries in the News Chronology section of *Chemical Weapons Convention Bulletin*, number 5, August 1989, Harvard Sussex Programme, pp. 7 and 10.
61. Analysis was undertaken of all OPCW documents publicly available on the OPCW website (http://www.opcw.org) as of 2nd August 2015. For further discussion see Chapter 5 of this publication.
62. Neill, D. *Riot Control and Incapacitating Chemical Agents Under the Chemical Weapons Convention*, Defence R&D Canada, Centre for Operational Research and Analysis, technical memorandum DRDC CORA TM 2007–2022, Ottawa, 2007, p. 6.
63. Sunshine Project, *Backgrounder 8, Non-Lethal Weapons Research in the US: Calmatives and Malodorants*, July 2001, p. 3.
64. Bickford, L., Bowie, D., Collins, K., Salem, H. and Dalton, P. Odorous Substances for Non-Lethal Application, presentation at NDIA Non-Lethal Defense IV, Tysons Corner, Virginia, 20–22nd March 2000; Sunshine Project, Backgrounder 8 (July 2001) *op.cit.*, p. 4.
65. Boguski, T., Breuer, L. and Erickson, L. Environmental Issues Associated with Malodorants, presentation to the Non-Lethal Technology and Academic Research Symposium, Kansas State University, 9th November 2001. Some

candidate compounds included isovaleric acid (which produces a "sweaty, putrid, swine odour" reminiscent of rancid cheese); skatole ("putrid, faecal"); n-caproic acid ("sharp, sour, rancid, goat"); and t-butyl mercaptan ("skunk, sulphurous"). As cited by Neill, D. (2007), *op.cit.*, endnote 23.
66. Lyon, D., Johnson, R. and Domanico, J. Design and Development of an 81mm Non-Lethal Mortar Cartridge. Presentation to Non-Lethal Defense IV, National Defense Industrial Association (NDIA), US, 20th–22nd March 2000. For further discussion see Crowley, M. Drawing the Line (April 2013) *op.cit.*, p. 30.
67. For discussion of the XM1063 155mm malodorant projectile, see Chapter 4 of this publication.
68. Hymes, K. Non-Lethal Weapons in Escalation of Force, Slide 11. *Proceedings of the 5th European Symposium on Non-Lethal Weapons*, Ettlingen, Germany, 11–13th May 2009; Hambling, D. US Military Malodorant Missiles Kick Up a Stink, *New Scientist*, issue 2867, 4th June 2012.
69. US Office of Naval Research, Joint Non-Lethal Weapons Program, Fiscal Year 2015 Non-Lethal Weapons Technologies, Broad Agency Announcement, ONRBAA14-008, 5th June 2014, p. 6.
70. 40mm Non-Lethal ammunition, malodorant, Security Devices International Inc., undated, available at http://6ukij4ety822xjv0j17p6pzv.wpengine.netdna-cdn.com/wp-content/uploads/2014/08/BIP-MO-Malodorant-One-Pager.pdf (accessed 2nd July 2015).
71. Skunk malodorant systems for crowd control, Mistral Inc. undated, available at http://www.mistralinc.com/portals/mistralinc/Images/product-photos/Final%20-%20Skunk%20Product%20Brochure%204.13.2015.pdf (accessed 2nd July 2015); Skunk, Mistral Inc., undated, available at http://www.mistralinc.com/Technologies/Non-Lethal-Technologies/Skunk (accessed 2nd July 2015). Although Mistral Inc. calls its "non-lethal" malodorant *"skunk"* and described it as a *"water based, biodegradable, vile smelling liquid"*, it is not known whether the US product is the same as the *"skunk"* developed in Israel.
72. See Odortec Ltd, available at http://www.skunk-skunk.com/121755/About-Us (accessed 2nd July 2015).
73. Material Safety Data Sheet (MSDS) Skunk, Odortec Ltd. Original MSDS 12th September 2004, Revised 9th September 2008, available at http://www.skunk-skunk.com/image/users/121755/ftp/my_files/MSDS_Skunk.pdf?id=3225191 (accessed 2nd July 2015).
74. Odortec Ltd MSDS (9th September 2008) *op.cit.*
75. *Ibid.*
76. Making a stink, Ben-Simhon, C./Haaretz, 4th September 2008, available at http://www.haaretz.com/making-a-stink-1.253207 (accessed 2nd July 2015).
77. See Odortec Ltd, available at http://www.skunk-skunk.com/121755/About-Us (accessed 2nd July 2015).
78. Crowd Control Israel's Use of Crowd Control Weapons in the West Bank, Michaeli, S./B'Tselem [The Israeli Information Center for Human Rights in the Occupied Territories], January 2013, available at http://www.btselem.org/download/201212_crowd_control_eng.pdf (accessed 29th May 2015), p. 35.
79. A whiff from hell: Skunk, a high-tec Israeli weapon against stone-throwers, *The Economist*, 6th June 2015
80. Law enforcement, Skunk TA03074, Tar Ideal Concepts, undated, available at http://www.tarideal.com/prdPics/prdFiles/TA03074_cp_attr_file_1.pdf (accessed 2nd July 2015).

81. Michaeli, S./B'Tselem (January 2013) *op.cit.*, p. 35
82. Stoppford, W. and Olajos, E. Issues/Concerns, Needs, Emerging Concepts/Trends, and Advances in Riot Control Agents, in Olajos, E. and Stoppford, W. (eds), *Riot Control Agents: Issues in Toxicology, Safety, and Health*. Boca Raton, FL: CRC Press, 2004, p. 323.
83. Dimitroglou, Y., Rachiotis, G. and Hadjichristodoulou, C. Exposure to the Riot Control Agent CS and Potential Health Effects: A Systematic Review of the Evidence, *International Journal of Environmental Research and Public Health*, volume 12, January 2015, pp. 1397–1411; Haber, L., et al., Human effectiveness and risk characterization of OC and P AVA handheld devices. Air Force Research Laboratory, DTIC Technical Report No. ADA476262, 1st May 2007; Hilmas, C., Poole, M., Katos, A. and Williams, Riot Control Agents, in Gupta, R. (ed.), *Handbook of Toxicology of Chemical Warfare Agents*. London: Academic Press, 2009, pp. 153–175; Levin, R and Mershon, M. Contact Sensitization to CS, A Riot Control Agent, Edgewood Arsenal, Aberdeen Proving Ground, Maryland, November 1973; Salem, H et al. (2008) *op.cit.*
84. Dimitroglou, Y., et al. (2015) *op.cit.*; Hilmas et al. (2009) *op.cit.*; Worthington, E. and Nee, P. CS Exposure-clinical Effects and Management, *Journal of Accident Emergency Medicine*, volume 168, 1999, pp. 168–170; U.K. Department of Health 1999 Annual Report of the Committees on Toxicity Mutagenicity Carcinogenicity of Chemicals in Food, Consumer Products and the Environment
85. Mendelson, J., Tolliver, B., Delucchi, K., Baggott, M., Flower, K., Harris, W., Galloway, G. and Berger, P. Capsaicin, an Active Ingredient in Pepper Sprays, Increases Lethality of Cocaine, *Forensic Toxicology*, volume 28, number 1, January 2010, pp. 33–37.
86. US National Institute of Standards and Technology, NIST keeps Users from Getting Burned by Bad Pepper Sprays, NIST Tech-Beat (May–June 2002), available at http://www.nist.gov/public_ affairs/techbeat/tb2002_0506.html (accessed 2nd April 2015).
87. Rappert, B. Health and Safety in Policing: Lessons from the Regulation of CS Sprays in the UK, *Social Science & Medicine*, volume 56, 2003, p. 1273.
88. Holopainen, M., Moilanen J., Hack, T. and Tervo, T. Toxic Carriers in Pepper Sprays may Cause Corneal Erosion, *Toxicology and Applied Pharmacology*, volume 186, 2003, pp. 155–162.
89. According to the UK Health Protection Agency, trichloroethylene is *"highly toxic by inhalation, ingestion and skin contact"*. Inhalation can "cause irritation of the respiratory tract, and sudden death due to cardiac arrhythmias". It is considered to be *"probably carcinogenic and mutagenic"* and presents a *"possible risk to the unborn child"*. [Foxall, K., Trichloroethylene, UK Health Protection Agency, 2008] See also Vesaluoma, M., Muller, L., Gallar, J., Lambiase, A., Moilanen, J., Hack, T., Belmonte, C. and Tervo, T. Effects of Oleoresin Capsicum Pepper Spray on Human Corneal Morphology and Sensitivity, *Investigative Ophthalmology and Visual Science*, volume 41, number 8, July 2000, pp. 2138–2147.
90. See for example: Personal Safety Corporation, Material Safety Data Sheet PE14 & PE17 OC FOGGER 8% 2MM, available at http://www.pepperenforcement.com/pe14_17.html (accessed 9th July 2015). The MSDS lists trichloroethylene but does not provide details of quantity or composition percentage.
91. Sutherland, R. (2008) *op.cit.*, p. 23.

92. Zarc International Incorporated, Consumer Alert: DuPont Cautions Against Use of HCFC Dymel in Pepper Sprays, 20 August 1993; as cited in: Sutherland, R. (2008) *op.cit.*, p. 23.
93. Hay, A., Giacaman, R., Sansur, R. and Rose, S. Skin Injuries Caused by New Riot Control Agent Used Against Civilians on the West Bank, *Medicine, Conflict and Survival*, volume 22, number 4, 2006, pp. 283–291.
94. Stoppford, W. and Olajos, E. (2004) *op.cit.*, p. 323.
95. Omega Research Foundation (2000) *op.cit.*, p. 36.
96. Hill, A., Silverberg, N., Mayorga, D. and Baldwin, H. Medical Hazards of the Tear Gas CS: A Case of Persistent, *Multisystem, Hypersensitivity Reaction and Review of the Literature, Medicine*, volume 79, 2000, pp. 234–240; Geneva Academy, Riot Control Agents, Weapons Law Encyclopaedia, available at http://www.weaponslaw.org/ (accessed 2nd July 2015).
97. Hu, H., Fine, J. and Epstein, P. et al. (1989) *op.cit.*
98. Karagama, Y., Newton, J. and Newbegin, C. Short Term and Long Term Physical Effects of Exposure to CS Spray, *Journal of the Royal Society of Medicine*, volume 96, number 4, 2003, pp. 172–174.
99. Sutherland, R. (2008) *op.cit.*, p. 12.
100. Physicians for Human Rights, Weaponizing Tear Gas: Bahrain's Unprecedented Use of Toxic Chemical Agents Against Civilians, August 2012, p. 15.
101. See, e.g., Karagama, Y. et al. (2003) *op.cit.*; Vesaluoma, M. et al. (2000) *op.cit.*
102. Arbak, P., Başer, I., Kumbasar, O., Ülger, F., Kılıçaslan, Z. and Evyapan, F. Long Term Effects of Tear Gases on Respiratory System: Analysis of 93 Cases, *The Scientific World Journal*, volume 2014, Article ID 963638.
103. See Physicians for Human Rights, The Casualties of Conflict: Medical Care and Human Rights in the West Bank and Gaza Strip, 1988; Physicians for Human Rights, The Use of Tear Gas in the Republic of Korea, 1987; Physicians for Human Rights (2012) *op.cit.*
104. See Chasseaud, L., Bunter, B., Robinson, W. and Barry, D. Suppression of Sebaceous Gland Non-Specific Esterase Activity by Electrophilicaβ-Unsaturated Compounds, *Experientia*, volume 31, number 10, pp. 1196–1197; Barry, D., Chasseaud, L., Hunter, B., and Robinson, W. The Suppression of Non-specific Esterase Activity in Mouse Skin Sebaceous Gland by CS Gas, *Nature*, volume 240, 1972, pp. 560–561. For further discussion see Hu, H., Fine, J. and Epstein, P. et al. (1989) *op.cit.*; Physicians for Human Rights (2012) *op.cit.*
105. Hu, H., Fine, J. and Epstein, P. et al. (1989) *op.cit.*, p. 660.
106. Carron, P. and Yerson, B. Clinical Review: Management of the Effects of Exposure to Tear Gas, *British Medical Journal*, volume 338, number 2283, 2009.
107. Casey-Maslen, S., Corney, N. and Dymond-Bass, A. Review of Weapons Under IHL and Human Rights Law, in Casey-Maslen, S. (ed.), *Weapons Under International Human Rights Law*. Cambridge University Press, 2014.
108. Geneva Academy, Riot Control Agents, Weapons Law Encyclopaedia, undated, available at http://www.weaponslaw.org/ (accessed 2nd July 2015).
109. See for example, Hu, H. et al. (1989) *op.cit.*; Amnesty International, the Pain Merchants: Security equipment and its use in torture and other ill-treatment, ACT 40/008/2003, December 2003; Amnesty International Israel and Occupied Palestinian Territories: Trigger-happy: Israel's use of excessive force in the West Bank, MDE 15/002/2014, 27th February 2014; Amnesty International, Gezi Park Protests: Brutal denial of the right to peaceful assembly in Turkey,

EUR 44/022/2013, October 2013; Physicians for Human Rights (2012) *op.cit.* See chapter four of this publication for further discussion and illustrative case studies.
110. Federal Laboratories (date unknown) Riot Control, Federal Laboratories, as cited in: Geneva Academy, Riot control agents, Weapons Law Encyclopaedia, available at http://www.weaponslaw.org/ (accessed 2nd April 2015).
111. According to Danto, *"If the CN Cartridges Are Too Old and the Agent Does Not Vaporize Adequately, Small Solid Particles May Strike the Cornea and Cause Damage"*. Danto, B. Medical Problems and Criteria Regarding the Use of Tear Gas by Police, *American Journal of Forensic Medical Pathology*, volume 8, December 1987, pp. 317–322; see also Physicians for Human Rights (2012) *op.cit.*
112. Geneva Academy, Riot Control Agents (undated) *op.cit.*
113. STR/Getty (2012) Getty Image ID: 143178781, as cited in Geneva Academy, Riot Control Agents (undated) *op.cit.*
114. Expired tear gas used on protesters during Cairo demonstrations, *Egypt Independent*, 28th January 2011; see also Elsadeq, K. (2011) via Twitter, as cited in Geneva Academy, Riot Control Agents (undated) *op.cit.*
115. Proud to be an Indian (2012) via Facebook, as cited in Geneva Academy, Riot Control Agents (undated) *op.cit.*
116. ACRI, ACRI to Israel Police: Issawiya Is Under Attack, 30th October 2014, available at http://www.acri.org.il/en/2014/10/30/issawiya-attack/ (accessed 9th July 2015); Wadi Hilweh Information Center (2011) Israel using Expired Tear Gas on Palestinians of East Jerusalem (with pictures), as cited in Geneva Academy, Riot Control Agents (undated) *op.cit.*
117. Physicians for human rights, Contempt for Freedom: State Use of Tear Gas as a Weapon and Attacks on Medical Personnel in Turkey, September 2013; See also Brown Moses (2013) Imgur Collection, as cited in Geneva Academy, Riot Control Agents (undated) *op.cit.*
118. Report for the Committee against Torture during its revision of the combined 3rd and 4th periodic reports submitted by the Bolivarian Republic of Venezuela (53rd period of sessions, 3–28 November 2014) Coalition of non-governmental organisations, academic institutions and organised civil society, October, 2014; see also Omega Research Foundation Archives (2009), as cited in Geneva Academy, Riot Control Agents (undated) *op.cit.*
119. Kasinof, L. and Goodman, D. Yemen Police and Protesters Clash as Deal Is Sought to End Political Crisis, *The New York Times*, 19th April 2011.
120. There is no data available in the public domain recording the frequency of RCA employment by law enforcement officials nor of the amounts of such chemical agents employed. Consequently, the relative frequency of reported misuse of RCAs could not be established. Instead the analysis is intended to indicate the broad types of reported RCA misuse and the range of countries where such misuse has reportedly occurred. For further information concerning the reported severity, frequency, circumstances and nature of the specific cases of misuse recorded, the reader should consult the primary documents cited.
121. The public documentation of relevant UN bodies and mechanisms including the UN Human Rights Council, UN Commission on Human Rights, UN Special Rapporteur on Torture and UN Special Rapporteur on Extra-Judicial Executions; and regional human rights bodies and mechanisms including the European Committee for the Prevention of Torture and Inhuman or Degrading Treatment or Punishment (CPT) and the Special Rapporteur for Freedom of Expression of

the Organization of American States, covering reported human rights violations from the start of 2009 till the end of 2013, was analysed.
122. The public documentation of AI, Article 19, HRW, International Federation for Human Rights (FIDH), and the World Organisation Against Torture (OMCT), covering reported human rights violations from the start of 2009 till the end of 2013, was analysed.
123. It should be noted that this review gives only a partial "snapshot" of reported misuse of RCAs by law enforcement personnel during a discrete time frame. Analysis of respected media, academic or other non-governmental information sources may well provide further cases of reported misuse of RCAs by law enforcement officials in these and additional countries.
124. For details of a previous analysis of alleged inappropriate use of RCAs by law enforcement officials from January 2004 till December 2008, see Crowley, M. Dangerous Ambiguities: Regulation of Riot Control Agents and Incapacitants Under the Chemical Weapons Convention, Bradford Non-Lethal Weapons Research Project, October 2009, pp. 41–48.
125. United Nations, Report on the visit of the Subcommittee on Prevention of Torture and Other Cruel, Inhuman or Degrading Treatment or Punishment (SPT) to Brazil, SPT, Optional Protocol to the Convention against Torture and Other Cruel, Inhuman or Degrading Treatment or Punishment (OPCAT), 5th July 2012.
126. OPCAT, SPT (5th July 2012) *op.cit.*, paragraph 127.
127. *Ibid.*, paragraph 128.
128. Maldives: The other side of paradise: A human rights crisis in the Maldives, Amnesty International, ASA 29/005/2012, 5th September 2012.
129. See B'Tselem, available at http://www.btselem.org/, and in particular: Crowd Control Israel's Use of Crowd Control Weapons in the West Bank, B'Tselem, January 2013.
130. See ACRI, available at http://www.acri.org.il/en/ for example: Spraying "skunk" at homes – violation of the right to dignity and freedom of demonstration, ACRI, 23rd January 2013 [original in Hebrew], https://www.acri.org.il/he/25560 (accessed 2nd July 2015).
131. Israel and Occupied Palestinian Territories: Trigger-happy: Israel's Use of Excessive Force in the West Bank, Amnesty International, MDE 15/002/2014, 27th February 2014.
132. Amnesty International (27th February 2014) *op.cit.*, p. 44.
133. *Ibid.*
134. B'Tselem (January 2013) *op.cit.*, pp. 36–37.
135. Amnesty International (27th February 2014) *op.cit.*, p. 44.
136. ACRI letter to Rav Nitzav Yohanan Danino Chief Commission of Israel Police, 24th November 2014, available at http://www.acri.org.il/en/wp-content/uploads/2015/05/EJ-Skunk-Spray-Letter-Nov-14.pdf; see also End the Use of Skunk Spray in East Jerusalem, 27th November 2014, available at http://www.acri.org.il/en/2014/11/27/skunk-jerusalem/; ACRI to Israel Police: Issawiya is Under Attack, 30th October 2014, available at http://www.acri.org.il/en/2014/10/30/issawiya-attack/ (accessed 2nd July 2015).
137. Illegal acts by Israeli authorities on the rise in the occupied West Bank – UN human rights expert, United Nations Office of the High Commissioner for Human Rights, 14th January 2011, available at http://www.ohchr.org/EN/NewsEvents/Pages/DisplayNews.aspx?NewsID=10633&LangID=E (accessed 8th August 2015).

138. Palestinian dies after inhaling gas at W Bank protest, 1st January 2011, BBC news, available at: http://www.bbc.co.uk/news/world-middle-east-12103825. (accessed 8th August 2015).
139. United Nations Office of the High Commissioner for Human Rights (14th January 2011) *op.cit.*
140. Tunisia in revolt, State violence during anti-government protests, Amnesty International, MDE 30/011/2011, February 2011.
141. United Nations CAT/OP/HND/1, Optional Protocol to the Convention against Torture and Other Cruel, Inhuman or Degrading Treatment or Punishment, 10th February 2010, Subcommittee on Prevention of Torture, Report on the Visit of the Subcommittee on Prevention of Torture and Other Cruel, Inhuman or Degrading Treatment or Punishment to Honduras, see in particular, paragraph 54.
142. Inter-American Commission on Human Rights, Honduras: Human Rights and the Coup D'etat, OEA/Ser.L/V/II. Doc. 55, 30th December 2009, see in particular, paragraphs 286–289.
143. IACHR (30th December 2009) *op.cit.*, paragraph 289.
144. Extrajudicial, Summary or Arbitrary Executions, Report of the Special Rapporteur, Philip Alston, Addendum, Summary of Cases Transmitted to Governments and Replies, UN Commission on Human Rights, 27th March 2006, pp. 317–318.
145. Amnesty International, Zimbabwe: Ten Dead Following Police Misuse of Tear Gas, Press release, 22nd September 2004; Amnesty International, Another Death at Porta Farm: 11 People Dead Following Police Misuse of Tear Gas, 4th October 2004; and Amnesty International and Zimbabwe Lawyers for Human Rights Zimbabwe, Shattered Lives: The Case of Porta Farm, AFR 46/004/2006, 31st March 2006.
146. On 31st August 2004, residents of Porta Farm obtained a High Court order (Case No. HC 10671/04) prohibiting the Minister of Local Government, Public Works and National Housing or anyone acting on his behalf from evicting people from Porta Farm unless and until the Government offered them suitable alternative accommodation. See Amnesty International and Zimbabwe Lawyers for Human Rights Zimbabwe, Zimbabwe: Shattered Lives: The Case of Porta Farm, *op.cit.*, p. 7.
147. Amnesty International (4th October 2004) *op.cit.*
148. Quoted from the affidavit of Christine, K., 22nd September 2004, Harare, as cited in: Amnesty International and Zimbabwe Lawyers for Human Rights Zimbabwe (31st March 2006), p. 8.
149. Amnesty International phone interviews and email communication with doctors and medical experts in Harare, London, September 2004. See Amnesty International and Zimbabwe Lawyers for Human Rights Zimbabwe, Zimbabwe: Shattered Lives: The Case of Porta Farm, *op.cit.*, p. 8.
150. Kingsley, P. How Did 37 Prisoners Come to Die at Cairo Prison Abu Zaabal? *The Guardian*, 22nd February 2014.
151. Kingsley, P. *The Guardian* (22nd February 2014) *op.cit.*
152. BBC News, Egypt Police Convicted Over Detainee Tear-Gas Deaths, 18th March, available at http://www.bbc.co.uk/news/world-middle-east-26626367 (accessed 2nd April 2015).
153. Kingsley, P. *The Guardian* (22nd February 2014) *op.cit*; BBC News, Egypt Police Convicted Over Detainee Tear-Gas Deaths, 18th March.

154. Amnesty International, Egypt: Security Forces Must Show Restraint After Reckless Policing of Violent Protest, 23rd August 2013, available at https://www.amnesty.org/press-releases/2013/08/egypt-security-forces-must-show-restraint-after-reckless-policing-violent-p/ (accessed 9th July2015).
155. Kingsley, P. *The Guardian* (22nd February 2014) *op.cit.*
156. *Ibid.*
157. BBC News, Egypt Police Convicted Over Detainee Tear-Gas Deaths, 18th March.
158. *Ibid.*; Kingsley, P. *The Guardian* (22nd February 2014) *op.cit.*
159. Human Rights Watch, *World Report 2015*, Egypt chapter, 29th January 2015; Reuters, Egypt Court Cancels Jail Sentence against Policeman Linked to 39 Deaths, 7th June 2014. Reuters, Egypt's high court orders re-trial of policemen linked to 37 deaths, 22nd January 2015.
160. See for example: La Rue, F. United Nations, Human Rights Council, Report of the Special Rapporteur on the Promotion and Protection of the Right to Freedom of Opinion and Expression, A/HRC/17/27/Add.1, 27th May 2011.
161. See for example: Amnesty International, Bloodied but Unbowed: Unwarranted State Violence against Bahraini Protesters, MDE 11/009/2011, March 2011; Human Rights Watch News, Bahrain: End Deadly Attacks on Peaceful Protesters: At Least 5 Dead, Hundreds Injured; Latest Assault Appeared Planned at Senior Levels, 17th February 2011; Human Rights Watch News, Bahrain: Hold Police Accountable in Teargas Episode, Direct Hit by Canister Injures Activist, 4th July 2012.
162. La Rue, F., UN Human Rights Council (27th May 2011) *op.cit.*, paragraph 175.
163. Human Rights Watch (17th February 2011) *op.cit.*
164. Amnesty International (March 2011) *op.cit.*
165. Human Rights Watch (17th February 2011) *op.cit.*
166. Amnesty International (March 2011) *op.cit.*
167. Correspondence to the Government of Bangladesh from the UN Special Rapporteur on the promotion and protection of the right to freedom of opinion and expression; the Special Rapporteur on the rights to freedom of peaceful assembly and of association; the Special Rapporteur on extrajudicial, summary or arbitrary executions; and the Special Rapporteur on torture and other cruel, inhuman or degrading treatment or punishment, 20th February 2013, REFERENCE: AL G/SO 214 (67-17) Assembly & Association (2010-1) G/SO 214 (33-27) G/SO 214 (53-24) BGD 4/2013. See also Méndez, J. E. United Nations, Human Rights Council, Report of the Special Rapporteur on Torture and Other Cruel, Inhuman or Degrading Treatment or Punishment, A/HRC/25/60/Add.2, 11th March 2013, paragraph 22.
168. UN Special Rapporteur on the promotion and protection of the right to freedom of opinion and expression et al. (20th February 2013) *op.cit.*, p. 2.
169. *Ibid.*
170. United Nations, Office of the High Commissioner for Human Rights, Pillay Urges Turkey's Government and Civil Society to Take Urgent Action to Defuse Tensions, 18th June 2013.
171. See for example: Amnesty International, Gezi Park Protests: Brutal Denial of the Right to Peaceful Assembly in Turkey, EUR 44/022/2013, October 2013; Human Rights Watch, Turkey: End Police Violence at Protests Government Intolerance of Free Speech, Assembly Triggers Demonstrations, press release, 1st June 2013 and 3rd June 2013 update; Human Rights Watch News, Turkey: A Weekend of Police Abuse, Many Protesters Arrested; Hospital Targeted, 18th June 2013.

172. United Nations, Office of the High Commissioner for Human Rights (18th June 2013) *op.cit.*
173. Amnesty International (October 2013), p. 19.
174. *Ibid.*, p. 20.
175. Human Rights Watch News, Turkey: A Weekend of Police Abuse, Many Protesters Arrested; Hospital Targeted, 18th June 2013.
176. Human Rights Watch (8th June 2013) *op.cit.*
177. Report of the International Commission of Inquiry mandated to establish the facts and circumstances of the events of 28th September 2009 in Guinea, UN doc. S/2009/693, 18th December 2009, paragraph 198.
178. Amnesty International, Guinea: "You Did Not Want the Military, So Now We Are Going to Teach You a Lesson", AFR 29/001/2010, London, February 2010.
179. Human Rights Watch, Guinea: September 28 Massacre Was Premeditated, In-Depth Investigation Also Documents Widespread Rape, 27th October 2009.
180. Human Rights Watch (27th October 2009) *op.cit.*, pp. 27–29.
181. *Ibid.*
182. Human Rights Watch interview (name withheld), Conakry, 15th October 2009 in Human Rights Watch (27th October 2009) *op.cit.*, p. 29.
183. United Nations (18th December 2009) *op.cit.*, paragraph 70.
184. Human Rights Watch (27th October 2009) *op.cit.*, pp. 27–29.
185. Amnesty International (February 2010) *op.cit.*, p. 27.
186. Stockholm International Peace Research Institute/Perry Robinson, J. and Leitenberg, M. *The Problem of Chemical and Biological Warfare, volume 1: The Rise of CB Weapons*, Stockholm: Almqvist and Wiksell, 1971; Perry Robinson, J., Hedén, C. and von Schreeb, H./Stockholm International Peace Research Institute, *The Problem of Chemical and Biological Warfare, Volume 2: CB Weapons Today*, Stockholm: Almqvist and Wiksell, 1973; Furmanski, M. Historical Military Interest in Low-Lethality Biochemical Agents, in Pearson, A., Chevrier, M. and Wheelis, M. (eds), *Incapacitating Biochemical Weapons*, Lanham, MD: Lexington Books, 2007, pp. 35–66; Verwey, W. *Riot Control Agents and Herbicides in War*, Leiden: Brill, 1977; Spiers, E. *A History of Chemical and Biological Weapons*, London: Reaktion Books, 2010; See also Perry Robinson, J. *Disabling Chemical Weapons: A Documented Chronology of Events, 1945–2003*, unpublished monograph, October 2003, copy given to author.
187. Fry, J. Contextualized Legal Reviews for the Methods and Means of Warfare: Cave Combat and International Humanitarian Law, *Colombia Journal of Transnational Law*, 28th February 2006.
188. Fry, J. (2006) *op.cit.*, p. 507.
189. *Ibid.*
190. *Ibid.*
191. Bahmanyar, M. *Afghanistan Cave Complexes 1979–2004: Mountain Strongholds of the Mujahideen, Taliban and Al Qaeda*, Osprey Publishing, Oxford, UK, 2004, p. 32.
192. *Ibid.*
193. See US Department of the Army, Operational Report – Lessons Learned, Headquarters United States Army Vietnam, Period Ending 31 July 1967 (U), Office of the Adjutant General. Washington, DC. 10310 14th December 1967, available at http://www.dtic.mil/dtic/tr/fulltext/u2/386396.pdf (accessed 9th July 2015); Encyclopedia of the Vietnam War, A Political, Social, and Military History (ed.), Tucker, S., ABC-CLIO.LLC. 2011, p. 1152.
194. Hon. Donald Rumsfeld, Secretary of Defense, Testimony Before the 108th Congress House Armed Services Committee, 5th February 2003.

195. Report from Kennzeichen D (ZDF, German TV), 27th October 1999. As cited in News Chronology, *CBW Conventions Bulletin*, number 46, December 1999, Harvard Sussex Program, p. 41; Sunshine Project, *Country Study No. 3, A Survey of Biological and Biochemical Weapons Related Research Activities in Turkey*, 2004, p. 15.
196. Report from Kennzeichen D (ZDF, German TV), 27th October 1999. As cited in Harvard Sussex Program (1999) *op.cit.*, p. 41; Sunshine Project (2004) *op.cit.*, p. 15.
197. According to the Sunshine Project Report, although the chemical analysis was conclusive, there was no independent proof that the shrapnel provided to the laboratory was removed from the cave in Sirnak, hence there is some uncertainty about this incident. Sunshine Project (2004) *op.cit*, p. 15.
198. Report from Kennzeichen D (ZDF, German TV), 27th October 1999. As cited in Harvard Sussex Program (1999) *op.cit.*, p. 41; Sunshine Project (2004) *op.cit.*, p. 15.
199. Anatolia News Agency (Ankara), 28th October 1999, as translated in BBC-SWB, 30th October 1999, EE/D3679/B. Spokesperson denies manufacture, use of chemical weapons. 991028, Harvard Sussex Program Events Database, retrieved 7th July 2009; Harvard Sussex Program (1999) *op.cit.*, p. 41.
200. Turkish training video, broadcast on 8 May 2004 on channel TRT 1 in the programme TSK Saati [stills in Sunshine Project (2004) *op.cit.*, p. 16].
201. Fry, J. Gas Smells Awful: UN Forces, Riot Control Agents and the Chemical Weapons Convention, *Michigan Journal of International Law*, volume 31, number 3, 2010, pp. 475–559. Fry defines the term "UN forces" as comprising three types of military operations falling under the auspices of the UN: UN-controlled peacekeeping operations, UN-controlled enforcement operations and UN-authorized enforcement operations. According to his definition, forces such as the Kosovo Force (KFOR) and the SFOR in Bosnia and Herzegovina are included among UN forces because the Security Council created them. See Fry, J. (2010) *op.cit.*, pp. 477–478. There has been no similar publicly available study undertaken by any author of RCA use by regional and pluri-lateral peacekeeping bodies.
202. Fry, J. (2010) *op.cit.*, p. 19.
203. *Ibid.*
204. Knickerbocker, B. Can New Arms Cut Casualties?, *Christian Scientist Monitor*, 11th March 2003; Walker, T. U.N. Troops Tear-Gas Bosnian Serbs, Mobs Smash Peacekeeping Vehicles in Karadzic's Towns as Revolt Spreads, *Globe & Mail* (Canada), 29th August 1997, both cited in Fry, J. (2010) *op.cit.*, p. 487.
205. World Briefs: Liberia: Students Confront U.N. Peacekeepers, *Miami Herald*, 21st October 2004; Paye–Layleh, J. Muslim–Christian Violence Erupts in Liberian Capital, Head of State Orders Round-the-Clock Curfew, *Associated Press*, 29th October 2004, both cited in Fry, J. (2010) *op.cit.*, p. 492.
206. Rizvi, H. "Massacre" Charged in U.N. Raid on Slum, *IPS-Inter Press Service*, 14th July 2005 (asserting that at least 25 people were killed in this operation, including a street vendor and a mother with her two children); Lacey, M. U.N. Troops Fight Haiti's Gangs One Battered Street at a Time, *New York Times*, 10th February 2007, both cited in Fry, J. (2010) *op.cit.*, p. 490.
207. Cameron, G., Pate, J., McCauley, D. and DeFazio, L. (1999) WMD Terrorism Chronology: Incidents Involving Sub-National Actors and Chemical, Biological, Radiological, and Nuclear Materials, *The Non-Proliferation Review*, volume 7, number 2, Summer 2000 pp. 157–174; Pate, J., Ackerman, G. and

McCloud, K. (2000) WMD Terrorism Chronology: Incidents Involving Sub-National Actors and Chemical, Biological, Radiological, and Nuclear Materials, April 2001, available from the website of the Monterey Institute's Center for Nonproliferation Studies, available at http://cns.miis.edu/reports/cbrn2k.html (accessed 1st December 2014). Although further annual WMD terrorism chronologies were published by CNS, tear gas was excluded from the subsequent scope of the research.
208. Cases occurred in: Belgium (1), Colombia (1), Denmark (1), France (1), Israel (1), Japan (3), Jordan (1), the Netherlands (1), Panama (1), Russia (1), South Africa (2), Taiwan (1), the UK (1), the US (9). Compiled from data in Cameron, G., Pate, J., McCauley, D. and DeFazio, L. (2000) *op.cit.*
209. Cases occurred in: Austria (1), Benin (1), Bosnia-Herzegovina (2), Canada (3), Colombia (1), Fiji (1), Haiti (1), Iran (1), Israel (3), Japan (1), Lebanon (1), Macedonia (1), Nigeria (1), Peru (1), Poland (1), Russia (1), Senegal (1), South Africa (1), Turkey (1), the United States (7), Former Yugoslavia (1), Zimbabwe (1). Compiled from data in: Pate, J., Ackerman, G. and McCloud, K. (2001) *op.cit.*
210. For example on 29th April 2000 in Veles, Macedonia, an unknown assailant(s) threw several tear-gas ampoules into a group of primary school pupils, injuring 73 people. Pate, J., Ackerman, G. and McCloud, K. (2001) *op.cit.*
211. For example on 24th March 2000, a tear gas canister was released in a nightclub in Durban, South Africa. In the stampede to escape the tear gas, youths trampled each other, and a brick wall collapsed on several people. Thirteen died, many due to severe suffocation, according to doctors. As many as 150 people were injured, including those from tear gas inhalation. Pate, J., Ackerman, G. and McCloud, K. (2001) *op.cit.*
212. On 12th December 1999, the FARC reportedly utilized RCAs against a naval base and police station in Jurado, Colombia. The FARC reportedly used tear gas in combination with other weapons (presumably firearms) and completely destroyed the police garrison. Cameron, G., Pate, J., McCauley, D. and DeFazio, L. (2000) *op.cit.*, p. 173.
213. On 7th October 2000, Hezbollah guerrilla fighters fired bullets and tear gas canisters at Israeli soldiers. No indication of the scale of the attack is given. Pate, J., Ackerman, G. and McCloud, K. (2001) *op.cit.*
214. Pate, J., Ackerman, G. and McCloud, K. (2001) *op.cit.*
215. Internet search of leading English language media outlets, covering 2000–2014, was undertaken by the author during April 2015.
216. LTTE used CS gas to attack soldiers, *Lanka Daily News*, 18th September 2008.
217. Sri Lankan Ministry of Defence, Army Ready for Any Type of LTTE Attack – Commander, 18th September 2008, available at http://www.defence.lk/PrintPage.asp?fname=20080918_02 (accessed 2nd April 2015); see also Troops 5 kms from Kilinochchi City Limits: Tigers Pull Out the Gas, 22nd September 2008, Permanent Mission of Sri Lanka to the UN Office at Geneva, previously available at http://www.lankamission.org/content/view/882/9/ (last accessed 13th December 2011, now removed).
218. See for example: Avant, D., *The Market for Force: The Consequences of Privatizing Security*, Cambridge: Cambridge University Press, 2005; Singer, P., *Corporate Warriors: The Rise of the Privatizes Military Industry*, Ithaca, NY: Cornell University Press, 2003.
219. 2005 Use of Gas by Blackwater Leaves Questions, *New York Times*, 10th January 2008.

220. *New York Times* (10th January 2008) *op.cit.*
221. *Ibid.*
222. Human Rights Watch, Gold's Costly Dividend, Human Rights Impacts of Papua New Guinea's Porgera Gold Mine, 1st February 2011.
223. Human Rights Watch interview #39, Porgera, 9th May 2010, in Human Rights Watch (February 2011) *op.cit.*, p. 54.
224. See for example: Chiara-Gillard, E. Business Goes to War: Private Military/Security Companies and International Humanitarian Law, *International Review of the Red Cross*, volume 88, number 863, September 2006, pp. 525–572.
225. See for example: Droege, C. *Private Military and Security Companies and Human Rights: A Rough Sketch of the Legal Framework*, Swiss Initiative on PMCs/PSCs Workshop in Küsnacht, 16th–17th January 2006, available at http://www.eda.admin.ch/psc (accessed 1st December 2014); Kontos, A. "Private" Security Guards: Privatized Force and State Responsibility Under International Human Rights Law, *Non-State Actors and International Law*, number 4, 2005, pp. 228–237.
226. Montreux Document on Pertinent International Legal Obligations and Good Practices for States related to Operations of Private Military and Security Companies during Armed Conflict, United Nations, General Assembly Sixty-Third Session, Agenda item 76, Status of the Protocols Additional to the Geneva Conventions of 1949 and relating to the protection of victims of armed conflicts, A/63/467–S/2008/636, 6th October 2008.
227. For further information see Federal Department of Foreign Affairs, Switzerland, International Code of Conduct, available at https://www.eda.admin.ch/eda/en/fdfa/foreign-policy/international-law/international-humanitarian-law/private-military-security-companies/international-code-conduct.html (accessed 2nd April 2015); Geneva Academy, Academy Briefing No. 4, The International Code of Conduct for Private Security Service Providers, August 2013.

4 Means of Delivering or Dispersing Riot Control Agents

1. Salem, H. Gutting, B. and Kluchinsky, T. et al. (2008) *op.cit.*, pp. 444, 453, and 458.
2. Symons, M. et al. Less Lethal Technologies, Review of Commercially Available and Near-market Products for the Association of Chief Police Officers, UK Home Office Scientific Development Branch, June 2008, p. 28; Janes Police and Homeland Security Equipment, 2009–2010, McBride, M. (ed), 2009, IHS Janes, p. 374; Geneva Academy, Riot Control Agents, Weapons Law Encyclopaedia, undated, available at http://www.weaponslaw.org/ (accessed 2nd July 2015).
3. Geneva Academy, Weapons Law Encyclopaedia, undated, *op.cit.*
4. Symons, M. et al. (June 2008) *op.cit.*, p. 30.
5. Geneva Academy, Weapons Law Encyclopaedia (undated) *op.cit.*
6. Geneva Academy, Weapons Law Encyclopaedia (undated) *op.cit.*
7. Geneva Academy, Weapons Law Encyclopaedia (undated) *op.cit.*
8. See Pepperball Technologies website, available at http://www.pepperball.com/.
9. See FN Herstal website, available at http://www.fnherstal.com/.
10. Algeria urged to allow peaceful protests, Amnesty International, News, 11th February 2011.
11. Bahrain: Reform shelved, repression unleashed, Amnesty International, MDE 11/062/2012, November 2012.

12. Blood on the Streets, The Use of Excessive Force During Bangladesh Protests, Human Rights Watch, 1st August 2013.
13. Report of the office of the Special Rapporteur for freedom of expression, Dr. Catalina Botero, Special Rapporteur for Freedom of Expression, OEA/Ser.L/V/II.147, Doc. 1, 5th March 2013.
14. Egypt: police continues to use lethal weapons during demonstrations, 7th March 2013, FIDH.
15. Greece: A law unto themselves: A culture of abuse and impunity in the Greek police, Amnesty International, EUR 25/005/2014, 3rd April 2014.
16. Haiti: Submission to the UN Human Rights Committee: 112th Session of the UN Human Rights Committee, 7–31 October 2014, Amnesty International, AMR 36/012/20147, October 2014.
17. Human Rights Watch, After the Coup: Ongoing Violence, Intimidation and Impunity in Honduras, 20th December 2010.
18. United Nations, Human Rights Council, Report of the Special Rapporteur on the promotion and protection of the right to freedom of opinion and expression, Frank La Rue, A/HRC/17/27/Add.1, 27th May 2011.
19. UN, Human Rights Council, Report of the Special Rapporteur on Extrajudicial, Summary or Arbitrary Executions, Philip Alston, A/HRC/14/24/Add.1, 18th June 2010.
20. Israel and Occupied Palestinian Territories: Trigger-happy: Israel's use of excessive force in the West Bank, Amnesty International, MDE 15/002/2014, 27 February 2014.
21. Malaysia: FIDH condemns violent and massive crackdown on peaceful protesters, Paris-Bangkok, FIDH, 11th July 2011.
22. Newsletter: Freedom of Expression in Eastern Africa, Article 19, 5th October 2012.
23. Human Rights Watch, Descent into Chaos: Thailand's 2010 Red Shirt Protests and the Government Crackdown, May 2011.
24. Annual report 2011, Amnesty International, Togo entry.
25. Tunisia in revolt, State violence during anti-government protests, Amnesty International, MDE 30/011/2011, February 2011.
26. Turkey: End Incorrect, Unlawful Use of Teargas Dozens Injured as Police Fired Teargas Canisters Directly at Protesters, 17th July 2013.
27. UN, Human Rights Council, A/HRC/16/52/Add.3, 11th October 2010, paragraph 58.
28. High Stakes: Political Violence and the 2013 Elections in Kenya, Human Rights Watch, 8th February 2013.
29. Unuvar, U., Yilmaz, D., Ozyildirim I., Kutlu, L. and Fiancani, S.B., Medical Evaluation of Gezi Cases, Human Rights Foundation of Turkey, December 2013.
30. Human Rights Watch, Turkey: End Incorrect, Unlawful Use of Teargas, Dozens Injured as Police Fired Teargas Canisters Directly at Protesters, 17th July 2013.
31. Human Rights Watch (17th July 2013) *op.cit.*
32. UN, Human Rights Council, Report of the Special Rapporteur on Extrajudicial, Summary or Arbitrary Executions, Philip Alston, A/HRC/14/24/Add.1, 18th June 2010.
33. Amnesty International, Iran: Election Contested, Repression Compounded, MDE 13/123/2009, 10th December 2009; see also United Nations, A/HRC/14/24/Add.1 (18th June 2010) *op.cit.*, p. 146.
34. Amnesty International (December 2009) *op.cit.*, p. 34.

35. Turkey: Gezi, one year on, International Federation for Human Rights (FIDH), the Human Rights Association (IHD) and the Human rights foundation of Turkey (HRFT), May 2014, available at https://www.fidh.org/IMG/pdf/turkey_avril_2014_uk_web.pdf. (accessed 11th August 2015), pp. 8 and 21.
36. FIDH/IHD/HRFT (May 2014) op.cit., pp. 8 and 21.
37. Crowley, M. The Use of Riot Control Agents in Law Enforcement, in Casey-Maslen, S. (ed), (2014) op.cit., p. 346.
38. Amnesty International and Omega Research Foundation, The Human Rights Impact of Less Lethal Weapons and Other Law Enforcement Equipment, ACT 30/1305/2015, April 2015.
39. Agence France Presse, 30th January 1999, Kenya. Dozens Injured as Kenyan Police Tackle Eco-Protestors.
40. Human Rights Watch, Malaysia: Investigate Use of Force Against Peaceful Rally, 14th November 2008.
41. Amnesty International, Gezi Park Protests: Brutal Denial of the Right to Peaceful Assembly in Turkey, EUR 44/022/2013, October 2013.
42. Amnesty International, Côte d'Ivoire: Excessive Use of Force to Repress Banned Demonstration, AFR 31/003/2004, 26th March 2004.
43. Amnesty International, South Korea: Call for Unimpeded Access to Food, Water and Necessary Medical Treatment for Ssanyong Striking Workers, ASA 25/007/2009, 31st July 2009.
44. Amnesty International, Peru: Bagua, Six Months On: "Just Because We Think and Speak Differently, They Are Doing This Injustice to Us", AMR 46/017/2009, 2nd December 2009.
45. Human Rights Watch, Diamonds in the Rough, Human Rights Abuses in the Marange Diamond Fields of Zimbabwe, June 2009.
46. Symons et al. (June 2008) op.cit., pp. 20–26; Amnesty International and Omega Research Foundation (April 2015) op.cit., p. 26; Geneva Academy, Weapons Law Encyclopaedia (undated) op.cit.
47. Geneva Academy, Weapons Law Encyclopaedia (undated) op.cit.
48. Amnesty International and Omega Research Foundation, The Human Rights Impact of Less Lethal Weapons and Other Law Enforcement Equipment, ACT 30/1305/2015, April 2015.
49. Ibid.
50. Amnesty International, Gezi Park Protests: Brutal Denial of the Right to Peaceful Assembly in Turkey, EUR 44/022/2013, October 2013.
51. Amnesty International (October 2013) op.cit., p. 19.
52. NTV, Vali Mutlu: İlaçlı su, kimyasal değil, 16th June 2013, available at http://www.ntvmsnbc.com/id/25449362/, as cited in Amnesty International (October 2013) op.cit.
53. Hürriyet Daily News, Substance in Water Cannons in Gezi Park Protests Harmful and Criminal, Experts Say, 18th June 2013.
54. Amnesty International (October 2013) op.cit., p. 19.
55. Ibid.
56. Human Rights Watch (June 2009) op.cit., p. 32.
57. Ibid., p. 31.
58. Ibid., p. 32.
59. Crowley, M. Drawing the Line: Regulation of "Wide Area" Riot Control Agent Delivery Mechanisms Under the Chemical Weapons Convention, Bradford Non-Lethal Weapons Project and Omega Research Foundation, April 2013; BNLWRP, ISS and ORF, Destruction by Turkey of all remaining 120mm mortar munitions

containing CS: A briefing note for CWC States Parties, 12th September 2011; BNLWRP, ISS and ORF, The production and promotion by a Russian Federation company of a range of munitions containing chemical irritants: A briefing note for CWC States Parties, 12th September 2011.
60. Stockholm International Peace Research Institute/Perry Robinson, J. and Leitenberg, M., *The Problem of Chemical and Biological Warfare, volume 1: The Rise of CB Weapons*, Stockholm, 1971.
61. United Nations Monitoring and Verification Committee [UNMOVIC], Compendium, The Chemical Weapons Programme, undated.
62. Henan Weida Military and Police Equipment Co., Ltd, Product Manual, undated, distributed at Asia Pacific China Police Expo 2014, China National Convention Center, Beijing, China, 20th–23rd May 2014, pp. 44–45 (copy of publication held by the Omega Research Foundation). Material promoting the WDTC-Q38 is also available on the company's website at: http://www.henanweida.com/cpzs_xx.asp?id=929 (accessed 8th July 2015).
63. Henan Weida Military and Police Equipment Co., Ltd, (undated) *op.cit.*, pp. 44–45.
64. Henan Weida Military and Police Equipment Co., Ltd, (undated) *op.cit.*, pp. 44–45.
65. Henan Weida Military and Police Equipment Co., Ltd, (undated) *op.cit.*, pp. 44–45; Henan Weida Military and Police Equipment Co., Ltd website, WDTC-Q18.
66. MSI Delivery Systems Inc., The AB2K-MMADS™, available at http://msideliverysystems.com/ (accessed 8th July 2015).
67. MSI Delivery Systems Inc., Mission Capabilities, available at http://msideliverysystems.com/mission-specs (accessed 8th July 2015). The company further stated that it "only provides the non-toxic training smoke. Additives for 'Irritants' are provided by the customer through their local suppliers." MSI Delivery Systems Inc., Mission Capabilities (undated) *op.cit.*
68. MSI Delivery Systems Inc. The AB2K-MMADS™ (accessed 8th July 2015) *op.cit.*
69. *Ibid.*
70. MSI Delivery Systems Inc., AB2K Capabilities, available at http://msideliverysystems.com/ab2k-mmads-variants (accessed 8th July 2015).
71. *Ibid.*
72. *Ibid.*
73. MSI Delivery Systems Inc., Multi-Mission Aerosol Delivery System press release, 10th January 2010, available at http://ab2kmmads.com/ab2kdocs/Press%20Release%20Intro-11112013.pdf (accessed 7th July 2015).
74. IronFist, NonLethal Technologies, Undated Brochure, Distributed at IDEX 2013, Abu Dhabi, the United Arab Emirates, 17th–21st February 2013.
75. *Ibid.*
76. *Ibid.*
77. *Ibid.*
78. *Ibid.*
79. IronFist, NonLethal Technologies, Revised But Undated, Brochure, available from Nonlethal Technologies, Inc. website, available at http://www.nonlethaltechnologies.com/pdf/NLT-BROCHURE.pdf (accessed 2nd April 2015).
80. IronFist, NonLethal Technologies, Product Brochure (Revised But Undated) *op.cit.*
81. IDEX 2015 was held in Abu Dhabi, the United Arab Emirates, from 22nd to 26th February 2015. The IDEX website provides information about NonLethal

Technologies participation, and includes a link to its product brochure which includes details of IronFist and Cobra, available at http://www.idexuae.ae/page.cfm/Link=1/t=m/goSection=1 (accessed 2nd April 2015).
82. Combined Systems Inc. Product Catalogue, Venom Launching Systems, Undated, Catalogue Distributed at IDEX 2013, Abu Dabhai, United Arab Emirates, 17th–21st February 2013. Also available at https://www.combinedsystems.com/userfiles/pdfs/CSI_MIL_Product_Catalog_2010.pdf (accessed 14th February 2013), p. 4.; Combined Systems Inc. Non-lethal Tube Launched Munition Systems, Venom, V-10, undated, p. 4.
83. Combined Systems Inc. Product Catalogue (undated) *op.cit.*, p. 4.
84. See Combined Systems Inc. Product Catalogue (undated) *op.cit.*, p. 4; Non-lethal Tube Launched Munition Systems, Venom, V-10, Combined Systems, Inc. (undated) *op.cit.*, p. 4.
85. Combined Systems Inc, Venom MC Launching System, available at https://www.combinedsystems.com/_pdf/ SpecSheets/SmallVENOM%20MC%20Flyer-JAN2013.pdf (accessed 2nd April 2015).
86. Combined Systems Inc, Venom MC Launching System (January 2013) *op.cit.*
87. *Ibid.*
88. *Ibid.*
89. Combined Systems Awarded VENOM System Contract, News Release, 16th June 2011, available at https://www.combinedsystems.com/userfiles/pdfs/NL_TLMS_VENOM_Press_ Release_6_15_2011.pdf (accessed 14th February 2013).
90. Combined Systems Inc, product catalogue, available at https://www.combinedsystems.com/userfiles/CTS_ LL_Catalog_2011.pdf (accessed 2nd April 2015).
91. B'TSelem, Crowd Control: Israel's Use of Crowd Control Weapons in the West Bank, January 2013.
92. Non-Lethal Munitions Section, *Volume 12 Ordnance and Munitions (English language version), Russia's Arms and Technologies. The XXI Century Encyclopedia,* version 2006.1eng, Arms and Technologies Publishing House, 5th May 2009 (copy held by the author).
93. Arms and Technologies Publishing House (2009) *op.cit.*
94. Information taken from China Ordnance Equipment Research Institute, Undated Catalogue, Distributed at Asia Pacific China Police Expo 2012. (Information is from an unofficial translation of the Chinese original on file with the Omega Research Foundation.)
95. China Ordnance Equipment Research Institute (undated catalogue) *op.cit.*, pp. 8 and 13.
96. *Ibid.*
97. *Ibid.*, p. 8.
98. *Ibid.*, p. 13.
99. *Ibid.*, p. 13.
100. Arms and Technologies Publishing House (2009) *op.cit.*
101. *Ibid.*
102. Department of Scientific and Technical Development, Chinese People's Armed Police Force, Brochure, copy distributed Asia Pacific China Police Expo 2002.
103. *Ibid.*
104. PP87 82mm Tear Gas Mortar Munition, China Ordnance Industry Group, State-Owned No. 672 Factory, Undated Brochure, Distributed at China (Beijing) International Exhibition and Symposium on Police and Anti-Terrorism Technology and Equipment (CIPATE 2011) (Information is from an unofficial translation of the Chinese original held by the Omega Research Foundation.)

105. Arms and Technologies Publishing House (2009) *op.cit.*
106. Non-Lethal Munitions Section, *Volume 12 Ordnance and Munitions (English language version), Russia's Arms and Technologies. The XXI Century Encyclopedia,* version 2006.1eng, Arms and Technologies Publishing House, 5th May 2009.
107. Foss, C. Turkey Details 120mm Automatic Mortar, *Jane's Defence Weekly*, 12th November 2003.
108. Turkish Defence Industry Catalogue, 2007, previously available at http://www2.ssm.gov.tr/katalog2007/data/24509/9/uruning/uruning34.html (last accessed 30th September 2010, subsequently removed).
109. MKEK was promoting the CS mortar round till at least mid-November 2009 on its website, previously available at http://mkekexport.com/ammunition.html (last accessed 16th November 2009, subsequently removed).
110. The 7th International Defense Industry Fair was held by the Turkish Armed Forces Foundation under the auspices of the Turkish Ministry of Defense at the Ankara Hippodrome between 27th and 30th September 2005. Over 400 companies from 49 countries exhibited their goods and services at IDEF, of which 108 were from Turkey.
111. AAD 2010 was held at Air Force Base Ysterplaat, Cape Town, South Africa, from 21st to 25th September 2010. For further information see AAD 2010 website, available at http://www.aadexpo.co.za/ (accessed 28th September 2010).
112. Correspondence from Ambassador Dogan, Permanent Representative of Turkey to the OPCW, to BNWLRP, ORF and ISS, 25th February 2011 (copy on file with the authors); correspondence from Mr Utkan, Counsellor, Permanent Representation of Turkey to the OPCW, to BNLWRP, ORF and ISS, 8th July 2011 (copy on file with the authors).
113. US Army, Picatinny Centre, Non-Lethal Artillery Structural Firing (FYO4) Purchase Order Contract in Support of the FY04 155mm Non-Lethal Artillery Projectile Program, Contract Number W15QKN-04-M-0328, 14th September 2004, previously available at http://www.sunshine-project.org/incapacitants/jnlwdpdf/XM1063.pdf (accessed 14th February 2013).
114. McCormick, J., Presentation on 155mm XM1063 Non-Lethal Personnel Suppression Projectile, General Dynamics OTS, National Defense Industrial Association, 42nd Annual Armament Systems: Gun and Missile Systems Conference and Exhibition, 23rd–26th April 2007, Charlotte, North Carolina, available at http://www.dtic.mil/ndia/2007gun_ missile/ GMTuePM2/ McCormickPresentation.pdf (accessed 14th February 2013), p. 4.
115. McCormick, J. (2007) *op.cit.*, p. 4.
116. For information on the M864 projectile see Globalsecurity.org, M864 Base Burn DPICM, available at http://www.globalsecurity.org/military/systems/munitions/m864.html (accessed 14th February 2013).
117. McCormick, J. 155mm XM1063 Non-Lethal Personnel Suppression Projectile. Presentation to the 41st Annual Armament Systems: Guns and Missile Systems, Conference & Exhibition, National Defense Industrial Association (NDIA), Sacramento, US, 27th–30th March 2006, available at http://www.dtic.mil/ndia/2006garm/tuesday/mccormick.pdf (accessed 14th February 2011), p. 12.
118. McCormick, J. (2006) *op.cit.*
119. NLOS-C Non-Lethal Personnel Suppression, US Army ARDEC Brochure, 2005, as cited in Davison, N. (2007) *op.cit.*, p. 34.

120. Hambling, D., US Weapons Research Is Raising a Stink, *The Guardian*, 10th July 2008, available at http://www.guardian.co.uk/science/2008/jul/10/weaponstechnology.research (accessed 14th February 2013).
121. US Army ARDEC, Solicitation, R – 155mm XM1063 Non-Lethal Artillery Engineering Support Contract, Solicitation Number: W15QKN-04-X-0819, 30th September 2004, available at http://www.fbodaily.com/archive/2004/10-October/02-Oct-2004/FBO-00687159.html (accessed 14th February 2013).
122. McCormick, J. (2006) *op.cit.*, p. 7.
123. US Army ARDEC (2004) Solicitation (Modification) R –155mm XM1063 Non-Lethal Artillery Engineering Support Contract (Ref: W15QKN-04-X-0819). *FBO Daily*, 30th September 2004, as cited in Davison, N. (2007) *op.cit.*, p. 34.
124. Joint Non-Lethal Weapons Directorate, Non-Lethal Weapons (NLW) Reference Book, 30th June 2011, available at http://publicintelligence.net/dod-non-lethal-weapons-2011/ (accessed 14th February 2013), p. viii.
125. Hambling, D. (2008) *op.cit.*
126. *Ibid.*
127. General Dynamics Website, available at http://www.gd-ots.com/agent_dispensing.html (accessed 22nd November 2013).
128. Hambling, D. US Military Malodorant Missiles Kick Up a Stink, *New Scientist*, issue 2867, 4th June 2012, available at http://www.newscientist.com/article/mg21428676.800-us-military-malodorant-missiles-kick-up-a-stink.html (accessed 14th February 2013).
129. Arms and Technologies Publishing House (2009) *op.cit.*
130. *Ibid.*
131. *Ibid.*
132. Skunk Riot Control Copter, Desert Wolf website, available at http://www.desert-wolf.com/dw/products/unmanned-aerial-systems/skunk-riot-control-copter.html (accessed 2nd April 2015).
133. Desert Wolf website (Undated) *op.cit.*
134. *Ibid.*
135. *Ibid.*
136. Kelion, L. African Firm Is Selling Pepper-Spray Bullet Firing Drones, BBC News, 18th June 2014, available at http://www.bbc.co.uk/news/technology-27902634 (accessed 2nd April 2015).
137. BBC News (18th June 2014) *op.cit.*
138. Martin G. Desert Wolf establishing production facilities in Oman for Skunk UAV, other products, 9th April 2015, available at http://www.defenceweb.co.za/index.php?option=com_ content&view=article&id=38716 (accessed 7th July 2015); Martin, G. Defenceweb, Desert Wolf Aiming to Build 1000 Skunk Riot Control UAVs a Month, 16th September 2014, available at http://www.defenceweb.co.za/index.php?option=com_ content&view=article&id=36281:desert-wolf-aiming-to-build-1000-skunk-riot-control-uavs-a-month&catid=35:Aerospace&Itemid=107 (accessed 2nd April 2015).
139. Technorobot, Riotbot Applications, available at http://www.technorobot.eu/en/riotbot.html (accessed 2nd April 2015).
140. Technorobot, Riotbot, specifications, http://www.technorobot.eu/en/riotbot_specifications.htm (accessed 14th February 2013). According to marketing materials published by Pepperball Technologies Inc., which manufacture the TAC 700 and related munitions, the launcher averages 700 rounds per minute in full

automatic mode with up to 60 ft. target accuracy and up to 150 ft. accuracy for saturating an area with pepper. The TAC 700 utilizes the 3-gram PAVA pepper projectile. See PepperBall Technologies Inc., PepperBall Products, Military, previously available at http://www.pepperball.com/mil/products.html#proj (last accessed 18th January 2013); see also PepperBall Technologies Inc., The TAC 700 Launcher: PepperBall Tactical Automatic Carbine, Brochure, available at http://www.technorobot.eu/en/pdf/tac700.pdf (accessed 8th July 2015).
141. Technorobot, Riotbot, Specifications, available at http://www.technorobot.eu/en/riotbot_specifications.html (accessed 8th August 2015).
142. *Ibid.*; and Technorobot, Riotbot Brochure, undated, http://www.technorobot.eu/en/pdf/dip_ing.pdf (accessed 2nd April 2015).
143. See Technorobot, Riotbot Applications, *Ibid.*
144. *Ibid.*
145. QinetiQ North America, MAARS Advanced Armed Robotic System, available at http://www.qinetiq-na.com/products/unmanned-systems/maars/ (accessed 2nd April 2015).
146. QinetiQ North America, Modular Advanced Armed Robotic System (MAARS), Data Sheet, Undated, available at https://www.qinetiq-na.com/wp-content/uploads/data-sheet_ maars.pdf (accessed 2nd April 2015), p. 1.
147. QinetiQ North America, *Ibid.*
148. QinetiQ North America, Unmanned Robotic Systems, edition 2, Undated, available at https://www.qinetiq-na.com/wp-content/uploads/catalog_ unmanned-systems.pdf (accessed 2nd April 2015), p. 12.
149. QinetiQ North America, Unmanned Robotic Systems, edition 2 (undated) *op.cit.*, p. 12.
150. QinetiQ North America, MAARS Data sheet (undated) *op.cit.*, p. 2.
151. Dubiel, J. Robots Can Stand in for Soldiers During Risky Missions, 11th August 2008, available at http://www.army.mil/article/11592/robots-can-stand-in-for-soldiers-during-risky-missions/ (accessed 2nd April 2015).
152. QinetiQ North America (2nd April 2015) *op.cit.*
153. QinetiQ North America, QinetiQ North America Ships First MAARS Robot, 5th June 2008, available at http://www.qinetiq.com/news/pressreleases/Pages/qna-ships-first-maars-robot.aspx (accessed 2nd April 2015).
154. QinetiQ North America, MAARS Modular Advanced Armed Robotic System, Warfighter Controls the Escalation of Force, available at http://www.qinetiq-na.com/wp-content/uploads/2011/12/pdf_ maars.pdf (last accessed 22nd January 2013); QinetiQ North America, MAARS Data sheet (undated) *op.cit.*, p. 2; QinetiQ North America, MAARS Product Overview *op.cit.*, p. 2.
155. QinetiQ North America, QinetiQ North America Ships First MAARS Robot, 5th June 2008, available at http://www.qinetiq.com/news/pressreleases/Pages/qna-ships-first-maars-robot.aspx (accessed 2nd April 2015).
156. Markoff, J. War Machines: Recruiting Robots for Combat, *The New York Times*, 27th November 2010, available at http://www.nytimes.com/2010/11/28/science/28robot.html?pagewanted=all&_ r=0 (accessed 2nd April 2015).
157. UGV Models Face Off Over Firepower, Load Carrying, Armytimes.com, 12th October 2013, available at http://www.armytimes.com/article/20131012/NEWS/310140003/UGV-models-face-off-over-firepower-load-carrying (accessed 2nd April 2015).
158. Stewart, J. Marine Corps Considers New Unmanned Tank, Micro-Drones, *The Marine Times*, 30th January 2015, available at http://www.marinecorpstimes.

com/story/military/pentagon/2015/01/30/new-unmanned-vehicle-has-tracks-machine-gun/22537723/ (accessed 2nd April 2015).

5 Application of the Chemical Weapons Convention to ICA Weapons, Riot Control Agents and Related Means of Delivery

1. Organisation for the Prohibition of Chemical Weapons (OPCW), Convention on the Prohibition of the Development, Production, Stockpiling and Use of Chemical Weapons and on Their Destruction (Chemical Weapons Convention [CWC]), 1993. The CWC text, details of States Parties and all OPCW documents are available from OPCW website: http://www.opcw.org. (accessed 19th August 2015).
2. An additional Signatory State (Israel) has signed the CWC, thus rendering political support to the objectives and principles of the CWC and committing itself to not undermining its objectives. Only four Non-Signatory States (Angola, DPRK, Egypt and South Sudan) have not taken relevant action on the CWC.
3. OPCW, About the OPCW, available at http://www.opcw.org (accessed 19th August 2015).
4. OPCW, CWC (1993) *op.cit.*, Articles VIII.1 to VIII.8.
5. *Ibid.*, Articles VIII.37 to VIII.47.
6. *Ibid.*, Articles VIII.37 to VIII.47.
7. *Ibid.*, Article I.1.b.
8. *Ibid.*, Article I.1.a.
9. Under Article X, States are permitted to "conduct research into, develop, produce, acquire, transfer or use means of protection against chemical weapons, for purposes not prohibited under this Convention". In order to increase "transparency of national programmes related to protective purposes", under Article X each State Party is obliged to "provide annually to the Technical Secretariat information on its programme". [See OPCW, CWC (1993) *op.cit.*, Article X.]
10. According to Krutzsch and Trapp: "Chemical defence programmes sometimes include threat assessment studies related to new, hitherto unknown, agents (new lethal agents, new incapacitants). When such assessments include research work that borders on what would be undertaken in chemical weapons development, they are no longer consistent with the object and purpose of the Convention." Krutzsch, W. and Trapp, R. Article II: Definitions and Criteria, pp. 73–104, in Krutzsch, W., Myjer, E. and Trapp, R. (eds), *The Chemical Weapons Convention: A Commentary*, Oxford: Oxford University Press, 2014, p. 93.
11. OPCW, CWC (1993) *op.cit.*, Article I.1.c.
12. *Ibid.*, Article I.1.d.
13. *Ibid.*, Article I.3.
14. *Ibid.*, Article I.4.
15. Krutzsch, W. Article I: General Obligations, in Krutzsch, W., Myjer, E. and Trapp, R. (2014) *op.cit.*, p. 64.
16. Krutzsch, W. (2014) *op.cit.*, p. 64.
17. OPCW, CWC (1993) *op.cit.*, Article XXII.
18. *Ibid.*, Article II.2.
19. Krutzsch, W. and Trapp, R. Article II: Definitions and Criteria, in Krutzsch, W., Myjer, E. and Trapp, R. (eds), (2014) *op.cit.*, p. 85.
20. *Ibid.*,

21. OPCW, CWC (1993) *op.cit.*, Article II.2.
22. *Ibid.*, Annex on Chemicals, Guidelines for Schedules of Chemicals.
23. *Ibid.*, Annex on Chemicals, Schedule 1.
24. *Ibid.*, Annex on Chemicals, Schedule 2.
25. *Ibid.*, Annex on Chemicals, Schedule 3.
26. *Ibid.*, Article II.2.
27. *Ibid.*, Article II.1.
28. Krutzsch, W. and Trapp, R. (2014) *op.cit.*, p. 77.
29. Meselson, M. and Perry Robinson, J. New Technologies and the Loophole, Editorial, *Chemical Weapons Convention Bulletin* 23 March 1994, Harvard Sussex Program, p. 1.
30. Meselson, M. and Perry Robinson, J. (1994) *op.cit.*, p. 1.
31. *OPCW, CWC (1993) op.cit.*, Article II.9.d.
32. *Ibid.*, Article II.1.a.
33. Additional terms that were left undefined or inadequately defined in the CWC, and have not subsequently been adequately addressed by the OPCW policy-making organs, include "method of warfare", "temporary incapacitation", "life processes", "munitions and devices" and "toxicity".
34. Meselson, M. and Perry Robinson, J. (1994) *op.cit.*, p. 2.
35. UN, Vienna Convention on the Law of Treaties, 23rd May 1969, Vienna, 1155 UN Treaty Series (UNTS) 331.
36. UN Treaty Collection website, available at https://treaties.un.org/pages/ViewDetailsIII.aspx?src=TREATY&mtdsg_ no=XXIII-1&chapter=23&Temp=mtdsg3&lang=en (accessed 9th July 2015). There are a further 45 Signatory States.
37. UN, Vienna Convention on the Law of Treaties (1969) *op.cit.*, Article 31.1.
38. *Ibid.*, Article 31.3.a–c.
39. *Ibid.*, Article 32.
40. Analysis was undertaken of all OPCW documents publicly available on the OPCW website (http://www.opcw.org) since the organization's inception up until 19th August 2015.
41. OPWC, CWC (1993) *op.cit.*, Article IX.
42. *Ibid.*, Articles IX.8 to IX.25.
43. *Ibid.*, Articles IX.3 to IX.7, IX.23.
44. For example, see Kelle, A. The CWC After Its First Review Conference: Is the Glass Half Full or Half Empty? *Disarmament Diplomacy*, issue 71, June–July 2003, pp. 31–40; Krutzsch, W. Never Under Any Circumstances: The CWC in the Third Year After Its First Review-Conference, *CBW Conventions Bulletin* 68, 2005, pp. 1, 6–12.; Perry Robinson, J. NonLethal Warfare and the Chemical Weapons Convention, Further HSP Submission to the OPCW Open-Ended Working Group on Preparations for the Second CWC Review Conference, 24th October 2007.
45. Krutzsch, W. and Dunworth, T. Article VIII: The Organization, in Krutzsch, W., Myjer, E. and Trapp, R. (ed.), (2014) *op.cit.*, pp. 258–260; Scott, D. Logjam in the OPCW: Time to Limit Consensus? CWC Special Paper Number 1, December 2002, p. 4, Acronym Institute; Krutzsch, W. Ensuring True Implementation of the CWC, *CBW Conventions Bulletin* 76/77, 2007, Harvard Sussex Program, p. 16.
46. OPCW, Director General, Report of the [Ekeus] Advisory Panel on Future Priorities of the Organisation for the Prohibition of Chemical Weapons, S/951/2011, 25th July 2011, paragraph 48.
47. Feakes, D. Workshop Report: Pugwash Study Group on the Implementation of the Chemical and Biological Weapons Conventions CWC Implementation:

Balancing Confidentiality and Transparency, 15th–17th May 1998; Tucker, J. Introduction, in Tucker, J. (ed.), *The Chemical Weapons Convention: Implementation Challenges and Solutions*, Washington DC: Monterey Institute of International Studies, April 2001, pp. 1–8; Moodie, M. Issues for the First CWC Review Conference, in Tucker, J. (ed.), (2001) *op.cit.*, pp. 59–66; Sands, A. and Pate, J. CWC Compliance Issues, in Tucker, J. (ed.), (2001) *op.cit.*, pp. 17–22; Kelle, A. (2002) *op.cit.*; Krutzsch, W. (2005) *op.cit.*; Krutzsch, W. (2007) *op.cit.*

48. OPCW, Note by the Technical Secretariat, The OPCW in 2025: Ensuring a world free of chemical weapons, Office of Strategy and Policy, S/1252/2015, 6th March 2015; See also OPCW, Ekeus Report (25th July 2011) *op.cit.*
49. OPCW, CWC (1993) *op.cit.*, Article II.7.
50. OPCW, Report of the twentieth session of the Scientific Advisory Board SAB-20/1, 10th–14th June 2013, 14th June 2013, Annex 4, Director-General's Request to the Scientific Advisory Board to consider which riot control agents are subject to declaration under the Chemical Weapons Convention.
51. OPCW, Technical Secretariat, Office of Strategy and Policy, S/1177/2014, 1st May 2014, Note by the Technical Secretariat, Declaration of riot control agents: advice from the Scientific Advisory Board. See also OPCW, Report of the twentieth session of the Scientific Advisory Board (14th June 2013) *op.cit.*, Annex 4.
52. Neill, D. *Riot Control and Incapacitating Chemical Agents Under the Chemical Weapons Convention*, Defence R&D Canada: Centre for Operational Research and Analysis, technical memorandum DRDC CORA TM 2007–2022, Ottawa, 2007, p. 6.
53. Neill, D. (2007) *op.cit.*, p. 6.
54. *Ibid.*; Sutherland, R. *Chemical and Biochemical Non-lethal Weapons, Political and Technical Aspects, SIPRI Policy Paper 26*. Stockholm, Sweden: SIPRI, 2008 , p. 20.
55. Analysis was undertaken of all OPCW documents publicly available on the OPCW website (http://www.opcw.org) as of 19th August 2015.
56. Hambling D. US Military Malodrant Missiles Kick Up a Stink, *New Scientist*, 4th June 2012, p. 2.
57. Hambling D. (June 2012) *op.cit.*, p. 2.
58. US Office of Naval Research, Joint Non-Lethal Weapons Program, Fiscal Year 2015 Non-Lethal Weapons Technologies, Broad Agency Announcement, ONRBAA14–008, 5th June 2014, p. 6.
59. US Office of Naval Research (5th June 2014) *op.cit.*, p. 6.
60. OPCW, CWC (1993) *op.cit.*, Article I.5.
61. The term "method of warfare" refers to the way or ways in which a weapon is used by parties to an armed conflict in the conduct of hostilities. Use of a weapon for law enforcement purposes is not referred to as a "method of warfare". For further discussion see: Method of warfare, weapons law encyclopedia, Geneva Academy (undated) *op.cit.*
62. President Gerald Ford, Executive Order 11850 – Renunciation of Certain Uses in War of Chemical Herbicides and Riot Control Agents, 8th April 1975. 40 FR 16187 CGR, 1971–1975 Comp., 980.
63. President Gerald Ford, Executive Order 11850 (1973) *op.cit.* When the US Senate ratified the CWC in 1997, further extensions to the use of RCAs were adopted. The Senate resolved that the US is not restricted by the CWC in its use of RCAs, including use against combatants who are parties to a conflict in three cases: (1) where the US is not a party to the conflict, (2) consensual

Chapter VI peacekeeping operations and (3) Chapter VII peacekeeping operations. Additionally, the Senate imposed a condition that "[t]he President shall take no measure, and prescribe no rule or regulation, which would alter or eliminate Executive Order 11850 of April 8, 1975". [See 143 US Congressional Record. S3657 (daily ed. 24th April 1997), paragraph 26 (A) and (B).]
64. United States, Law of War Manual, Office of General Counsel, US Department of Defense, June 2015, Section 6.16.
65. See for example: US, *The Commander's Handbook on the Law of Naval Operations*, US Department of the Navy, July 2007, paragraph 10.3.2.1. For further examples of US military manuals and relevant policy statements see ICRC, Customary IHL, Practice Relating to Rule 75. Riot Control Agents.
66. See Chapter 3 for case studies detailing reported use of RCAs in conjunction with firearms by Turkish armed forces against Kurdish armed fighters in April 1999; reported employment by US military of RCAs as part of its cave-clearing techniques in Afghanistan.
67. OPCW, CWC (1993) *op.cit.*, Article II.9(d).
68. See written response by Ambassador Ledogar to the US Senate Committee on Foreign Relations. [102d US Congress, 2nd Session, Senate, Committee on Foreign Relations, hearing, 1st May 1992, Chemical Weapons Ban Negotiations Issues, S.Hrg.102–719, 1992, pp. 34–35.]
69. Editorial, New Technologies and the Loophole in the Convention, *Chemical Weapons Convention Bulletin*, number 23, March 1994, Harvard Sussex Program, p. 1.
70. See Fidler, D. The Meaning of Moscow: "Non-Lethal" Weapons and International Law in the Early 21st Century, *International Review of the Red Cross*, volume 87, number 859, September 2005, pp. 525–552; Fidler, D. Incapacitating Chemical and Biochemical Weapons and Law Enforcement Under the Chemical Weapons Convention, in Pearson, A., Chevrier, M. and Wheelis, M. (eds), *Incapacitating Biochemical Weapons*, Lanham, MD: Lexington Books, 2007, pp. 171–194; Neill, D. (2007) *op.cit.*; Von Wagner, A. Toxic Chemicals for Law Enforcement Including Domestic Riot Control Purposes Under the Chemical Weapons Convention, in Pearson, A., Chevrier, M. and Wheelis, M. (eds), *Incapacitating Biochemical Weapons*, Lanham, MD: Lexington Books, 2007, pp. 195–208.
71. Such activities would also need to be in full accordance with international humanitarian law and/or international human rights law, as appropriate.
72. *Commentary on Geneva Convention III Relative to the Treatment of Prisoners of War*, ICRC, Geneva, 1960, p. 247, as cited in Fidler, D. (2005) *op.cit.*, p. 545.
73. Von Wagner, A. (2007) *op.cit.*, p. 202.
74. On 3rd August 2007, the Multi-National Forces in Iraq reportedly used tear gas against rioting inmates at the Badoush detention centre outside of Mosul. See Ballard, K. Convention in Peril? Riot Control Agents and the Chemical Weapons Ban, *Arms Control Today*, September 2007.
75. According to Fidler, non-traditional military operations would be legitimate under international law if they were conducted pursuant to: (a) a request for peacekeeping forces from a sovereign State; (2) the authorization of peacekeeping operations by the UN Security Council Under Chapter VII of the UN Charter. Fidler, D. (2007) *op.cit.*, p. 181.
76. *Ibid.*, pp. 180–183.
77. Von Wagner, A. (2007) *op.cit.*, p. 202.

78. See in particular: Fry, J. Gas Smells Awful: UN Forces, Riot Control Agents and the Chemical Weapons Convention, *Michigan Journal of International Law*, volume 31, number 3, 2010, pp. 475–559.
79. Fry, J. (2010) *op.cit.*, p. 4.
80. *Ibid.*, p. 19.
81. Krutzsch, W. and Trapp, W. (2014) *op.cit.*, p. 103.
82. Law enforcement could be defined in ordinary parlance as "the activity of making certain that the laws of an area are obeyed". It is generally understood as referring to the exercise of police powers, especially with respect to the powers of arrest and detention, and may involve the lawful use of force. According to Chayes and Meselson: The term "law enforcement" in Art. II (9) (d) means actions taken within the scope of a nation's "jurisdiction to enforce" its national law, as that term is understood in international law. When such actions are taken in the context of law enforcement or riot control functions under the authority of the UN, they must be specifically authorized by that organization. No act is one of "law enforcement" if it otherwise would be prohibited as a "method of warfare" under Article II (9) (c).
   [Chayes, A. and Meselson, M. Proposed Guidelines on the Status of Riot Control Agents and Other Toxic Chemicals Under the Chemical Weapons Convention, *Chemical Weapons Convention Bulletin*, volume 35, March 1997, p. 15.]
83. UN, Vienna Convention on the Law of Treaties (1969) *op.cit.*, Article 31 (3) (a)–(c).
84. Law enforcement officials are considered to include:
   "all officers of the law, whether appointed or elected, who exercise police powers, especially the powers of arrest or detention.... In countries where police powers are exercised by military authorities, whether uniformed or not, or by State security forces, the definition of law enforcement officials shall be regarded as including officers of such services."
   United Nations, UN Code of Conduct for Law Enforcement Officials adopted by General Assembly resolution 34/169 of 17th December 1979, Article 1.
85. Professor Julian Perry Robinson in interview with author, 7th July 2009.
86. Krutzsch, W. and Trapp, W. (2014) *op.cit.*, p. 96.
87. *Ibid.*
88. *Ibid.*
89. Analysis was undertaken of all OPCW documents publicly available on the OPCW website up to 19th August 2015.
90. OPCW, CWC (1993) *op.cit.*, Article III(1) (e).
91. *Ibid.*, Article III(1) (e).
92. *Ibid.*, Article III(1) (e).
93. Krutzsch, W. (2014) *op.cit.*, p. 72.
94. OPCW, CWC (1993) *op.cit.*, Article III(1) (e).
95. *Ibid.*, Article III.6.
96. As with any other declaration information, this data is provided by the Technical Secretariat to States Parties on request. Author's interview of former OPCW official, 1st September 2008.
97. See for example: OPCW, Report of the OPCW on the implementation of the Convention on the Prohibition of the Development, Production, Stockpiling and Use of Chemicals and Their Destruction in 2013, Nineteenth Session of Conference of States Parties, C-19/4, 3rd December 2014.
98. OPCW, CWC (1993) *op.cit.*, Article II.1.

99. *Ibid.*, Article I.5.
100. *Ibid.*, Article II.9.
101. Krutzsch, W. (2014) *op.cit.*, p. 71.
102. *Ibid.*
103. For example, see Chayes, A. and Meselson, M. (March 1997) *op.cit.*
104. Neill, D. (2007) *op.cit.*, p. 12.
105. NATO, Research and Technology Organisation, AC/323(HFM-073)TP/65, The Human Effects of Non-Lethal Technologies, August 2006, Chapter 6, p. 9.
106. OPCW, CWC (1993) *op.cit.*, Article II.9 and Article I.5.
107. Crowley, M. Drawing the Line: Regulation of "Wide Area" Riot Control Agent Delivery Mechanisms Under the Chemical Weapons Convention, April 2013, BNLWRP & ORF, pp. 59–64.
108. Crowley (2013) *op.cit.*, pp. 59–64.
109. *Ibid.*
110. Krutzsch, W. (2014) *op.cit.*, p. 72.
111. *Ibid.*
112. See for example: Crowley, M. Drawing the Line: Regulation of "Wide Area" Riot Control Agent Delivery Mechanisms Under the Chemical Weapons Convention, April 2013, BNLWRP & ORF; BNLWRP, ISS and ORF, Destruction by Turkey of all remaining 120mm mortar munitions containing CS: A briefing note for CWC States Parties, 12th September 2011, available at http://www.omegaresearch foundation.org/assets/downloads/publications/Briefing%20Paper%20Turkey001.pdf (accessed 9th July 2015); BNLWRP, ISS and ORF, The production and promotion by a Russian Federation company of a range of munitions containing chemical irritants: A briefing note for CWC States Parties, 12th September 2011, available at http://www.brad.ac.uk/acad/nlw/publications/russia_ crowley.pdf (accessed 9th July 2015). These briefing papers were distributed by BNLWRP, ISS and ORF to the Director General, Executive Council and all CWC States Parties. See also Crowley, M. (October 2009) *op.cit.*, pp. 107–109; Schneidmiller, C. Danger of "Nonlethal" Agents Grows Amid States' Inaction, Report Says, *Global Security Newswire*, 6th November 2009; Crowley, M. Toxic Traps: Weaknesses of the Chemical Control Regime, *Jane's Intelligence Review*, December–January 2009; Hatch Rosenberg, B. Riot Control Agents and the Chemical Weapons Convention, Open Forum on Challenges to the Chemical Weapons Ban, 1st May 2003.
113. OPCW, Report of the Scientific Advisory Board on Developments in Science and Technology, RC-3/DG.1, Third Review Conference, 29th October 2012, paragraph 56.
114. Correspondence from Ambassador Dogan, Permanent Representative of Turkey to the OPCW, to BNLWRP, ORF and ISS, 25th February 2011.
115. Correspondence from Mr Utkan, Counsellor, Permanent Representation of Turkey to the OPCW, to BNLWRP, ORF and ISS, 8th July 2011. Turkey's position was subsequently reaffirmed by Ambassador Dogan, Permanent Representative of Turkey to the OPCW, to BNLWRP and ORF, 29th March 2013.
116. Correspondence from Ambassador Dogan (25th February 2011) *op.cit.*; correspondence from Mr Utkan (8th July 2011) *op.cit.*
117. Although the CWC Schedules currently list only one ICA: BZ (Schedule 2.a.), and two of its immediate precursors, 3-Quinuclidinol and Benzilic Acid (both Schedule 2.b.), [See OPCW, CWC (1993) *op.cit.*, Annex on Chemicals, B. Schedules of

Chemicals, Schedule 2.], all toxic chemicals promoted as ICAs fall within the CWC's ambit.
118. OPCW, CWC (1993) *op.cit.*, Article II.9.
119. *Ibid.*, Article III.1.(a–c).
120. *Ibid.*, Article I.2 and I.4; Article IV and Article V.
121. *Ibid.*, Article II.9(d).
122. Chayes, A. and M. Meselson (March 1997) *op.cit.*, pp. 13–18.
123. Krutzsch, W. Non-Lethal Chemicals for Law Enforcement, BITS Research Note 03.2, April 2003; Krutzsch, W. and Von Wagner, A. Law Enforcement Including Domestic Riot Control: The Interpretation of Article II, Paragraph 9(d), available at http://cwc2008.files.wordpress.com/2008/04/krutzsch-von-wagner-law-enforcement.pdf (accessed 11th January 2015).
124. Krutzsch, W. and Trapp, R. Article II: Definitions and Criteria, in Krutzsch, W., Myjer, E. and Trapp, R (eds), *The Chemical Weapons Convention: A Commentary*, Oxford: Oxford University Press, 2014, pp. 73–104.
125. Von Wagner, A. (2007) *op.cit.*
126. The only possible exceptions to this restriction recognized are those toxic chemicals used for judicially sanctioned execution, provided such chemicals are not on the CWC's Schedule 1 list. See, for example, A. Chayes and M. Meselson (March 1997) *op.cit.*, pp. 17–18.
127. *Ibid.*, p. 17.
128. Krutzsch, W and Trapp, R. (2014) *op.cit.*, p. 102.
129. Fidler, D. (2007) *op.cit.*, p. 174.
130. Neill, D. Riot Control Agents and Incapacitating Chemical Agents Under the Chemical Weapons Convention, Defence R&D Canada, Technical Memorandum DRDC CORA TM 2007–2022, June 2007.
131. Fidler, D. (2007) *op.cit.*, p. 175.
132. *Ibid.*
133. Fidler, D. (2007) *op.cit.*, p. 185.
134. *Ibid.*
135. US Department of the Navy, Office of the Judge Advocate General, International & Operational Law Division, Preliminary legal review of proposed chemical-based non-lethal weapons, 30th November 1997.
136. Office of the Judge Advocate General, (30th November 1997) *op.cit.*, p. 21.
137. *Ibid.*, p. 20.
138. *Ibid.*, p. 22.
139. OPCW, Statement by Ambassador Robert P. Mikulak, United States Delegation to the Seventy-Ninth Session of the Executive Council, 7th July 2015.
140. For full discussion of this incident, see Chapter 2 of this publication.
141. See *CBW Conventions Bulletin*, number 59, Harvard Sussex Program, March 2003, p. 16.
142. See Yesterday in Parliament, *The Guardian*, 29th October 2002; Jones, G. Moscow's Handling of Siege Backed by Blair, *Daily Telegraph*, 29th October 2002; BBC News, West Backs Russia Over Rescue Tactics, 28th October 2002.
143. BBC News, Putin: Foreign Support but Also Concern, 28th October 2002; Russian Counts Cost of Deadly Siege, *Japan Today*, 27th October 2002.
144. NATO (August 2006) *op.cit.*, Annex M: Medical Aspects of the Moscow Theatre Hostage Incident, p. 136.
145. NATO (August 2006) *op.cit.*, chapter 6: Human effects issues affecting NLW development, testing and acceptance, p. 55.

146. For a range of divergent argumentation on this issue see Fidler, D. The Meaning of Moscow: Non-Lethal Weapons and International Law in the 21st Century, *International Review of the Red Cross*, volume 87, number 859, September 2005, pp. 525–552; Fidler, D. (2007) *op.cit.*; Von Wagner, A. (2007) *op.cit.*; Koplow, D. (2006) *op.cit.*; Neill, D. (2007) *op.cit.*
147. Wheelis, M. Will the New Biology Lead to New Weapons? *Arms Control Today*, July/August 2004.
148. Interview with Professor Mark Wheelis, by email, correspondence dated 13th July 2010.
149. Author's interview with Western Government expert, 28th April 2010.
150. Interview with senior US State Department official, 27th May 2010. [Interviewee speaking in private capacity.]
151. *Ibid.*
152. Interview with Professor Mark Wheelis, by email, correspondence dated 13th July 2010.
153. UN Human Rights Committee, Concluding Observations of the Human Rights Committee: Russian Federation, UN Doc. CCPR/CO/79/RUS, 2003, paragraph 14.
154. UN Commission on Human Rights, Report of the Special Rapporteur, Asma Jahangir, submitted pursuant to Commission on Human Rights resolution 2002/36, E/CN.4/2003/3, 13th January 2003.
155. Amnesty International (October 2003) *op.cit.*; Amnesty International. *The Pain Merchants: Security Equipment and Its Use in Torture and Other Ill-Treatment*, December 2003.
156. Human Rights Watch, press release: *Independent Commission of Inquiry Must Investigate Raid on Moscow Theater: Inadequate Protection for Consequences of Gas Violates Obligation to Protect Life*, 30th October 2002.
157. *Statement of the International Committee of the Red Cross*, First Special Session of the Conference of the States Parties to Review the Operation of the Chemical Weapons Convention, The Hague, 28th April–9th May 2003.
158. On 2nd May 2015, an analysis was undertaken of all OPCW documents publicly available on the OPCW website (http://www.opcw.org).
159. Address by Mr Rogelio Pfirter, Director General of the OPCW, to the Fourth National Dialogue Forum, Russian Implementation of the Chemical Weapons Convention: Status and Perspectives as of Year End 2002, 11th November 2002.
160. BBC News, Russia Names Moscow Siege Gas, 31st October 2002.
161. Von Twickel, N. Unmasking Dubruvka's Mysterious Gas, *The Moscow Times*, 23rd October 2007.
162. O'Brien, M. 4th November 2002, Response to Parliamentary Question, *Hansard*, HC Deb 04 November 2002 vol. 392 c74W, available at http://hansard.millbanksystems.com/written_ answers/2002/nov/04/moscow-theatre-siege (accessed 12th July 2015).
163. Interview by author of senior US State Department official, 27th May 2010. [Interviewee speaking in private capacity.]
164. On 19th August 2015, an analysis was undertaken of all OPCW documents publicly available on the OPCW website (http://www.opcw.org).
165. UN, Vienna Convention on the Law of Treaties (1969) *op.cit.*, Article 31.3.b.
166. Fidler, D. (2005) *op.cit.*
167. UK Government Official, correspondence with author, 10th November 2008.

168. Pearson, A. Incapacitating Chemical Weapons and the Chemical Weapons Convention, *Presentation by the Center for Arms Control and Non-Proliferation and the Scientists Working Group on Biological Chemical Weapons to the Open-Ended Working Group for the preparation of the Second Review Conference of the Chemical Weapons Convention*, 18th November 2007, available at http://archive.armscontrolcenter.org/issues/biochem/articles/incapacitating_chemical_weapons_and_cwc/ (accessed 9th July 2015).
169. Pearson, A. (November 2007) *op.cit.*
170. Former OPCW official, interview with author, 1st September 2008.
171. See for example: OPCW, Statement by Ambassador Dominik M. Alder, Permanent Representative of Switzerland to the OPCW, Second Review Conference of the Chemical Weapons Convention, General Debate, The Hague, Netherlands, 8th April 2008.
172. OPCW, Conference of States Parties, Switzerland: Statement by Markus Börlin, Permanent Representative of Switzerland to the OPCW, General Debate, Statement at the Third Special Session of the Conference of the States Parties to Review the Operation of the Chemical Weapons Convention, 8th April 2013.
173. The text of all National Papers and Statements to the Third CWC Review Conference area available from the OPCW website at https://www.opcw.org/documents-reports/conference-states-parties/third-review-conference/national-statements/ (accessed 2nd May 2015).
174. OPCW, Conference of States Parties, Germany: Statement by Ambassador Rolf Wilhelm Nikel, Commissioner of the Federal German Government for Disarmament and Arms Control, at the Third Review Conference, Third Review Conference RC-3/NAT.28, 8th–19th April 2013, 9th April 2013.
175. *Ibid.*
176. In the UK, ICA weapon research reportedly occurred in the 1950s and 1960s, but this had apparently ended by the early 1970s. In contrast, in the US, ICA weapon research was reported to have continued in the 1990s and into the 2000s. See Chapter 2 of this publication for further discussion.
177. OPCW, Conference of States Parties, United States of America: Statement by Rose E. Gottemoeller, Acting Under Secretary for Arms Control and International Security, at the Third Review Conference, Third Special Session RC-3/NAT.45, 8th–19th April 2013, 9th April 2013.
178. *Ibid.*
179. OPCW, Conference of States Parties, United Kingdom: Statement by Mr Alistair Burt, Parliamentary Under Secretary of State for Foreign and Commonwealth Affairs, Third Review Conference, RC-3/NAT.22, 8th–19th April 2013, 9th April 2013.
180. *Ibid.*
181. *Ibid.*
182. *Ibid.*
183. OPCW, Conference of States Parties, Australia, Weaponisation of Central Nervous System Acting Chemicals for Law Enforcement, Nineteenth Session C-19/NAT.1, 1st–5th December 2014, 14th November 2014, paragraph 5.
184. During the Third CWC Review Conference, certain States Parties i.e. Germany and Switzerland explicitly declared that only RCAs can be employed in their countries for law enforcement.
185. OPCW, Conference of States Parties, Australia (14th November 2014) *op.cit.*, paragraph 6.

186. OPCW, Conference of States Parties, Australia (14th November 2014) *op.cit.*
187. OPCW, Canada, Statement by H.E. Ambassador James Lambert, Permanent Representative of Canada to the OPCW at the Seventy Third Session of the Executive Council, EC-73/NAT.23, 16th July 2013; OPCW, Canada, Statement by H.E. Ambassador James Lambert, Permanent Representative of Canada to the OPCW at the Seventy Eighth Session of the Executive Council, EC-78/NAT., 17th March 2015.
188. OPCW, Conference of States Parties, Third Review Conference, Germany (9th April 2013) *op.cit.*
189. OPCW, Malaysia, Statement by H.E. Dr Fauziah Mohamad Taib, Permanent Representative of Malaysia to the OPCW at the Seventy Seventh Session of the Executive Council, EC-77/NAT.29, 7th October 2014.
190. OPCW, Statement by H.E. Ambassador Dr Jan Borkowski, Permanent Representative of Poland to the OPCW at the Seventy Eighth Session of the Executive Council, EC-78/NAT.14, 17th March 2015.
191. OPCW, Conference of States Parties, Third Review Conference, Switzerland (8th April 2013) *op.cit.*
192. OPCW, Conference of States Parties, Third Review Conference, United Kingdom (9th April 2013) *op.cit.*
193. OPCW, United States of America: Statement by Ambassador Robert P. Mikulak, United States Delegation to the OPCW, at the Seventy-Second Session of the Executive Council, EC-72/NAT.8, 6th May 2013.
194. Czech Republic, National Authority of the CWC, Reply to the University of Bradford, Re: Request for information concerning research potentially related to incapacitating chemical agents, 14th July 2014 (copy on file with author).
195. Correspondence to Dr M. Crowley, BNLWRP, from Dr H. Farajvand, Secretary of the National Authority for the CWC, Ministry of Foreign Affairs of the Islamic Republic of Iran, 26th July 2014.
196. OPCW, Conference of States Parties, Australia (14th November 2014) *op.cit.*, paragraph 7.
197. OPCW, Australian Intervention in the General Debate, Seventy Eighth Session of the Executive Council, 17th March 2015.
198. OPCW, Canada, Executive Council (16th July 2013) *op.cit.*; OPCW, Canada, Executive Council (17th March 2015) *op.cit.*
199. OPCW, Germany: Statement by H. E. Ambassador Dr Christoph Israng Permanent Representative of Germany to the OPCW at the Seventy-Eighth Session of the Executive Council, EC-78/NAT.16, 17th March 2015.
200. OPCW, Malaysia, Executive Council (7th October 2014) *op.cit.*
201. OPCW, New Zealand: Statement by H.E. Ambassador Janet Lowe Permanent Representative of New Zealand to the OPCW at the Seventy-Eighth Session of the Executive Council, EC-78/NAT.25, 17th March 2015.
202. OPCW, Norway: Statement by H.E. Ambassador Anniken R. Krutnes, Permanent Representative of Norway to the OPCW, at the Seventy-Second Session of the Executive Council, EC-72/NAT.6, 6th May 2013; OPCW, Norway, Statement by Mr Thomas Mosberg-Stangeby, Charge D' Affaires A.I, Deputy Permanent Representative of Norway to the OPCW, EC-73/NAT.8, 17th July 2013.
203. OPCW, Switzerland: Statement by H.E. Ambassador Urs Breiter Permanent Representative of Switzerland to the OPCW at the Seventy-Eighth Session of the Executive Council, EC-78/NAT.28, 17th March 2015.

204. OPCW, UK, Statement by Mr John Foggo, National Authority of the UK, Seventy-Ninth Session of the OPCW Executive Council, undated.
205. OPCW, United States of America: Statement by H.E. Ambassador Robert P. Mikulak, Permanent Representative of the United States of America to the OPCW at the Seventy-Eighth Session of the Executive Council, EC-78/NAT.10, 17th March 2015.
206. OPCW, Switzerland, Statement by H.E. Ambassador Urs Breiter, Permanent Representative of Switzerland to the OPCW at the Seventy-Ninth Session of the Executive Council, EC-79/NAT 23, 7th July 2015.
207. OPCW, Statement by Ambassador Robert P. Mikulak, United States Delegation to the Seventy-Ninth Session of the Executive Council, 7th July 2015.
208. See: OPCW, Note by the Technical Secretariat, The OPCW in 2025: Ensuring A World Free of Chemical Weapons, Office of Strategy and Policy, S/1252/2015, 6th March 2015. (Of particular relevance is paragraph 21.b which states that: "The Organisation will also require enhanced capabilities to monitor the full spectrum of relevant toxic chemicals falling within its mandate, ranging from toxic industrial chemicals to chemicals used for example in medicine or law enforcement, including those acting on the central nervous system.")
209. OPCW, Russian Federation, Statement by the Delegation of the Russian Federation at the Seventy-Ninth Session of the Executive Council, EC-79/NAT 15, 8th July 2015.

6 Arms Control and Disarmament Agreements Applicable to ICA Weapons and Riot Control Agents

1. Protocol for the Prohibition of the Use in War of Asphyxiating, Poisonous or Other Gases, and of Bacteriological Methods of Warfare (Geneva Protocol), 17th June 1925, available from ICRC, Treaties and States Parties to such treaties, available at https://www.icrc.org/ihl/INTRO/280?OpenDocument (accessed 2nd August 2015).
2. Details of the States Parties and Signatory States, available from United Nations Office for Disarmament Affairs website at http://disarmament.un.org/treaties/t/1925 (accessed 2nd August 2015).
3. Geneva Protocol (1925) op.cit.
4. United Nations General Assembly (UNGA) Resolutions 2162 B (XXI) of 5th December 1966, 2454 A (XXIII) of 20 December 1968, 2603 B (XXIV) of 16th December 1969, and 2662 (XXV) of 7th December 1970.
5. UNGA Resolution 2603 A (XXIV) of 16th December 1969.
6. For discussion see Henckaerts, J. and Doswald-Beck, L. *Customary International Humanitarian Law, Volume I: Rules*, Cambridge: Cambridge University Press, 2005, pp. 263–265; Henckaerts, J. and Doswald-Beck, L. *Customary International Humanitarian Law, Volume II: Practice*, Cambridge: Cambridge University Press, 2005, pp. 1742–1762.
7. See Henckaerts, J. and Doswald-Beck, L. (2005) Volume II, *op.cit.*, pp. 1750–1753.
8. *Ibid.*, pp. 1742–1762.
9. United States President G. Ford, Executive Order 11850 – Renunciation of certain uses in war of chemical herbicides and riot control agents, 8th April 1975, 40 Federal Register 16187, 3 CFR, 1971–1975 Compilation, p. 980.
10. Von Wagner, A. Toxic Chemicals for Law Enforcement Including Domestic Riot Control Purposes Under the Chemical Weapons Convention, in Pearson, A.,

Chevrier, M. and Wheelis, M. (eds), *Incapacitating Biochemical Weapons*. Lanham, MD: Lexington Books, 2007, p. 204; see also Henckaerts, J. and Doswald-Beck, L. (2005) Volume I, *op.cit.*, pp. 263–265.
11. See Nuclear Threat Initiative website at http://www.nti.org/treaties-and-regimes/protocol-prohibition-use-war-asphyxiating-poisonous-or-other-gasses-and-bacteriological-methods-warfare-geneva-protocol/ (accessed 2nd August 2015).
12. Kelle, A., Nixdorff, K. and Dando, M. *Controlling Biochemical Weapons: Adapting Multilateral Arms Control for the 21st Century*, Basingstoke: Palgrave Macmillan, 2006, p. 16; Perry Robinson, J. Item 456, Near-Term Development of the Governance Regime for Biological and Chemical Weapons, 2006, Appendix, p. 21.
13. See for example, UN, Final Document of the Seventh BTWC Review Conference, 22nd December 2011, Article VIII, paragraph 44.
14. Henckaerts, J. and Doswald-Beck, L. (2005) *op.cit.*, For a discussion see Rule 74. The use of chemical weapons is prohibited, and Rule 75. The use of riot control agents as a method of warfare is prohibited, pp. 259–265.
15. Convention on the Prohibition of the Development, Production and Stockpiling of Bacteriological (Biological) and Toxin Weapons and on Their Destruction (BTWC), 1972. For text of the Convention, as well as additional understandings and agreements, documents from Review Conferences and Meetings of States Parties, details of membership, etc, see the website of the UN Office at Geneva available at http://www.unog.ch/bwc/docs (accessed 2nd August 2015).
16. Central African Republic, Côte d'Ivoire, Egypt, Haiti, Liberia, Nepal, Somalia, Syria and Tanzania.
17. Angola, Chad, Comoros, Djibouti, Eritrea, Guinea, Israel, Kiribati, Micronesia, Namibia, Niue, Samoa, South Sudan and Tuvalu.
18. BTWC (1972) *op.cit.*, Preamble.
19. BTWC (1972) *op.cit.*, Article I.
20. United Nations, Final Document of the Fourth BTWC Review Conference, BWC/CONF. IV/9, 25th November–6th December (1996), Article I, paragraph 3.
21. United Nations, Final Document of the Sixth BTWC Review Conference, BWC/CONF.VI/6, 20th November–8th December, 2006, Geneva, Article I, paragraph 3.
22. United Nations, Final Document of the Seventh Review Conference, BWC/CONF.VII/7, 5th–22nd December 2011, Geneva, Article I, paragraph 3.
23. For further discussion see Kelle, A., Nixdorff, K. and Dando, M. (2006) *op.cit.*, p. 43; Kelle, A. Ensuring the Security of Synthetic Biology: Towards a 5P Governance Strategy, *Systems and Synthetic Biology*, volume 3, 2009, pp. 85–90.
24. Chevrier, M. and Leonard, J. Incapacitating Biochemicals and the Biological Weapons Convention, in Pearson, A., Chevrier, M. and Wheelis, M. (eds), *Incapacitating Biochemical Weapons*, Lanham, MD: Lexington Books, 2007, pp. 209–224, p. 211.
25. For example see World Health Organization (2002 and revised 2004), *Public Health Response to Biological and Chemical Weapons: WHO Guidance*, Annex 2: Toxins; Chevrier, M. and Leonard, J. (2007) *op.cit.*, p. 211.
26. Author's interview with Western Government expert, 28th April 2010.
27. Authors interview with Western Government expert, 20th April 2011.
28. United Nations, BTWC Seventh Review Conference Final Document (2011) *op.cit.*, Article I, paragraph 1.

29. United Nations, Seventh Review Conference of the States Parties to the Convention on the Prohibition of the Development, Production and Stockpiling of Bacteriological (Biological) and Toxin Weapons and on Their Destruction, New scientific and technological developments relevant to the Convention, Background information document submitted by the Implementation Support Unit, BWC/CONF.VII/INF.3, 10th October 2011.
30. United Nations, New scientific and technological developments relevant to the Convention, Addendum: Submissions from States Parties (November 2011) *op.cit.*, paragraph 125.
31. United Nations, New scientific and technological developments relevant to the Convention, Addendum: Submissions from States Parties (October 2011) *op.cit.*, paragraph 13.
32. Response of the Secretary of State for Foreign and Commonwealth Affairs, Fourth Report from the Foreign Affairs Committee, Session 2008–09, Global Security: Non-Proliferation, The Stationary Office, August 2009, p. 22.
33. USA, Department of the Navy, Office of the Judge Advocate General, *Legal Review of Oleoresin Capsicum (OC) Pepper Spray*, for Commander, Marine Corps Systems Command, Ser 103/353, 19th May 1998, p. 10.
34. US, Department of the Navy, Office of the Judge Advocate General (1998) *op.cit.*, p. 10.
35. According to the Joint Non-Lethal Weapons Directorate (JNLWD), legal reviews were prepared of Tiger Light Guardian OC Spray (3rd June 2008) and MK3 First Defense/Pepper Foam (17th April 2003). See JNLWD, *Non-Lethal Weapons (NLW) Reference Book*, 30th June 2011, p. ix.
36. United States, Law of War Manual, Office of General Counsel, US Department of Defense, June 2015, Section 6.9.
37. BTWC (1972) *op.cit.*, Article 1.
38. Goldblat, J. The Biological Weapons Convention: An Overview, *International Review of Red Cross, Volume 79*, article number 318, 30th June 1997; Sossai, M. Drugs as Weapons: Disarmament Treaties Facing the Advances in Biochemistry and Non-Lethal Weapons Technology, *Journal of Conflict & Security Law*, volume 15, 2010, p. 5.
39. Goldblat, J. (1997) *op.cit.*; Sossai, M. (2010) *op.cit.*, p. 5.
40. Goldblat, J. (1997) *op.cit.*
41. See Wheelis, M. and Dando, M. On the Brink: Biodefence, Biotechnology and the Future of Weapons Control, *CBW Convention Bulletin*, number 58, Harvard Sussex Programme, 2002, pp. 3–7; Wheelis M. and Dando, M. Back to Bioweapons? *Bulletin of the Atomic Scientist*, number 59, 2003, pp. 40–46; Hatch Rosenberg, B. Defending against Biodefence: The Need for Limits, *Disarmament Diplomacy*, issue 69, February–March 2003, available at www.acronym.org.uk/dd/dd69/69op03.html
42. Henckaerts, J. and Doswald-Beck, L. (2005) Volume I, *op.cit.*, p. 256.
43. Author's interview with Western Government expert, 20th April 2011.
44. Although there is no equivalent of an OPCW for the BTWC, in September 2006, the Sixth BTWC Review Conference decided to create and fund a (three-person) Implementation Support Unit (ISU) within the Office for Disarmament Affairs (UNODA) of the United Nations Office at Geneva. The ISU was launched in August 2007 and its mandate was renewed and extended by the Seventh BTWC Review Conference to run until 2016. [See UN, Final

Document of the Seventh BTWC Review Conference (2011) *op.cit.*, Decisions and recommendations, Implementation support unit, paragraph 31, p. 23.]
45. BTWC (1972) *op.cit.*, Article V.
46. Biological and Toxin Weapons Convention (1972) *op.cit.*, Article V.
47. Cuba requested a formal consultative meeting to address its allegation that, in October 1996, a US Government crop-dusting aircraft deliberately released Thrips palmi, an insect pest, over an area of Cuba in order to harm the country's agricultural economy. The overall "finding of fact" from the consultative meeting was officially inconclusive. For further discussion see Tucker, J. Strengthening Consultative Mechanisms Under Article V to Address BWC Compliance Concerns, Harvard Sussex Program, May 2011, available at http://www.sussex.ac.uk/Units/spru/hsp/occasional%20papers/HSPOP_ 1.pdf (accessed 10th July 2015), pp. 8–9.
48. Biological and Toxin Weapons Convention (1972) *op.cit.*, Article VI.
49. Hampson, F. International Law and the Regulation of Weapons, in Pearson, A., Chevrier, M. and Wheelis, M. (eds), *Incapacitating Biochemical Weapons*, Lanham, MD: Lexington Books, 2007, pp. 231–260.
50. UN, Final Document of the Seventh BTWC Review Conference (2011) *op.cit.*, Article 1, paragraph 2.
51. UN, Final Document of the Seventh BTWC Review Conference (2011) *op.cit.*, Decisions and recommendations, paragraph 23.
52. UNGA, Resolution 35/144C, 12th December 1980.
53. UNGA, Resolution 37/98D, 13th December 1982.
54. UNGA Resolution 42/37c, 30th November 1987.
55. Tucker, J. Putting Teeth in the Biological Weapons Convention, *Issues in Science and Technology*, Spring 2002, National Academy of Sciences/Institute of Medicine, University of Texas.
56. UNGA Resolution 46/35b, 9th December 1991.
57. For discussions of each investigation see Littlewood, J. Investigating Allegations of CBW Use: Reviving the UN Secretary-General's Mechanism, *Compliance Chronicles*, number 3, Canadian Centre for Treaty Compliance, 2006, pp. 10–19.
58. UNGA, A More Secure World, A/59/565, 2nd December 2004, paragraph 141, p. 46; UNGA, In Larger Freedom, A/59/2005, 21st March 2005, paragraph 104, p. 29; UNGA, A/60/825, 27th April 2006, paragraph 90, p. 18.
59. UNGA, Resolution A/RES/60/288, Annex: Plan of Action, Section II, Paragraph 11, 20th September 2006.
60. Kraatz-Wadsack, G. Implementing the UN Secretary-General's Mechanism on Alleged Use Investigations for Chemical, Biological and Toxin Weapons,* Second Global Conference of OIE Reference Laboratories and Collaborating Centres, Paris, 21st–23rd June 2010.
61. Memorandum of understanding between the World Health Organization and the United Nations, 31st January 2011, available at http://www.un.org/disarmament/WMD/Secretary-General_ Mechanism/UN_WHO_MOU_2011.pdf (accessed 2nd May 2015).
62. See Kraatz-Wadsack, G. (2010) *op.cit*; UN ODA, The Secretary General's Mechanism for the Investigation of Alleged Use of Chemical and Biological Weapons, Factsheet, June 2013.
63. Editorial note, Kurds Accuse Turkish Military of Using Chemical Weapons, *WMD Insights*, April 2006, available at http://cns.miis.edu/wmd_insights/WMDInsights_ 2006_04.pdf (accessed 10th July 2015).

64. Five permanent members of the UNSC: China, France, Russian Federation, UK and US.
65. UN Security Council, Note by the Secretary General, Report of the Mission dispatched by the Secretary-General to investigate an alleged use of chemical weapons in Mozambique, S/24065, 12th June 1992; also see Littlewood, J. (2006) *op.cit.*; Stock, T. Chemical and Biological Weapons: Developments and Proliferation, in *SIPRI Yearbook 1993*, Oxford: Oxford University Press, pp. 259–292. Gould, C. and Folb, P. The South African Chemical and Biological Warfare Program: An Overview, *The Non-Proliferation Review* (Fall/Winter 2000) volume 7, number 3, pp. 10–23; Gould, C. and Folb, P. *Project Coast: Apartheid's Chemical and Biological Warfare Programme*, UNIDIR, Geneva, 2002; Gould, C. South Africa's Chemical and Biological Warfare Programme 1981–1995, PhD thesis, Rhodes University, August 2005.
66. As documented in: UN, General Assembly, Chemical and Bacteriological (Biological) Weapons, Report of the Secretary General, A/37/259, 1st December 1982.
67. UNGA, Resolution 42/37c, 30th November 1987.
68. See OPCW, CWC (1997) *op.cit.*, Article IX and Article X.
69. CWC (1993) *op.cit.*, Verification Annex, Part XI.
70. *Ibid.*
71. UNGA, Agreement Concerning the Relationship Between the United Nations and the Organisation for the Prohibition of Chemical Weapons, Resolution, A/Res/55/283, 7 September 2001.
72. Amnesty International, Egypt: Human Rights in Crisis: Systemic Violations and Impunity, MDE 12/034/2014, 1st July 2014.
73. Amnesty International, Israel and Occupied Palestinian Territories: Trigger-Happy: Israel's Use of Excessive Force in the West Bank, MDE 15/002/2014, 27th February 2014.
74. See for example: *Country Profiles, Syria, Chemical*, Nuclear Threat Initiative, available at http://www.nti.org/country-profiles/syria/chemical/ (accessed 11th August 2014).
75. Associated Press, Syrian Regime Makes Chemical Warfare Threat, *The Guardian*, 23rd July 2012.
76. Both the Syrian Government armed forces and the armed opposition forces have been accused of utilizing chemical weapons in Syria. As of August 2015, the unconfirmed allegations of ICA weapons use appear to have been confined to the Syrian Government armed forces.
77. Yezdani, I. Chemical Weapons Used Against Syrians, Says Defected Soldier, *Hürriyet Daily News*, 21 February 2012: cited and discussed in Perry Robinson, J. *Alleged Use of Chemical Weapons in Syria*, Harvard Sussex Program Occasional Paper No. 4, 26th June 2013, pp. 11–12.
78. *Ibid.*
79. Perry Robinson, J. (2013) *op.cit.*, p. 12.
80. *Ibid.*,
81. Rogin, J. *Exclusive: Secret State Department Cable: Chemical Weapons Used in Syria*, 15th January 2013, p. 1.
82. Rogin, J. (January 2013) *op.cit.*, p. 2.
83. US plays down media report that Syria used chemical weapons, Reuters, 16th January 2013.

84. Zaher Al-Saket, Interview Broadcast on Al-Arabiya TV on 27 April 2013: cited by Perry Robinson, J. (2013) *op.cit.*, pp. 7–8.
85. The UN Mission was established by the UN Secretary General based on his authority under UNGA Resolution 42/37 C and UN Security Council Resolution 620 (1988). The purpose of this Mission was to ascertain the facts related to the allegations of use of chemical weapons, to gather relevant data, to undertake the necessary analyses for this purpose, and to deliver a report to the Secretary General.
86. UN, *United Nations Mission to Investigate Allegations of the Use of Chemical Weapons in the Syrian Arab Republic: Report on the Alleged Use of Chemical Weapons in the Ghouta Area of Damascus on 21 August 2013*, Note by the Secretary-General, 16th September 2013, available at http://www.un.org/disarmament/content/slideshow/Secretary_ General_Report_of_ CW_Investigation.pdf (accessed 1st June 2014), p. 1.
87. *Ibid.*, paragraph 27.
88. *Ibid.*, paragraph 28.
89. United Nations, *United Nations Mission to Investigate Allegations of the Use of Chemical Weapons in the Syrian Arab Republic*, Final report, 12th December 2013, available at https://unoda-web.s3.amazonaws.com/wp-content/uploads/2013/12/report.pdf (accessed 1st June 2014).
90. OPCW, Press release: OPCW to Review Request from Syria, 13th September 2013; OPCW, Executive Council, *Decision: Destruction of Syrian Chemical Weapons*, EC-M-33/DEC.1, 27th September 2013; OPCW, Press release: *Syria's Accession to the Chemical Weapons Convention Enters into Force*, 14th October 2013.
91. See for example: Israeli Official Says Syria Has Used Chemical Incapacitants against Insurgents, *Jane's Defence Weekly*, 10th April 2014.
92. OPCW, News article: OPCW Fact Finding Mission: "Compelling Confirmation" that Chlorine Gas Used as Weapon in Syria, 10th September 2014, available at http://www.opcw.org/news/article/opcw-fact-finding-mission-compelling-confirmation-that-chlorine-gas-used-as-weapon-in-syria/ (accessed 18th September 2014).
93. OPCW, Third Report of the OPCW Fact-Finding Mission in Syria, Technical Secretariat, S/1230/2014, 18th December 2014, available at http://photos.state.gov/libraries/netherlands/328666/pdfs/THIRDREPORTOFTHEOPCWFACTFINDINGMISSIONINSYRIA.pdf (accessed 4th October 2015).
94. Interview by author with Dr Jez Littlewood, 29th March 2010.
95. United Nations Security Council, Resolution 1540 (2004). Adopted by the Security Council at Its 4956th Meeting, on 28th April 2004, S/RES/1540 (2004), available at http://www.un.org/en/ga/search/view_ doc.asp?symbol=S/RES/1540%20(2004) (accessed 10th July 2015).
96. *Ibid.*, Article 2.
97. *Ibid.*, Article 3.
98. United Nations Security Council, Resolution 1540 (2004) *op.cit.*, Article 4.
99. *Ibid.*, Article 4.
100. Kelle, A., Nixdorff, K. and Dando, M. (2006) *op.cit.*, pp. 165–166.
101. *Ibid.*, p. 166.
102. Arnold, A. UN Security Council Resolution 1540, Part 1: Resolution 1810: Progress since 1540, *WMD Insights*, August 2008, p. 1.
103. UNSC, Resolution 1540 (2004), paragraph 4.

104. UNSC, Resolution 1673, 27th April 2006; UNSC, Resolution 1810, 25th April 2008.
105. UNSC, Resolution 1977, 20th April 2011.
106. UNSC, Resolution 1977 (2011), paragraph 2.
107. *Ibid.*, paragraph 5.a.
108. *Ibid.*, paragraph 10.
109. *Ibid.*, paragraph 3.
110. UNSC, Resolution 2118, 27th September 2013.
111. UNSC, Resolution 2118 (2013) *op.cit.*, Article 14.

7 International Humanitarian Law Applicable to ICA Weapons and Riot Control Agents

1. UN, Vienna Convention on the Law of Treaties, 1969, Article 18. For further discussion see Hampson, F. International law and the Regulation of Weapons, in Pearson, A., Chevrier, M. and Wheelis, M. (eds), *Incapacitating Biochemical Weapons*, Lexington Books, Lanham, MD, 2007, pp. 231–260.
2. Convention (I) for the Amelioration of the Condition of the Wounded and Sick in Armed Forces in the Field. Geneva, 12th August 1949; Convention (II) for the Amelioration of the Condition of Wounded, Sick and Shipwrecked Members of Armed Forces at Sea. Geneva, 12th August 1949; Convention (III) relative to the Treatment of Prisoners of War. Geneva, 12th August 1949; Convention (IV) relative to the Protection of Civilian Persons in Time of War. Geneva, 12th August 1949 [Geneva Conventions (I–IV)]. As of August 2015 there were 196 States Parties to these Conventions. For text, commentaries and details of States Parties, see ICRC website at https://www.icrc.org/applic/ihl/ihl.nsf/vwTreaties1949.xsp. (accessed 10th August 2015).
3. Protocol Additional to the Geneva Conventions of 12th August 1949, and relating to the Protection of Victims of International Armed Conflicts [Protocol I], 8th June 1977; Protocol Additional to the Geneva Conventions of 12th August 1949, and relating to the Protection of Victims of Non-International Armed Conflicts [Protocol II], 8th June 1977. As of August 2015 there were 174 States Parties to Additional Protocol I and 168 States Parties to Additional Protocol II. For text and commentary of the Protocols and details of States Parties see ICRC website.
4. International Court of Justice, Statute of the International Court of Justice, 1945, Article 38(1)(b).
5. Henckaerts, J. M. and Doswald-Beck, L. (eds), *Customary International Humanitarian Law: Rules, Volume I*, Cambridge: Cambridge University Press, 2005, pp. xxxi–xxxii.
6. Henckaerts, J. M. and Doswald-Beck, L. (2005) *op.cit.* It should be noted that the scope of some of the alleged norms discussed in the ICRC study are controversial and contested.
7. See for example: Geneva Conventions (1949) *op.cit.*, Common Article 2.
8. Additional Protocol II (1977) *op.cit.*, Article 1, paragraphs 1–2. In addition, Common Article 3 of the Geneva Conventions can apply to armed conflicts where a State is not a party, i.e. conflicts between two or more organised armed groups.
9. For example, the Russian Federation denied that its military operations in Chechnya were part of an internal armed conflict and therefore covered by IHL, even though the Russian Constitutional Court had characterized the situation as coming within Additional Protocol II. See Judgement of the Constitutional

Court of the Russian Federation, 31st July 1995, on the constitutionality of the Presidential Decrees and the Resolutions of the Federal Government concerning the situation in Chechnya, European Commission for Democracy through Law of the Council of Europe, CDL-INL (96), 1, as cited by Hampson, F. (2007) *op.cit.*, p. 251.
10. See for example: Clapham, A. *Human Rights Obligations of Non-State Actors*, Oxford: Oxford University Press, 2006, p. 271; and Cassese, A. *International Law*, 2nd edn, Oxford: Oxford University Press, 2005, p. 125, as cited by Casey-Maslen, S. *Non-Kinetic-Energy Weapons Termed "Non-Lethal", A Preliminary Assessment Under International Humanitarian Law and International Human Rights Law*, Geneva Academy, October 2010, p. 13.
11. Hampson, F. (2007) *op.cit.*, p. 238.
12. Whilst IHL and international criminal law are distinct, elements of IHL can be enforced through international criminal law. See Chapter 9 of this publication for a discussion of the application of international criminal law to ICA weapons and RCAs.
13. International Criminal Tribunal for the former Yugoslavia, *Prosecutor v. Tadic*, Decision on the Defence Motion for Interlocutory Appeal on Jurisdiction (Appeals Chamber), 2nd October 1995, Case no. IT-94-1, paragraphs 119 and 127. However, the ruling did include an important qualification, elaborated in paragraph 126 which stated: "The emergence of the aforementioned general rules on internal armed conflicts does not imply that internal strife is regulated by general international law in all its aspects. Two particular limitations may be noted: (i) only a number of rules and principles governing international armed conflicts have gradually been extended to apply to internal conflicts; and (ii) this extension has not taken place in the form of a full and mechanical transplant of those rules to internal conflicts; rather, the general essence of those rules, and not the detailed regulation they may contain, has become applicable to internal conflicts..."
14. For a detailed discussion of this issue see Coupland, R. Incapacitating Biochemical Weapons: Risks and Uncertainties, in Pearson, A., Chevrier, M. and Wheelis, M. (eds), *Incapacitating Biochemical Weapons*, Lanham, MD: Lexington Books, 2007, pp. 225–230; Fidler, D. The meaning of Moscow: "Non-lethal" Weapons and International Law in the Early 21st Century, *International Review of the Red Cross*, volume 87, number 859, September 2005, pp. 525–552; Hampson, F. (2007) *op.cit.*; Herby, P. Protecting and Reinforcing Humanitarian Norms: The Way Forward, in Pearson, A., Chevrier, M. and Wheelis, M. (eds), *Incapacitating Biochemical Weapons*, Lanham, MD: Lexington Books, 2007, pp. 285–290.
15. Regulations concerning the Laws and Customs of War on Land, annexed to Convention (IV) respecting the Laws and Customs of War on Land, The Hague, 18th October 1907 [Hague Regulations (1907)], Article 22.
16. Additional Protocol I (1977) *op.cit.*, Article 35.1.
17. Convention on Cluster Munitions, Diplomatic Conference for the Adoption of a Convention on Cluster Munitions, Dublin, 19th–30th May 2008, 20th preambular paragraph.
18. Casey-Maslen, S. (2010) *op.cit.*, p. 13.
19. Geneva Conventions (1949) *op.cit.*, Common Article 3.1.
20. *Ibid.*, Common Article 3.1.a–c.
21. *Ibid.*, Common Article 3.2.

22. For discussion see Henckaerts, J. M. and Doswald-Beck, L. (2005) *op.cit.*, Rule 87, pp. 306–308; Rule 110, pp. 400–403; Rule 111, pp. 403–405.
23. Statement of the International Committee of the Red Cross, First CWC Review Conference, 28th April–9th May 2003.
24. *Ibid.*
25. Royal Society, Science Policy Centre, *Brain Waves Module 3, Neuroscience, Conflict and Security*, RS publications, London, February 2012, p. 19.
26. Royal Society (February 2012) *op.cit.*, p. 19. Specifically Under: Third Geneva Convention (1949) *op.cit.*, Article 17.
27. Additional Protocol I (1977) *op.cit.*, Article 11.1.
28. *Ibid.*, Article 11.2. (b).
29. *Ibid.*, Article 11.4.
30. Additional Protocol II (1977) *op.cit.*, Article 10.2.
31. Royal Society (February 2012) *op.cit.*, p. 19.
32. See for example: Additional Protocol I (1977) *op.cit.*, Article 35.
33. See for example, Convention on Certain Conventional Weapons, 1980, third preambular paragraph.
34. Rome Statute of the International Criminal Court, A/CONF.183/9, 17th July 1998, Article 8(2)(b)(xx).
35. Henckaerts, J. M. and Doswald-Beck, L. (2005) *op.cit.*, Rule 70, pp. 237–244.
36. *Ibid.*, Rule 70, p. 242.
37. For a discussion from a perspective of those opposed to the project, see Verchio, D. Just Say No! The SIrUS Project: Well-Intentioned, But Unnecessary and Superfluous, *Air Force Law Review*, volume 51, number 183, 2001, pp. 183–228.
38. WMA Resolution on the SIrUS Project, Adopted by the 50th World Medical Assembly Ottawa, Canada, October 1998.
39. Coupland, R. (ed.), The SIRUS Project, Towards a Determination of Which Weapons Cause "Superfluous Injury or Unnecessary Suffering", ICRC, Geneva, 1997, p. 23, available at http://www.loc.gov/rr/frd/Military_Law/pdf/SIrUS-project.pdf (accessed 10th August 2015).
40. Herby, P. (2007) *op.cit.*, p. 286.
41. *Ibid.*, p. 286.
42. Additional Protocol I (1977) *op.cit.*, Articles 48 and 51.
43. Henckaerts, J. M. and Doswald-Beck, L. (2005) *op.cit.*, Rules 1 [pp. 3–8], 7 [pp. 25–29], 9, 10 [pp. 25–34], 11 and 12 [pp. 37–43].
44. *Ibid.*
45. International Court of Justice, Nuclear Weapons case, Advisory Opinion, 1996, as cited by Henckaerts, J. M. and Doswald-Beck, L. (2005) *op.cit.*, Rule 1, p. 5.
46. See for example: Fenton, G. Current and Prospective Military and Law Enforcement Use of Chemical Agents for Incapacitation, in Pearson A., Chevrier, M. and Wheelis, M. (eds), *Incapacitating Biochemical Weapons*, Lanham, MD, Lexington Books, 2007, pp. 103–120; Mayer, C. Non-Lethal Weapons and Non-Combatant Immunity: Is it Permissible to Target Noncombatants?, *Journal of Military Ethics*, volume 6, number 3, 2007, pp. 221–231.
47. Herby, P. (2007) *op.cit.*, p. 286.
48. Under US Presidential Executive Order 11850, use of RCAs is permitted: "(b) In situations in which civilians are used to mask or screen attacks and civilian casualties can be reduced or avoided". [G. Ford, Executive Order 11850 – Renunciation of Certain Uses in War of Chemical Herbicides and Riot Control

Agents, 8th April 1975. 40 FR 16187 CGR, 1971–1975 Comp., 980]. See also: United States, Law of War Manual, Office of General Counsel, U.S. Department of Defense, June 2015, section 6.16.
49. Geneva Conventions (1949) *op.cit.*, Common Article 1; Additional Protocol I (1977) *op.cit.*, Article 1.
50. Royal Society (February 2012) *op.cit.*, p. 20.
51. For example of this principle, see the Statute of the International Criminal Court (1998) *op.cit.*, Article 31, paragraphs 1(a) and (b). [Cited in Royal Society (2012) *op.cit.*, p. 20.]
52. See for example, Additional Protocol 1(1977) *op.cit.*, Article 1(2).
53. The International Court of Justice, Legality of the Threat or Use of Nuclear Weapons, Advisory Opinion, 8th July 1996, paragraph 84.
54. Additional Protocol I (1977) *op.cit.*, Article 1(2).
55. ICRC, *A Guide to the Legal Review of New Weapons, Means and Methods of Warfare, Measures to Implement Article 36 of Additional Protocol 1 of 1977*, January 2006, p. 17.
56. The International Court of Justice (1996) *op.cit.*, paragraph 87.
57. *Ibid.*, paragraph 78.
58. Herby, P. (2007) *op.cit.*, p. 288.
59. For a range of positions on this issue, see Cassese, A. The Martens Clause: Half a Loaf or Simply Pie in the Sky?, *European Journal of International Law*, volume 11, 2000, pp. 187–216; Crawford, E. The Modern Relevance of the Martens Clause, *Legal Studies Research Paper Number 11/27*, Sydney Law School, University of Sydney, May 2011; Meron, T. The Martens Clause, Principles of Humanity, and Dictates of Public Conscience, *American Journal of International Law*, volume 91, number 1, January 2000, pp. 78–89; Ticehurst, R. The Martens Clause and the Laws of Armed Conflict, *International Review of the Red Cross*, number 317, 30th April 1997, pp. 125–134.
60. The phrase "grave breaches" is a technical term which is defined in each of the Geneva Conventions and Additional Protocol I. For example Article 50 of the 1st Geneva Convention states: "Grave breaches to which the preceding Article relates shall be those involving any of the following acts, if committed against persons or property protected by the Convention: wilful killing, torture or inhuman treatment, including biological experiments, wilfully causing great suffering or serious injury to body or health, and extensive destruction and appropriation of property, not justified by military necessity and carried out unlawfully and wantonly."
61. Geneva Convention I (1949) *op.cit.*, Article 49; Geneva Convention II (1949) *op.cit.*, Article 50; Geneva Convention III (1949) *op.cit.*, Article 129; Geneva Convention IV (1949) *op.cit.*, Article 146.
62. Geneva Convention I (1949) *op.cit.*, Article 52; Geneva Convention II (1949) *op.cit.*, Article 53; Geneva Convention III (1949) *op.cit.*, Article 132; Geneva Convention IV (1949) *op.cit.*, Article 149.
63. Additional Protocol I (1977) *op.cit.*, Article 90.2.(c).(i).
64. Additional Protocol I (1977) *op.cit.*, Article 90.2.(c).(ii).
65. The potential application of international criminal law, in certain circumstances, is explored in Chapter 9 of this publication. In addition in terms of IHRL, the UN Human Rights Council has established Commissions of Inquiry in certain cases. Although such Commissions can establish the facts of the case, this rarely leads to accountability.
66. Henckaerts, J. M. and Doswald-Beck, L. (2005) *op.cit.*, Rule 73, p. 256.

67. *Ibid.*, Rule 74, p. 259.
68. *Ibid.*, Rule 75, p. 263.
69. Henckaerts, J. M. Study on Customary International Humanitarian Law, *International Review of the Red Cross*, volume 87, number 857, March 2005, p. 205.
70. Additional Protocol I (1977) *op.cit.*, Article 36.
71. ICRC, *A Guide to the Legal Review of New Weapons, Means and Methods of Warfare, Measures to Implement Article 36 of Additional Protocol I of 1977*, January 2006, p. 4.
72. ICRC (2006) *op.cit.*, p .9.
73. Daoust, I., Coupland, R. and Ishoey, R. New Wars, New Weapons? The Obligation of States to Assess the Legality of Means and Methods of Warfare, *International Review of the Red Cross*, June 2002, volume 84, number 846, pp. 345–363.
74. ICRC (2006) *op.cit.*, p. 23.
75. *Ibid.*, p. 24.
76. See for example: McClelland, J. The Review of Weapons in Accordance with Article 36 of Additional Protocol 1, *International Review of the Red Cross*, volume 85, number 850, June 2003, p. 411; International Committee of the Red Cross. *A Guide to the Legal Review of New Weapons, Means and Methods of Warfare, Measures to Implement Article 36 of Additional Protocol I of 1977*, Geneva, January 2006, p. 24.
77. Boothby, W. *Weapons and the Law of Armed Conflict*, Oxford University Press, Oxford, May 2009, pp. 345–346.
78. Casey-Maslen, S. (2010) *op.cit.*, p. 22.
79. Lawand, K. Reviewing the Legality of New Weapons, Means and Methods of Warfare, *International Review of the Red Cross*, volume 88, number 864, December 2006, p. 929.
80. ICRC (2006) *op.cit.*, pp. 18–19.
81. Final Goal 2.5 of the Agenda for Humanitarian Action Adopted by the 28th International Conference of the Red Cross and Red Crescent, Geneva, 2nd–6th December 2003, paragraph 2.5.2.
82. ICRC (2006) *op.cit.*, pp. 18–19.
83. *Ibid.*, p. 10.
84. *Ibid.*, p. 10.
85. ICRC Commentary on the Additional Protocols, paragraph 1469, as cited in International Committee of the Red Cross (2006) *op.cit.*, p. 10.
86. Fry, J. Contextualized Legal Reviews for the Methods and Means of Warfare: Cave Combat and International Humanitarian Law, *Colombia Journal of Transnational Law*, 28th February 2006, volume 44, number 2, pp. 470–471. As part of his analysis demonstrating how legality might shift depending on the setting in which weapons were employed, Fry incorporated a case study of the US military's alleged use of RCAs in caves and other contained spaces, and argued that such alleged use appeared to breach the CWC and also the Hague Regulations, the Hague Gas Declaration and the Geneva Gas Protocol. [See in particular Fry, J. (2006) *op.cit.*, pp. 506–509.]
87. ICRC (2006) *op.cit.*, p. 26.
88. *Ibid.*, p. 20.
89. Danish Red Cross, Reviewing the Legality of New Weapons, December 2000.
90. Daoust, I., Coupland, R. and Ishoey, R. (2002) *op.cit.*, pp. 354–360.

91. See for example: Daoust, I. Coupland, R. and Ishoey, R. (2002) *op.cit.*, pp. 352–353; Lawland, K. (2006) *op.cit.*, p. 929; 28th International Conference of the Red Cross and Red Crescent (2003) *op.cit.*, paragraph 2.5.1.
92. See ICRC, Commentary on the Additional Protocols, undated, paragraph 1469. [Available from ICRC website].
93. ICRC, Commentary on the Additional Protocols, (undated) *op.cit.*, paragraphs 1469 and 1481.
94. The scope and nature of such potential information provision regarding prohibited weapons would have to address issues of commercial confidentiality as well as the risks of it being misused by less scrupulous States for their own R&D activities.
95. Section 21, Final Goal 1.5 of the Plan of Action for the years 2000–2003 adopted by the 27th International Conference of the Red Cross and Red Crescent, Geneva, 31st October to 6th November 1999.
96. 28th International Conference of the Red Cross and Red Crescent (2003) *op.cit.*, paragraph 2.5.3.
97. Final Declaration of the Second Review Conference of the States Parties to the Convention on Certain Conventional Weapons, Geneva, 11th–21st December 2001, CCW/CONF.II/2, p. 11.
98. NATO Research and Technology Organisation, Non-Lethal Weapons and Future Peace Enforcement Operations, RTO-TR-SAS-040, December 2004, Annex C.
99. See for example: 30th International Conference of the Red Cross and Red Crescent, Geneva, 26th–30th November 2007, Paragraph 19, Resolution 3.
100. Australia, Belgium, Canada, Denmark, Germany, the Netherlands, Norway, Sweden, the UK and the US. See Fry (2006) *op.cit.*, p. 474.
101. *Ibid.*
102. Lawand, K. (2006) *op.cit.*, p. 926.

8 Human Rights Law Applicable to ICA Weapons and Riot Control Agents

1. International Court of Justice, Legal Consequences of the Construction of a Wall in the Occupied Palestinian Territory, Advisory Opinion, 9th July 2004, paragraphs 107–112.
2. UN, Human Rights Committee, General Comment 31 on The Nature of the General Legal Obligation on the States Parties to the International Covenant on Civil and Political Rights, CCPR/C/21/Rev.1/Add.13, 26th May 2004, paragraph 11.
3. Hampson, F. International Law and the Regulation of Weapons, in Pearson, A., Chevrier, M. and Wheelis, M. (eds), *Incapacitating Biochemical Weapons*, Lanham, MD: Lexington Books, 2007, pp. 244–245.
4. See, e.g., Universal Declaration of Human Rights, adopted and proclaimed by UN General Assembly Resolution 217 A (III), 10th December 1948, Article 3; International Covenant on Civil and Political Rights, adopted on 16th December 1966, Article 6.
5. See, e.g., European Convention on the Protection of Human Rights and Fundamental Freedoms, signed on 4th November 1950, Article 2. The obligations deriving from such regional instruments are binding upon those States party to the relevant agreements, and their application has subsequently been clarified following judgements made by the relevant regional legal institutions.

6. UN, ICCPR (1966) *op.cit.*, Article 6(1).
7. Human Rights Committee, General Comment No. 6: The right to life, 16th Session, 30th April 1982, paragraph 1.
8. *Ibid.*, paragraph 3.
9. *Ibid.*, paragraph 1.
10. Basic Principles on the Use of Force and Firearms by Law Enforcement Officials, 7th September 1990, adopted by the Eighth United Nations Congress on the Prevention of Crime and the Treatment of Offenders, Havana, Cuba, 27th August–7th September 1990.
11. Code of Conduct for Law Enforcement Officials, adopted by United Nations General Assembly Resolution 34/169 of 17th December 1979.
12. Melzer, N. *Targeted Killings*, Oxford: Oxford Monographs in International Law, Oxford University Press, 2009, pp. 199–201.
13. UNBP (1990) *op.cit.*, Principle 4.
14. *Ibid.*, Principle 5 (a)–(c).
15. *Ibid.*, Principle 2.
16. ICRC (September 2012) *op.cit.*, p. 3.
17. Aceves, W. Human Rights Law and the Use of Incapacitating Biochemical Weapons, in Pearson, A., Chevrier, M. and Wheelis, M. (eds), *Incapacitating Biochemical Weapons*, Lanham, MD: Lexington Books, 2007, pp. 261–284, 286.
18. Fidler, D. Incapacitating Chemical and Biochemical Weapons and Law Enforcement Under the Chemical Weapons Convention, in Pearson, A., Chevrier, M. and Wheelis, M. (eds), *Incapacitating Biochemical Weapons*, Lanham, MD: Lexington Books, 2007, pp. 171–194.
19. See Chapter 2 of this publication, for further discussion of this case.
20. Leading human rights NGOs have also raised concerns about this use of an ICA weapon by the Russian Federation. See, e.g., Human Rights Watch, press release: *Independent Commission of Inquiry Must Investigate Raid on Moscow Theater: Inadequate Protection for Consequences of Gas Violates Obligation to Protect Life*, 30th October 2002; Amnesty International, *Rough Justice: The Law and Human Rights in the Russian Federation*, EUR 46/054/2003, October 2003.
21. UN, Commission on Human Rights, Report of the Special Rapporteur, Asma Jahangir, submitted pursuant to Commission on Human Rights Resolution 2002/36, UN doc. E/CN.4/2003/3, 13th January 2003, p. 15, paragraph 34.
22. UN, Human Rights Commission, Concluding observations of the Human Rights Committee: Russian Federation, UN doc. CCPR/CO/79/RUS, 6th November 2003, p. 4, paragraph 14.
23. *Ibid.*
24. UN Human Rights Commission (6th November 2003) *op.cit.*, p. 4, paragraph 14.
25. See European Court of Human Rights (ECtHR), *Finogenov and Others v. Russia*, application numbers 18299/03 and 27311/03, Judgement, 20th December 2011.
26. *Ibid.*, paragraph 201.
27. *Ibid.*, paragraph 202.
28. *Ibid.*, paragraph 202.
29. *Ibid.*, paragraph 203.
30. This term refers to the space for manoeuvre that the Strasbourg organs are willing to grant national authorities, in fulfilling their obligations under the ECHR.
31. ECtHR (20th December 2011) *op.cit.*, paragraph 213.
32. *Ibid.*, paragraph 232.

33. *Ibid.*, paragraph 236.
34. *Ibid.*, paragraph 266. [For further discussion, see paragraphs 243–262.]
35. *Ibid.*, paragraph 282. [For further discussions, see paragraphs 277, 279 and 281.]
36. Kelle, A. The Message From Strasbourg, *Bulletin of the Atomic Scientists*, 23rd February 2012; Kelle, A. Legally Incapacitated, Politically Outmaneuvered, *Bulletin of the Atomic Scientists*, 7th June 2012.
37. ICRC, Toxic Chemicals as Weapons for Law Enforcement, A Threat to Life and International Law?, ICRC, Geneva, September 2012, p. 3.
38. See in particular, ECtHR (20th December 2011) *op.cit.*, paragraphs 162–164 and 228–229.
39. ECtHR (20th December 2011) *op.cit.*, paragraph 229.
40. See ECtHR, Press release issued by the Registrar of the Court, ECHR 270 (2012), 27th June 2012.
41. In the case of *Finogenov and Others v. Russia*, the Court ruled that Russia was to pay all 64 applicants a total award – as regards non-pecuniary damage – of €1,254,000, and €30,000, jointly, for costs and expenses. See ECtHR (20th December 2011) *op.cit.*, paragraphs 285–296.
42. The African Court on Human and Peoples' Rights, the European Court of Human Rights, and the Inter-American Courts of Human Rights.
43. The Human Rights Committee can also consider a case raised through individual petition, but can only reach non-binding conclusions in such instances. See Hampson, F. (2007) *op.cit.*, p. 243.
44. *Ibid.*
45. There is, however, an ongoing attempt, initiated by Italy in 2007, to establish an international moratorium on the use of the death penalty with a view to progressing towards abolition. A number of UNGA Resolutions passed in 2007, 2008 and 2010, 2012 and 2014, supported by an increasing number of Member States, call upon all States that still maintain the death penalty to progressively restrict the use of the death penalty, reduce the number of offences for which it may be imposed, and establish a moratorium on executions with a view to abolishing the death penalty altogether. See, for example: UNGA Resolution A/RES/69/186 of 18th December 2014.
46. UN, ICCPR (1966) *op.cit.*, Article 6(2).
47. *Ibid.*, Article 6(5).
48. Amnesty International, Death Sentences and Executions in 2014, ACT 50/001/2015; April 2015.
49. Amnesty International, Execution by Lethal Injection, ACT 50/007/2007, October 2007, p. 38.
50. Amnesty International, Death Sentences and Executions in 2013, ACT 50/001/2014; March 2014.
51. *Ibid.*
52. Amnesty International, Death Sentences and Executions in 2009, AI Index ACT 50/001/2010 (March 2010), p. 6.
53. *Ibid.* (March 2010), p. 6.
54. Amnesty International, Lethal Injection Looms for 117 Prisoners, Urgent Action, UA 161/13, 24th June 2013; Amnesty International, ACT 50/001/2015 (April 2015) *op.cit.*
55. Amnesty International, ACT 50/001/2015 (April 2015) *op.cit.*
56. See for example: Amnesty International, ACT 50/001/2014, (March 2014) *op.cit.*

57. Criminal Procedure Law of the People's Republic of China, 1st July 1979, Article 212, as cited in: Cornell University Law School, Death Penalty Worldwide website, http://www.deathpenaltyworldwide.org/country-search-post.cfm?country=China (accessed 2nd April 2015).
58. China Makes Ultimate Punishment Mobile, *USA Today*, 14th June 2006; Segura, C. China Injects "Humanity" into Death Sentence, *Asia Times Online*, 16th December 2009, http://www.atimes.com/atimes/China/KL16Ad01.html (accessed 2nd April 2015).
59. Amnesty International, Execution by Lethal Injection, ACT 50/007/2007, October 2007, p. 17.
60. Chinese Courts Purchasing Mobile Execution Units, AFP, 18th December 2003.
61. China Makes Ultimate Punishment Mobile, *USA Today*, 14th June 2006.
62. Katyal, S. China to Swap Bullets for Lethal Injections, Reuters, 16th June 2009, http://www.reuters.com/article/idUSTRE55F0XT20090616 (accessed 2nd April 2015); Segura, C. China Injects "Humanity" into Death Sentence, *Asia Times Online*, 16th December 2009.
63. Segura, C. China Injects "Humanity" into Death Sentence, *Asia Times Online*, 16th December 2009.
64. Amnesty International, Execution by Lethal Injection, ACT 50/007/2007, October 2007, p. 38
65. Death Penalty Information Center, methods of execution, authorized methods, http://www.deathpenaltyinfo.org/methods-execution (accessed 2nd April 2015).
66. For further discussion see, for example, Amnesty International, Execution by Lethal Injection, ACT 50/007/2007, October 2007.
67. World Medical Association, Resolution on Physician Participation in Capital Punishment Adopted by the 34th World Medical Assembly Lisbon, Portugal, 28th September–2nd October 1981; International Council of Nurses, Torture, Death Penalty and Participation by Nurses in Executions, adopted 1998, revised 2006.
68. UN, Report of the Special Rapporteur, Asma Jahangir, submitted pursuant to Commission on Human Rights Resolution 2002/36, 13th January 2003, paragraph 12.
69. UN, Report of the Special Rapporteur on the Rights to Freedom of Peaceful Assembly and of Association, Maina Kiai, Human Rights Council, Twentieth Session, 21st May 2012, A/HRC/20/27, paragraph. 35.
70. CPT, CPT/Inf (2009) 13, 19th March 2009, paragraph 92.
71. See, e.g., Council of Europe, ECHR (1950) *op.cit.*, Articles 10 and 11.
72. See, e.g., UN, Universal Declaration of Human Rights (1948) *op.cit.*, Articles 19 and 20; UN, ICCPR (1966) *op.cit.*, Articles 19, 21 and 22.
73. UN, ICCPR (1966) *op.cit.*, Article 21.
74. UNBP (1990) *op.cit.*, Principle 13.
75. See for example: UN, Human Rights Council Resolution 15/21: The rights to freedom of peaceful assembly and of association, adopted on 30th September 2010.
76. See for example: UN, Human Rights Council, Resolution 25/38: Promotion and protection of human rights in the context of peaceful protests, adopted on 11th April 2014.
77. UN, Human Rights Council, Resolution 25/38 (11th April 2014) *op.cit.*, paragraph 3.

78. *Ibid.*, paragraph 9.
79. See, e.g., UN, Universal Declaration of Human Rights (1948) *op.cit.*, Article 3; and UN, ICCPR (1966) *op.cit.*, Article 9.
80. See, e.g., Council of Europe, ECHR (1950) *op.cit.*, Article 5.
81. UN, ICCPR (1966) *op.cit.*, Article 9(1).
82. Casey-Maslen, S. Non-Kinetic-Energy Weapons Termed "Non-Lethal", A Preliminary Assessment Under International Humanitarian Law and International Human Rights Law, Geneva Academy, 2010, p. 34.
83. *Ibid.*
84. NATO Standardisation Agency, Glossary of Terms and Definitions, AAP-6 (2010), p. 2-1-1, 22nd March 2010.
85. UN, Universal Declaration of Human Rights (1948) *op.cit.*, Article 5; UN, ICCPR (1966) *op.cit.*, Article 7; UN, Convention against Torture and Other Cruel, Inhuman or Degrading Treatment or Punishment, adopted by UNGA Resolution 39/46, 10th December 1984.
86. See for example: European Convention for the Prevention of Torture and Inhuman or Degrading Treatment or Punishment, Strasbourg, 26th November 1987.
87. See, e.g., UN, ICCPR (1966) *op.cit.*, Articles 4 and 7; UN, Human Rights Committee, General Comment Number 29: States of Emergency (Article 4), UN doc. CCPR/C/21/Rev.1/Add.11, 21 August 2001, paragraph 7.
88. UNCoC (1979) *op.cit.*, Article 5.
89. UN, Convention against Torture (1984) *op.cit.*, Article 1(1).
90. Rome Statute of the International Criminal Court, UN doc. A/CONF.183/9 of 17th July 1998. Article 8(2) (a) (ii), as cited in J. M. Henckaerts and L. Doswald-Beck (eds), *Customary International Humanitarian Law Study* (2005) *op.cit.*, Rule 90.
91. European Commission of Human Rights, Greek case (cited in Vol. II, Ch. 32, §1339), as cited in Henckaerts, J. M. and L. Doswald-Beck (eds), *Customary International Humanitarian Law Study* (2005) *op.cit.*, Rule 90.
92. Body of Principles for the Protection of All Persons Under Any Form of Detention or Imprisonment, adopted by UNGA Resolution 43/173 of 9th December 1988, Principle 6, Commentary.
93. Torture and other cruel, inhuman or degrading treatment, Report of the Special Rapporteur on the question of torture, Manfred Nowak, UN doc. E/CN.4/2006/6, 23rd December 2005, paragraph 38.
94. UN Commission on Human Rights, Study on the situation of trade in and production of equipment which is specifically designed to inflict torture or other cruel, inhuman or degrading treatment, its origin, destination and forms, submitted by Theo Van Boven, Special Rapporteur on torture, pursuant to resolution 2002/38 of the Commission on Human Rights, UN doc. E/CN.4/2003/69, 13th January 2003.
95. UN, Report of the Special Rapporteur on the question of torture, Theo Van Boven, UN doc. E/CN.4/2005/62, 15th December 2004, paragraph 13.
96. ECtHR, *Ali Güneş v. Turkey*, application number 9829/07, Judgement 10 April 2012. This judgement became final on 10 July 2012. See paragraphs 5–10.
97. ECtHR, *Ali Güneş v. Turkey* (April 2012) *op.cit.*, paragraph 43.
98. *Ibid.*, paragraph 43.
99. The applicant was also awarded € 1,500 for costs and expenses. ECtHR, *Ali Güneş v. Turkey* (April 2012) *op.cit.*, paragraphs 58 and 61.

100. Aceves, W. (2007) *op.cit.*, p. 271.
101. Royal Society, Science Policy Centre, *Brain Waves Module 3, Neuroscience, Conflict and Security*, Royal Society, London, February 2012, p. 24.
102. *Ibid.*, p. 24, note 78.
103. Fidler, D. (2007) *op.cit.*, p. 176.
104. *Ibid.*
105. UN, ICCPR (1966) *op.cit.*, Article 19.
106. Nowak, M., *U.N. Covenant on Civil and Political Rights: CCPR Commentary*, 2nd edn, Kehl: N. P. Engel, 2005, p. 340.
107. Human Rights Watch and Geneva Initiative on Psychiatry, *Dangerous Minds: Political Psychiatry in China Today and Its Origins in the Mao Era*, August 2002.
108. See, e.g., British Medical Association, *Medicine Betrayed: The Participation of Doctors in Human Rights Abuses*, London: Zed Books, 1992, pp. 64–72; Amnesty International, *Prisoners of Conscience in the USSR: Their Treatment and Conditions*, London: AI Publications, 1980; Bloch, S. and Reddaway, P. *Soviet Psychiatric Abuse: The Shadow Over World Psychiatry*, London: Gollancz, 1984.
109. *Ibid.*
110. UN, Human Rights Council, Seventh Session, *Report of the Special Rapporteur on torture and other cruel, inhuman or degrading treatment or punishment, Manfred Nowak, Addendum, Summary of information, including individual cases, transmitted to Governments and replies received*, A/HRC/7/3/Add.1, 19 February 2008, paragraph 123, p. 149.
111. Amnesty International, Viet Nam, Lawyer Detained in Mental Hospital, *The Wire*, volume 37, number 06, July 2007, AI Index: NWS 21/006/2007.
112. Human Rights Watch, *List of Imprisoned Uzbek Defenders and Activists, 11th December 2008*; Human Rights Watch, *Uzbekistan: Imprisoned Activists' Health in Danger UN Body Should Urge Their Immediate Release, End to Repression*, 11th December 2008.
113. Dando, M. and Furmanski, M. Midspectrum incapacitant programs, in Wheelis, M., Rózsa, L. and Dando, M. (eds), *Deadly Cultures: Biological Weapons since 1945*. Cambridge, MA: Harvard University Press, 2006, pp. 243–244.
114. Perry Robinson, J. Disabling Chemical Weapons: A Documented Chronology of Events, 1945–2003, Unpublished monograph, 1st November 2003, copy given to author. For US research see also US Senate, Committee on Intelligence and Human Resources Subcommittee on Health and Scientific Research, Joint Hearing, *Project MKULTRA, the CIA's Program of Research in Behavioral Modification*, 3rd August 1977.
115. Acharya, S. Is Narco Analysis a Reliable Science? Present Legal Scenario in India, *Legal Service India*, 19th February 2008, http://www.legalserviceindia.com/article/l176-Narco-Analysis.html (accessed 14th January 2011); Mumbai Attacks: Militant Kept in Underwear to Prevent Suicide, *Daily Telegraph*, 8th December 2008.
116. Amnesty International, *Turkmenistan: Individuals Continue to Be at Risk of Violations in Turkmenistan*, AI Index: EUR 61/001/2009, 12th February 2009; Human Rights Watch, *Turkmenistan: Open Letter from a Coalition of Human Rights Organizations*, 17th July 2006.
117. Human Rights Watch, *"We Need a Law for Liberation" Gender, Sexuality, and Human Rights in a Changing Turkey*, May 2008.
118. Warrick, J. Detainees Allege Being Drugged, Questioned. U.S. Denies Using Injections for Coercion, *Washington Post*, 22nd April 2008; UN Commission

on Human Rights, Situation of Detainees at Guantánamo Bay, 27th February 2006, UN doc. E/CN.4/2006/120. The report was prepared jointly by: the Chairperson-Rapporteur of the Working Group on Arbitrary Detention, the Special Rapporteur on the independence of judges and lawyers, the Special Rapporteur on torture and other cruel, inhuman or degrading treatment or punishment, the Special Rapporteur on freedom of religion or belief and the Special Rapporteur on the right of everyone to the enjoyment of the highest attainable standard of physical and mental health. For details of the US Government position on this issue, see United States Department of Defense, Deputy Inspector General for Intelligence, *Investigation of Allegations of the Use of Mind-Altering Drugs to Facilitate Interrogations of Detainees*, 23rd September 2009.
119. UN, Principles for the Protection of Detainees (9th December 1988) *op.cit.*, Principle 21.
120. UNGA Resolution 37/194, Principles of Medical Ethics, adopted on 18th December 1982.
121. UN, Principles of Medical Ethics (18th December 1982) *op.cit.*, Principle 3.
122. *Ibid.*, Principle 4.
123. *Ibid.*, Principle 6.
124. Supreme Court of India, Criminal Appellate Jurisdiction, *Selvi v. State of Karnataka & Anr.*, Criminal Appeal Number 1267 of 2004, 5th May 2010. In its ruling the Court did, however "leave room for the voluntary administration of the impugned techniques in the context of criminal justice, provided that certain safeguards are in place" and in very limited circumstances. The Court ruled that even when the subject had given consent to undergo narcoanalysis, "the test results by themselves cannot be admitted as evidence because the subject does not exercise conscious control over the responses during the administration of the test". However, the Court did rule that "any information or material that is subsequently discovered with the help of voluntary administered test results can be admitted". *Ibid.*, paragraph 223. Although this is a narrow loophole, it does appear that the practice of narcoanalysis where voluntary consent has been given continues. See for example: Lie Test Hints at Internal Plot Behind, *The Indian Express*, 18th July 2013; Murder Case: Accused to Seek Narco-Analysis Test, *The Hindu*, 22nd March 2015.
125. *Ibid.*, paragraph 195.
126. See US Supreme Court, *Townsend v. Sain*, 372 U.S. 293 (1963), pp. 307–309.
127. Organisation of American States, Inter-American Convention to Prevent and Punish Torture (1985) *op.cit.*, Article 2. It should, however, be noted that the US has neither signed nor acceded to this Convention
128. US Army, *Field Manual FM 2–22.3 (FM 34–52): Human Intelligence Collector Operations*, Headquarters, Department of the Army, September 2006, p. 102.
129. See also United Nations, Universal Declaration of Human Rights (1948) *op.cit.*, Article 25.
130. International Covenant on Economic, Social and Cultural Rights, adopted by UNGA Resolution 2200 (XXI), 16th December 1966, Article 12.
131. Committee on Economic, Social and Cultural Rights, General Comment No. 14: The Right to the Highest Attainable Standard of Health, UN doc. E/C.12/2000/4, 11th August 2000, paragraph 50.
132. *Ibid.*, paragraph 51.
133. Casey-Maslen, S. Non-Kinetic-Energy Weapons Termed "Non-Lethal", (2010) *op.cit.*, p. 35.

134. *Ibid.*, p. 36.
135. See, e.g., Euripidou, E., MacLehose, R. and Fletcher, A. An Investigation into the Short-Term and Medium-Term Health Impacts of Personal Incapacitant Sprays: A Follow-Up of Patients Reported to the National Poisons Information Service (London), *Emergency Medicine Journal*, volume 21, 2004, pp. 548–552; Hu, H., Fine, J., Epstein, P., Kelsey, K., Reynolds, P. and Walker, B. Tear Gas: Harassing Agent or Toxic Chemical?, *Journal of the American Medical Association*, volume 262, 1989, pp. 660–663.
136. See, for example, British Medical Association, The Use of Drugs as Weapons (May 2007) *op.cit.*
137. See UN, Report of the Special Rapporteur, Asma Jahangir, submitted pursuant to Commission on Human Rights Resolution 2002/36 (13th January 2003) *op.cit.*, paragraph 12.
138. ECtHR, *Oya Ataman v. Turkey*, no. 74552/01, Final Judgement, 5th March 2007, paragraph 18; ECtHR, *Ali Gunes v Turkey*, no. 9829/07, Final Judgement, 10th July 2012, paragraph 37.
139. See, in particular, CPT, The CPT Standards, CPT/Inf/E (2002) 1 – Rev. 2010, paragraph 38, p. 69.
140. Amnesty International, *Pain Merchants: Security Equipment and Its Use in Torture and Other Ill-Treatment*, ACT 40/008/2003, 2nd December 2003, pp. 68–73.
141. See CPT, CPT/Inf (2009) 25, 14th October 2009, paragraph 79; CPT, CPT/Inf (2009) 8, 5th February 2009, paragraph 46.
142. ECtHR, *Ali Güneş v. Turkey* (10th July 2012) *op.cit.*, paragraph 41.
143. Popular protest and challenges to freedom of assembly, media and speech, Resolution 1947 (2013) adopted by the Parliamentary Assembly of the Council of Europe at its 25th Sitting (27th June 2013).
144. Resolution 1947, Parliamentary Assembly of the Council of Europe (June 2013) *op.cit.*, paragraph 7.
145. *Ibid.*, paragraph 9.
146. *Ibid.*, paragraph 9.4.
147. See, e.g., British Medical Association, The Use of Drugs as Weapons (May 2007) *op.cit.*
148. See, e.g., *Ibid.*
149. Human Rights Watch, press release: *Independent Commission of Inquiry Must Investigate Raid on Moscow Theater* (2002) *op.cit.*
150. UN Commission on Human Rights, Situation of Detainees at Guantánamo Bay, 27th February 2006, UN doc. E/CN.4/2006/120.
151. UN Commission on Human Rights, Situation of Detainees at Guantánamo Bay (27th February 2006) *op.cit.*, paragraph 70.
152. UN Commission on Human Rights, Situation of Detainees at Guantánamo Bay (27th February 2006) *op.cit.*, paragraph 70. Although the Commission report did not elaborate on the nature of the drugs administered nor the purpose of such actions, a number of former detainees have alleged that injections of unknown drugs occurred before interrogation and were intended to coerce confessions (see Warrick, J. (22nd April 2008) *op.cit.*
153. UN Commission on Human Rights, Situation of Detainees at Guantánamo Bay (2006) *op.cit.*, paragraph 75.
154. UN, Standard Minimum Rules for the Treatment of Prisoners, adopted by the First United Nations Congress on the Prevention of Crime and the Treatment of Offenders, held at Geneva in 1955, and approved by the Economic and Social

Council by its Resolutions 663 C (XXIV) of 31st July 1957 and 2076 (LXII) of 13th May 1977.
155. UN, General Assembly, Israeli practices affecting the human rights of the Palestinian people in the Occupied Palestinian Territory, including East Jerusalem, Report of the Secretary-General, A/66/356, 13th September 2011.
156. UN, Human Rights Council, Report of the Special Rapporteur on the promotion and protection of the right to freedom of opinion and expression, Frank La Rue, Mission to Israel and the Occupied Palestinian Territory, A/HRC/20/17/Add.2, 11th June 2012.
157. UN, A/HRC/20/17/Add.2 (11th June 2012) *op.cit.*, paragraph 80.
158. *Ibid.*, paragraph 83.
159. ECtHR, case of *Abdullah Yaşa and Others v. Turkey*, application number 44827/08, judgement: 16th July 2013.
160. ECtHR, Disproportionate use of force by police to disperse a violent demonstration, Press release issued by the Registrar of the Court, ECHR 225 (2013) 16th July 2013.
161. ECtHR, *Abdullah Yaşa and Others v. Turkey* (October 2013) *op.cit.*, paragraph 48.
162. *Ibid.*, paragraph 50.
163. *Ibid.*, paragraph 51.
164. ECtHR, Death during a demonstration: Turkey must regulate the use of tear-gas grenades, press release issued by the Registrar of the Court, ECHR 227 (2014), 22nd July 2014. Concerning case of *Ataykaya v. Turkey*, application number 50275/08. This judgement is not yet final.
165. ECtHR (22nd July 2014) *op.cit.*, Principal facts, p. 2.
166. *Ibid.*, Decision of the Court, Article 2. p. 3.
167. *Ibid.*, Decision of the Court, Article 2. p. 3.
168. *Ibid.*, Decision of the Court, Article 2. p. 3.
169. Protocol Additional to the Geneva Conventions of 12th August 1949, and relating to the Protection of Victims of International Armed Conflicts [Protocol I], 8th June 1977, Article 36.
170. UN Basic Principles on the Use of Force and Firearms by Law Enforcement Officials (1990) *op.cit.*, Principle 3.
171. UN, Human Rights Council, Resolution 25/38 (11th April 2014) *op.cit.*, paragraph 15.
172. *Ibid.*, paragraph 14.
173. ICRC, *A guide to the legal review of new weapons, means and methods of warfare, Measures to implement Article 36 of Additional Protocol 1 of 1977*, January 2006.
174. UNBP (1990) *op.cit.*, Principle 1. Which reads: "Governments and law enforcement agencies shall adopt and implement rules and regulations on the use of force and firearms against persons by law enforcement officials. In developing such rules and regulations, Governments and law enforcement agencies shall keep the ethical issues associated with the use of force and firearms constantly under review."
175. UNBP (1990) *op.cit.*, Principle 20.
176. *Ibid.*, Article 7.
177. UN, Human Rights Council, Resolution 25/38 (11th April 2014) *op.cit.*, paragraph 12.
178. Hampson, F. (2007) *op.cit.*, p. 243.
179. *Ibid.*

9 International Criminal Law Applicable to ICA Weapons and Riot Control Agents

1. Cassese, A. *International Criminal Law*, Oxford: Oxford University Press, 2003, p. 23, as cited by Oñate, S., Exterkate, B., Tabassi, L. and van der Borght, E. Lessons Learned: Chemicals Trader Convicted of War Crimes, *Hague Justice Journal*, volume 2, number 1, 2007, p. 37.
2. Cassese, A. *International Criminal Law*, 2nd edn, Oxford: Oxford University Press, 2005, p. 199. Vienna Convention on the Law of Treaties (adopted 23rd May 1969, entered into force 27th January 1980) 1155 UNTS 331 Article 53, as cited by Oñate, S., Exterkate, B., Tabassi, L. and van der Borght, E. (2007) *op.cit.*, p. 37.
3. Oñate, S., Exterkate, B., Tabassi, L. and van der Borght, E. (2007) *op.cit.*, p. 38.
4. Hampson, F. International Law and the Regulation of Weapons, in Pearson, A., Chevrier, M. and Wheelis, M. (eds), *Incapacitating Biochemical Weapons*. Lanham, MD: Lexington Books, 2007, pp. 231–260.
5. International Criminal Court, Rome Statute of the International Criminal Court, 17th July 1998, A/CONF.183/9. This and other Court documents are available from the ICC website http://www.icc-cpi.int/Pages/default.aspx (accessed 10th August 2015).
6. International Criminal Court, About the Court, ICC website.
7. For full details of the States Parties and Signatories, see ICC website. Three of the 16 Signatory States – Israel, Sudan and the United States – subsequently informed the UN Secretary General that they no longer intended to be party to the treaty and have no legal obligations arising from their signatures.
8. Rome Statue of the International Criminal Court (1998) *op.cit.*, Article 5.
9. *Ibid.*, Articles 8 (2) (b) (xvii) and (xviii).
10. Allen, K. with Spence, S. and Leal, R. Chemical and Biological Weapons Use in the Rome Statute: A Case for Change, *VERTIC brief 14*, February 2011, p. 2.
11. International Criminal Court, Elements of Crimes, Official Records of the Assembly of States Parties to the Rome Statute of the International Criminal Court (ASP), First session, New York, 3rd–10th September 2002, ICC-ASP/1/3, part II.B.
12. *Ibid.*, Article 8(2) (b) (xviii), Elements.
13. International Criminal Court, ASP (2002) *op.cit.*, Article 8(2) (b) (xvii), Elements.
14. To date, however, there has been no such case brought before the ICC. For further discussion of ICC and chemical weapons, see Aceves, W. Human Rights Law and the Use of Incapacitating Biochemical Weapons, in Pearson, A., Chevrier, M. and Wheelis, M. (eds), *Incapacitating Biochemical Weapons*. Lanham, MD: Lexington Books, 2007, pp. 261–284; Hampson, F. (2007) *op.cit.*; Tabassi, L. Impact of the CWC: Progressive Development of Customary International Law and Evolution of the Customary Norm Against Chemical Weapons, *CBW Conventions Bulletin*, number 63, pp. 1–7, March 2004.
15. Tabassi, L. (2004) *op.cit.*, p. 2.
16. Allen, K. (2011) *op.cit.*, p. 10.
17. Rome Statue of the International Criminal Court (1998) *op.cit.*, Article 8(2) (a) (ii).
18. *Ibid.*, Article 8(2) (a) (iii).
19. *Ibid.*, Articles 11 and 12.
20. *Ibid.*, Article 17.
21. Tabassi, L. (2004) *op.cit.*, p. 2.
22. Tabassi, L. and van der Borght, E. Chemical Warfare as Genocide and Crimes against Humanity, *Hague Justice Journal*, volume 2, number 1, 2007, p. 5.

23. Tabassi, L. (2004) *op.cit.*, p. 2.
24. Tabassi, L. and van der Borght, E. (2007) *op.cit.*, p. 22.
25. *Ibid.*
26. International Criminal Court, Rome Statute amendment proposals, Elements of crimes corresponding to the proposed amendment contained in Annex III to Resolution ICC-ASP/8/Res.6.
27. Austria, Argentina, Bolivia, Bulgaria, Burundi, Cambodia, Cyprus, Germany, Ireland, Latvia, Lithuania, Luxembourg, Mauritius, Mexico, Romania, Samoa, Slovenia and Switzerland were the co-sponsoring States.
28. First Review Conference of the Rome Statute, 31st May–10th June 2010, Kampala, Uganda. For more information, see ICC website.
29. International Criminal Court, Review Conference of the Rome Statute concludes in Kampala, 12th June 2010, press release; International Criminal Court, Review Conference of the Rome Statute, Kampala, Uganda, 31st May–11th June 2010, Resolution RC/Res.5, Amendments to Article 8 of the Rome Statute, http://www.icc-cpi.int/iccdocs/asp_ docs/Resolutions/RC-Res.5-ENG.pdf (accessed 2nd April 2015).
30. Rome Statue of the International Criminal Court (1998) *op.cit.*, Article 121, paragraph 5.
31. For further discussion see: Zgonec-Rožej, M. Historical Development and the Establishment of the International Courts and Tribunals, in: *International Criminal Law Manual*, International Bar Association, May 2010.
32. For further information, see United Nations, International Criminal Tribunal for the Former Yugoslavia, available at http://www.icty.org/sections/AbouttheICTY (accessed 2nd April 2015).
33. For further information, see United Nations, International Criminal Tribunal for Rwanda, available at http://www.unictr.org/en (accessed 2nd April 2015).
34. For further discussion, see Zgonec-Rožej, M. (May 2010) *op.cit.*, pp. 65–76.
35. For further information, see International Bar Association, Special Panels for Serious Crimes (East Timor), available at http://www.ibanet.org/Committees/WCC_EastTimor.aspx (accessed 2nd April 2015).
36. For further information, see International Bar Association, Special Court for Sierra Leone, available at http://www.ibanet.org/Committees/WCC_SCSL.aspx (accessed 2nd April 2015).
37. For further information, see International Bar Association, Extraordinary Chambers in the Courts of Cambodia, available at http://www.ibanet.org/Committees/WCC_Cambodia.aspx (accessed 2nd April 2015).
38. For further information, see United Nations, Special Tribunal for Lebanon, available at http://www.stl-tsl.org/ (accessed 2nd April 2015).
39. For further information on the War Crimes Chamber in Bosnia-Herzegovina, see International Tribunal Spotlight, War Crimes Chamber: Court of Bosnia and Herzegovina, International Judicial Monitor, American Society of International Law and the International Judicial Academy, volume 2, number 2, July/August 2007, available at http://www.judicialmonitor.org/archive_0707/spotlight.html (accessed 10th July 2015).
40. Zgonec-Rožej, M. (2010) *op.cit.*, pp. 72–73.
41. For further information on the Iraqi High Tribunal, see International Bar Association, Iraqi High Tribunal, available at http://www.ibanet.org/Committees/WCC_IHT.aspx (accessed 2nd April 2015); Scharf, P. The Iraqi High Tribunal: Avai

lable Experiment in International Justice? *Journal of International Criminal Justice*, volume 5, number 2, 2007, pp. 258–263–6.
42. Hampson, F. (2007) *op.cit.*, p. 232. For further information on these bodies, see Rikhof, J. Fewer Places to Hide? The Impact of Domestic War Crimes Prosecutions on International Impunity, *Criminal Law Forum*, volume 20, number 1, 2009, pp. 4–8; Zgonec-Rožej, M. (May 2010) *op.cit.*, pp. 55–61 and 65–76.
43. For information and analysis of the Anfal trial, see International Centre for Transitional Justice, the Anfal trial and the Iraqi High Tribunal updates 1–3, 2006, available at http://ictj.org/publications (accessed 2nd April 2015); Trahan, J. A Critical Guide to the Iraqi High Tribunal's Anfal Judgement: Genocide Against the Kurds, *Michigan Journal of International Law*, volume 30, 13th March 2009, pp. 305–407; Tabassi, L. and van der Borght, E. Chemical Warfare as Genocide and Crimes Against Humanity, *Hague Justice Journal*, volume 2, number 1, 2007.
44. Saddam Hussein Majid Al-Tikriti, Ali Hassan Al-Majid Al-Tikriti, Sultan Hashem Ahmed, Sabir Abdul-Aziz Al-Douri, Hussein Rashid Al-Tikriti, Tahir Tawfiq al-A'ni, Farhan Mutlak Al-Joubori.
45. Saddam Accused of Genocide in New Charges, *Associated Press* (4th April 2006), as cited by Tabassi, L. and van der Borght, E. (2007) *op.cit.*
46. For a detailed analysis, see Trahan, J. (2009) *op.cit.*
47. Hampson, F. (2007) *op.cit.*, p. 232.
48. Oñate, S., Exterkate, B., Tabassi, L. and van der Borght, E. (2007) *op.cit.*, p. 38.
49. Convention on the Prevention and Punishment of the Crime of Genocide, New York, 9th December 1948. For text, see International Humanitarian Law – treaties and documents, International Committee of the Red Cross, available at http://www.icrc.org/ihl.nsf/full/357?OpenDocument (accessed 10th August 2015). As of August 2015, there were 146 States Parties and one Signatory State to this Convention. Further details are available at https://www.icrc.org/applic/ihl/ihl.nsf/States.xsp?xp_viewStates=XPages_NORMStatesParties&xp_treatySelected=357 (accessed 10th August 2015).
50. Conventions (I–IV), Geneva, 12th August 1949 [Geneva Conventions]. For text and commentaries, see ICRC, Treaties and States Parties to such Treaties, available at https://www.icrc.org/applic/ihl/ihl.nsf/ (accessed 2nd April 2015). As of August 2015 there were 196 States Parties to these Conventions. For further details, see https://www.icrc.org/applic/ihl/ihl.nsf/States.xsp?xp_viewStates=XPages_NORMStatesParties&xp_treatySelected=375 (accessed 10th August 2015).
51. Rikhof, J. (2009) *op.cit.*, p. 8.
52. Hankin, S. Overview of Ways to Import Core International Crimes into National Criminal Law, in Bergsmo, M., Hayashi, M. and Harlem, N. (eds), *Importing Core International Crimes into National Criminal Law*, 2nd edn, Oslo: Torkel Opsahl Academic EPublisher, 2010, pp. 20–28.
53. Rikhof, J. (2009) *op.cit.*, pp. 9–12.
54. For example, Australia, Jordan, Malta and the UK. See Rikhof, J. (2009) *op.cit.*, p. 9. See also Coalition for the International Criminal Court, Ratification and Implementation, available at http://www.coalitionfortheicc.org/?mod=ratimp (accessed 10th July 2015).
55. For example, Kenya, New Zealand, South Africa and Uganda. See Rikhof, J. (2009) *op.cit.*, p. 9; CICC website.
56. Gould, C. More Questions Than Answers: The Ongoing Trial of Dr Wouter Basson, *Disarmament Diplomacy*, number 52, November 2000, available at http://www.

acronym.org.uk/52trial.html (accessed 12th July 2015). See Chapter 12 of this publication for further discussion of Dr Basson's alleged alleged activities as head of Project Coast.

57. See, for example, trial by Dutch courts of Guus van Kouwenhoven, convicted of violating a United Nations arms embargo in Liberia: the verdict was subsequently overturned; and the Belgian court indictment of Ephrem Nkezabera for alleged activities in the Rwandan genocide related to financing and arming of the Interahamwe militia, as discussed in Rikhof, J. (2009), *op.cit*, p. 21 and pp. 22–23.
58. For a discussion of this case, see The Hague Justice Portal, Frans van Anraat, available at http://www.haguejusticeportal.net/eCache/DEF/6/411.html (accessed 2nd April 2015); Zwanenburg, M. and den Dekker, G. Prosecutor v. Frans van Anraat, *American Journal of International Law*, volume 104, number 1, January 2010, pp. 86–94.
59. Both the Prosecution and van Anraat appealed the decision. On 9th May 2007, the Appeal Chamber confirmed the decision of the District Court and increased Mr van Anraat's sentence to 17 years imprisonment, this was subsequently reduced by 6 months on further appeal to the Supreme Court in June 2009. [For further discussion, see The Hague Justice Portal, van Anraat accomplice to war crimes, available at http://www.haguejusticeportal.net/eCache/DEF/2/243.html (accessed 2nd April 2015).]; van der Wilt, Genocide, Complicity in Genocide and International v. Domestic Jurisdiction: Reflections on the van Anraat Case, *Journal of International Criminal Justice*, volume 4, 2006, pp. 239–257, as cited by Rikhof, J. (2009) *op.cit*.
60. The Hague Justice Portal, Frans van Anraat, available at http://www.haguejusticeportal.net/eCache/DEF/6/411.html (accessed 2nd April 2015); Zwanenburg, M. and den Dekker, G. (2010) *op.cit*., p. 87.
61. Oñate, S., Exterkate, B., Tabassi, L. and van der Borght, E. (2007) *op.cit*., p. 23.
62. The Hague Justice Portal, Frans van Anraat (2015) *op.cit*.
63. Oñate, S., Exterkate, B., Tabassi, L. and van der Borght, E. (2007) *op.cit*., p. 23.
64. Meselson, M. and Perry Robinson, J. A Draft Convention to Prohibit Biological and Chemical Weapons Under International Criminal Law, *Fletcher Forum of World Affairs*, volume 28, edition 1, Winter 2004, pp. 57–70, at p. 57.
65. *Ibid*.
66. Meselson, M. and Robinson, J. P. (2004) *op.cit*., p. 58.
67. *Ibid*.
68. Harvard Sussex Program, Draft Convention on the Prevention and Punishment of the Crime of Developing, Producing, Acquiring, Stockpiling, Retaining, Transferring or Using Biological or Chemical Weapons, 2009. Also see accompanying Draft Legal Commentary. Both available from the HSP website, available at http://www.sussex.ac.uk/Units/spru/hsp/Harvard-Sussex-Program-draft-convention-Further-Info.html (accessed 12th July 2015).
69. Harvard Sussex Program (2009) *op.cit*., Preamble.
70. *Ibid*., Preamble.
71. *Ibid*., Article III.
72. See for example: Dunworth, T. HSP Draft Convention: Some Thoughts from a Legal Perspective, *CBW Conventions Bulletin, Special Issue*, February 2011, pp. 31–24; Evidence by Simms, N. and Feakes, D. to the UK Foreign Affairs Committee. House of Commons, Foreign Affairs Committee, Global Security: Non-Proliferation, Fourth Report of Sessions 2008–2009, 14th June 2009.

73. Dando, M. Bringing Increased Biological and Chemical Weapons Provisions to the ICC, *Bulletin of Atomic Scientists*, 11th November 2009, available at http:thebulletin.org/bringing-increased-biological-and-chemical-weapons-provisions-icc (accessed 12th July 2015).
74. Perry Robinson, J. Criminalization of Biological and Chemical Armament, *CBW Conventions Bulletin*, Special Issue, February 2011, Harvard Sussex Program, p. 4.
75. The Draft HSP Convention, Implementation of the proposal, HSP website; see also Jefferson, C. The Harvard Sussex Draft Convention as a complement to Resolution 1540, Resolution 1540: At the crossroads, Stanley Foundation Civil Society Event, 1st October 2009.
76. See Memorandum from the Foreign and Commonwealth Office to the Foreign Affairs Committee, House of Commons papers 150, Session 2002–03; United Kingdom Foreign and Commonwealth Office, Strengthening the Biological and Toxin Weapons Convention: Countering the Threat from Biological Weapons, April 2002, CM 5484, p. 15.
77. Meselson, M. and Robinson, J. P. (2004) *op.cit.*, p. 61.
78. Perry Robinson, J. (2011) *op.cit.*, p. 5.
79. Rome Statute of the International Criminal Court (1998) *op.cit.*, Preamble.
80. Nuremberg IMT: Judgment and Sentence, *American Journal of International Law*, volume 41, 1947, p. 172, as cited by Dunworth, T. (2011) *op.cit.*
81. First Chautauqua Declaration, 29th August 2009, previously available at http://www.asil.org/chaudec/index_ files/frame.html (last accessed 1st May 2012, subsequently removed). Signed by the prosecutors of the Special Court for Sierra Leone, International Military Tribunal at Nuremberg, International Criminal Tribunal for Rwanda, International Criminal Court, Extraordinary Chambers in the Courts of Cambodia, International Criminal Tribunal for the Former Yugoslavia.

10 Mechanisms to Regulate the Transfer of ICA Weapons, Riot Control Agents and Related Means of Delivery

1. Organisation for the Prohibition of Chemical Weapons (OPCW), Convention on the Prohibition of the Development, Production, Stockpiling and Use of Chemical Weapons and on Their Destruction (Chemical Weapons Convention [CWC]), 1993, Article XI.2.(b).
2. *Ibid.*, Article I.1.(a).
3. *Ibid.*, Article I.1.(d).
4. *Ibid.*, Article VI and Verification Annex, Part VI, A.
5. *Ibid.*, Verification Annex, Part VI, B.
6. *Ibid.*, Verification Annex, Part VI, A.1.
7. *Ibid.*, Article VI and also Verification Annex, Part VII, C.
8. *Ibid.*, Article VI and also Verification Annex, Part VII.
9. *Ibid.*, Article VI and also Verification Annex, Part VIII.
10. *Ibid.*, Article XV (4–5).
11. BZ has been listed under Schedule 2.a
12. 3-Quinuclidinol and Benzilic Acid have been listed under Schedule 2.b.
13. RCAs are defined under the Convention as "Any chemical not listed in a Schedule..." [See OPCW, CWC (1993) *op.cit.*, Article II.7.]
14. OPCW, CWC (1993) *op.cit.*, Article VI.2.

15. OPCW, International Transfer of Scheduled Chemicals Under the Chemical Weapons Convention, available at http://www.opcw.org/our-work/non-proliferation/international-transfer-of-scheduled-chemicals/ (accessed 2nd May 2015).
16. Convention on the Prohibition of the Development, Production and Stockpiling of Bacteriological (Biological) and Toxin Weapons and on Their Destruction (BTWC), 1972, Article III.
17. *Ibid.*, Article I.(2).
18. *Ibid.*, Article X.
19. United Nations, Final Document of the Seventh Review Conference, BWC/CONF.VII/7, 5th–22nd December 2011, Geneva, Article III, paragraph 9.
20. *Ibid.*, paragraph 10.
21. BTWC (1972) *op.cit.*, Article I.(1).
22. *Ibid.*, Article I.(2).
23. See Australia Group, Objectives of the Group [all AG documents discussed are available from the AG website at: http://www.australiagroup.net/en/index.html (last accessed 10th August 2015).
24. *Ibid.*
25. See Australia Group, Introduction and Activities.
26. See Australia Group, Participants.
27. Australia Group, Introduction.
28. *Ibid.*
29. Australia Group, Introduction and Activities.
30. Australia Group, Export Control List: Chemical Weapons Precursors, September 2014.
31. Australia Group, Dual-use chemical manufacturing facilities and equipment and related technology and software, September 2014.
32. Australia Group, Dual-use biological equipment and related technology and software, September 2014.
33. Australia Group, List of human and animal pathogens and toxins for export control, January 2014.
34. Australia Group, Export Control List: Chemical Weapons Precursors, September 2014.
35. Australia Group, List of human and animal pathogens and toxins for export control, January 2014.
36. For further discussion, see Chapter 2.
37. Australia Group, Dual-use biological equipment and related technology and software, September 2014.
38. *Ibid.* Part I (Equipment), section 9.
39. Australia Group, Guidelines for Transfers of Sensitive Chemical or Biological Items, June 2012.
40. *Ibid.*, Article 1.
41. *Ibid.*, Article 2.
42. *Ibid.*, Article 4.
43. *Ibid.*, Article 4.a.
44. Australia Group, Guidelines for Transfers of Sensitive Chemical or Biological Items, June 2012, Article 4.b.
45. Australia Group, Guidelines for Transfers of Sensitive Chemical or Biological Items, further provisions applicable to Australia Group Participants: catch-all, June 2012.

46. *Ibid.*
47. Australia Group, Guidelines for Transfers of Sensitive Chemical or Biological Items, further provisions applicable to Australia Group Participants: no-undercut policy, June 2012.
48. *Ibid.*
49. Australia Group, Guidelines for Transfers of Sensitive Chemical or Biological Items, further provisions applicable to Australia Group Participants: catch-all, June 2012.
50. Brehm, M. Conventional Arms Transfers in the Light of Humanitarian and Human Rights Law, LLM thesis, University Centre for International Humanitarian Law, Geneva, February 2005, pp. 6–9.
51. UN Security Council Commission for Conventional Armaments Resolution adopted at its 13th meeting, 12th August 1948.
52. UN Guidelines for international arms transfers in the context of General Assembly resolution 46/36 H of 6 December 1991; UN Disarmament Commission, UNGA, Supplement No. 42 (A/51/42), 22nd May 1996, paragraph 8.
53. UN (1996) *op.cit.*, paragraph 7.
54. UNGA, Resolution A/RES/64/48, 12th January 2010, paragraph 4.
55. UNGA, Resolution 67/234, 2nd April 2013.
56. United Nations, Office for Disarmament Affairs, Arms Trade Treaty, http://www.un.org/disarmament/ATT/ (accessed 8th October 2015).
57. United Nations, Arms Trade Treaty, 2013, Articles 2 & 3. The text of the treaty, details of States Parties and related documents are available from the UN Office of Disarmament Affairs website at http://www.un.org/disarmament/ATT/ (accessed 8th October 2015).
58. Wassenaar Arrangement, Guidelines and Procedures, including the Initial Elements, 6th July 1996, last amended July 2014, Purposes, Article 1 [all Wassenaar Arrangement documents are available from the WA website at: http://www.wassenaar.org/ (last accessed 18th August 2015].
59. *Ibid.*
60. For further details see: Wassenaar Arrangement, Basic Documents, Compiled by WA Secretariat, January 2015.
61. Wassenaar Arrangement, Frequently Asked Questions.
62. Wassenaar Arrangement, Introduction.
63. Wassenaar Arrangement, List of Dual Use Goods and Technologies and Munitions List, WA-LIST (14) 2 Corr,* 25th March 2015.
64. Wassenaar Arrangement, List of Dual Use Goods and Technologies and Munitions List (2015) *op.cit.*, ML7.
65. OPCW, S/1177/2014 (1st May 2014) *op.cit.*
66. OPCW, SAB III/I (27th April 2000) *op.cit.*
67. Wassenaar Arrangement, List of Dual Use Goods and Technologies and Munitions List (2015) *op.cit.*, Definitions, Cat 1, ML7.
68. Wassenaar Arrangement, List of Dual Use Goods and Technologies and Munitions List (2015) *op.cit.*, ML 7.b.3.a.
69. Wassenaar Arrangement, List of Dual Use Goods and Technologies and Munitions List (2015) *op.cit.*, ML 7.a.
70. Wassenaar Arrangement, List of Dual Use Goods and Technologies and Munitions List (2015) *op.cit.*, ML 7.e.
71. Wassenaar Arrangement, List of Dual Use Goods and Technologies and Munitions List (2015) *op.cit.*, ML 18.
72. Wassenaar Arrangement, Introduction.

73. Wassenaar Arrangement, Frequently Asked Questions.
74. Wassenaar Arrangement, Guidelines and Procedures, including the Initial Elements (2014) *op.cit.*, section II (scope), paragraph 3.
75. Wassenaar Arrangement, Elements for Objective Analysis and Advice Concerning Potentially Destabilising Accumulations of Conventional Weapons, WA Plenary approved 3rd December 1998 and amended by the WA Plenary in December 2004 and December 2011, as included in: Wassenaar Arrangement, Basic Documents, Compiled by WA Secretariat, January 2015.
76. *Ibid.*
77. Council of the European Union, Common Position 2008/944/CFSP adopted on 8th December 2008, *Official Journal of the European Union L 335/99*, 13th December 2008.
78. European Union, EU Code of Conduct on Arms Exports, European Union, EU Council 8675/2/98, Rev. 2, Brussels, 5th June 1998.
79. European Union, 13th Annual Report According to Article 8(2) of Council Common Position 2008/944/CFSP Defining Common Rules Governing Control of Exports of Military Technology and Equipment, 2011/C 382/01, *Official Journal of the European Union*, 30th December 2011, p. 1.
80. EU NGO Coalition, *Taking Control: The Case for a More Effective European Union Code of Conduct on Arms Exports*, Saferworld, September 2004, pp. 10–12.
81. European Union, Council Common Position 2008/944/CFSP (2008) *op.cit.*, paragraph 3.
82. European Union, Council Common Position 2008/944/CFSP (2008) *op.cit.*, paragraph 4.
83. European Union, Council Common Position 2008/944/CFSP (2008) *op.cit.*, Article 1.
84. European Union, Council Common Position 2008/944/CFSP (2008) *op.cit.*, Article 2.
85. European Union, Council Common Position 2008/944/CFSP (2008) *op.cit.*, Article 2, Criterion 2.
86. European Union, Council Common Position 2008/944/CFSP (2008) *op.cit.*, Article 2, Criterion 2.
87. European Union, Council Common Position 2008/944/CFSP (2008) *op.cit.*, Article 2.2.
88. European Union, Council Common Position 2008/944/CFSP (2008) *op.cit.*, Article 2.2.c.
89. European Union, Council Common Position 2008/944/CFSP (2008) *op.cit.*, Article 4.
90. European Union, Council Common Position 2008/944/CFSP (2008) *op.cit.*, Article 8.
91. European Union, Council Common Position 2008/944/CFSP (2008) *op.cit.*, Article 12.
92. European Union, Council Common Position 2008/944/CFSP (2008) *op.cit.*, Article 1.
93. European Union, Common Military List of the European Union (equipment covered by Council Common Position 2008/944/CFSP defining common rules governing the control of exports of military technology and equipment), (CFSP) (2015/C 129/01), adopted by the Council on 9th February 2015. *Official Journal of the European Union C 129/1.* volume 58, 21st April 2015.
94. Article 6 of Common Position 2008/944/CFSP extended the scope of coverage to dual use goods and technologies covered by EC Regulation 1334/2000. For

scope of this coverage, see EC Regulation (EC) No. 1334/2000 of 22nd June 2000 setting up a Community regime for the control of exports of dual-use items and technology, *Official Journal of the European Communities*, EN L 159/1, 30th June 2000.
95. It should be noted that Article 12 of the EU Common Position stipulates that the EU Common Military List shall act as a reference point for Member States' national military technology and equipment lists, but shall not directly replace them.
96. European Union, EU Common Position, paragraph 16.
97. See European Union, Common Military List (2015) *op.cit.*
98. It should be noted, however, that Article 3 of the Common Position stated that: "This Common Position shall not affect the right of Member States to operate more restrictive national policies." A more inclusive list of ICA weapons, RCAs and delivery mechanisms may therefore be added to individual national control lists.
99. European Union, User's Guide to Council Common Position 2008/944/CFSP defining common rules governing the control of exports of military technology and equipment, 10858/15, COARM 172, CFSP/PESC 393, 20th July 2015.
100. European Union, Council Common Position 2008/944/CFSP (2008) *op.cit.*, Article 8.3.
101. For links to most EU Member State annual reports, see Council of the European Union, External Action, Arms Export Control, http://www.eeas.europa.eu/non-proliferation-and-disarmament/arms-export-control/index_ en.html (accessed 2nd April 2015).
102. European Union, Council Common Position (2008) *op.cit.*, Article 8.3.
103. For latest version, see European Union, 16th Annual Report (2015) *op.cit.*
104. For examples, see Amnesty International, Blood at the Crossroads: Making the Case for a Global Arms Trade Treaty, ACT 30/011/2008, 2008, pp. 11–12; Amnesty International, Arms Transfers to the Middle East and North Africa: Lessons for an Effective Arms Trade Treaty, ACT 30/117/2011, 19th October 2011; Amnesty International, Guinea: "You Did Not Want the Military, So Now We Are Going to Teach You a Lesson", AFR 29/001/2010, February 2010.
105. For further discussion and case studies, see chapters 2, 3, 4 and 8 of this publication.
106. Report of the Special Rapporteur on the question of torture, Theo Van Boven, Commission on Human Rights, E/CN.4/2005/62, 15th December 2004, paragraphs 14 and 37.
107. Amnesty International, Stopping the Torture Trade, ACT 40/002/2001, February 2001.
108. European Union, Council Regulation (EC) No. 1236/2005 of 27 June 2005 concerning trade in certain goods which could be used for capital punishment, torture or other cruel, inhuman or degrading treatment or punishment, *Official Journal of the European Union*, 30th July 2005.
109. European Union, Council Regulation (EC) No. 1236/2005 (2005) *op.cit.*, Article 3.
110. European Union, Council Regulation (EC) No. 1236/2005 (2005) *op.cit.*, Article 5.
111. European Union, Council Regulation (EC) No. 1236/2005 (2005) *op.cit.*, Article 6.

112. European Union, Council Regulation (EC) No. 1236/2005 (2005) *op.cit.*, Article 6.1.
113. European Union, Council Regulation (EC) No. 1236/2005 (2005) *op.cit.*, Article 17.
114. European Union, Council Regulation (EC) No. 1236/2005 (2005) *op.cit.*, Article 11.
115. European Union, Council Regulation (EC) No. 1236/2005(2005) *op.cit.*, Article 13.
116. European Union, Council Regulation (EC) No. 1236/2005 (2005) *op.cit.*, Article 17.
117. European Union, Council Regulation (EC) No. 1236/2005(2005) *op.cit.*, pre-ambulatory paragraph 23.
118. European Union, Council Regulation (EC) No. 1236/2005(2005) *op.cit.*, Article 12.2 and see also Article 15.
119. European Union, Council Regulation (EC) No. 1236/2005 (2005) *op.cit.*, Annex III.
120. This item does not control individual portable devices, even if containing a chemical substance when accompanying their user for the user's own personal protection. [See European Union, Council Regulation (EC) No. 1236/2005 (2005) *op.cit.*, Annex III, footnote.]
121. For further discussion, see Chapter 8 of this publication.
122. Death Penalty Information Centre (DPIC), Lethal Injection, available at http://www.deathpenaltyinfo.org/lethal-injection-moratorium-executions-ends-after-supreme-court-decision (accessed 5th March 2015); DPIC, State by State Lethal Injection, available at http://www.deathpenaltyinfo.org/state-lethal-injection (accessed 5th March 2015).
123. European Commission, Commission Implementing Regulation (EU) No. 1352/2011 of 20th December 2011 amending Council Regulation (EC) No. 1236/2005 concerning trade in certain goods which could be used for capital punishment, torture or other cruel, inhuman or degrading treatment or punishment, L338/34, 21st December 2011, Annex III, paragraph 4.
124. European Commission, Commission Implementing Regulation (EU) No. 1352/2011 (2011) *op.cit.*, Annex III, paragraph 4.1.
125. For full discussion see: Amnesty International & the Omega Research Foundation, Grasping the Nettle: ending Europe's trade in execution and torture technology, Amnesty International, EUR 01/1632/2015, May 2015, pp. 26–27.
126. For example, between 1953 and 1974, the US reportedly explored a wide range of potential ICA weapons including depressants (barbiturates and opiates), but rejected them as unsuitable. See, for example: Croddy, E. and Wirtz, J. (eds), *Weapons of Mass Destruction: An Encyclopedia of Worldwide Policy, Technology and History, volume 2*, Santa Barbara, CA: ABC-CLIO, 2005, p. 229.
127. As proposed in: Amnesty International and the Omega Research Foundation, No More Delays: Putting an End to the EU Trade in "Tools of Torture", ACT 30/062/2012, Amnesty International, June 2012, pp. 24–26.
128. European Commission, Commission Implementing Regulation (EU) No. 775/2014 of 16th July 2014 amending Council Regulation (EC) No. 1236/2005 concerning trade in certain goods, which could be used for capital punishment, torture or other cruel, inhuman or degrading treatment or punishment, *Official Journal of the European Union* L 210/1, volume 57, 17th July 2014.

129. European Commission, Commission Implementing Regulation (EU) No. 775/2014 (17th July 2014) *op.cit.*, paragraph 3.5.
130. European Commission, Commission Implementing Regulation (EU) No. 775/2014 (17th July 2014) *op.cit.*, paragraph 3.6.
131. European Commission, Proposal for a Regulation of the European Parliament and of the Council amending Council Regulation (EC) No. 1236/2005 concerning the trade in certain goods which could be used for capital punishment, torture or other cruel, inhuman or degrading treatment or punishment, COM (2014) 1 final, 2014/0005 (COD), 14th January 2014.
132. Amnesty International and the Omega Research Foundation (May 2015) *op.cit.*
133. UK Department of Business, Enterprise and Regulatory Reform, Export Control Act 2002. Review of export control legislation (2007): Government's initial response to the public consultation (February 2008).
134. European Parliament resolution of 17th June 2010 on implementation of Council Regulation (EC) No. 1236/2005 concerning trade in certain goods, which could be used for capital punishment, torture or other cruel, inhuman or degrading treatment or punishment, P7_ TA-PROV(2010)0236, Trade in goods used for torture (B7-0360, 363, 365, 368 and 0369/2010), paragraph 16.
135. For further discussion see: Amnesty International and Omega Research Foundation (May 2015) *op.cit.*, pp. 30–31.
136. UNGA, Resolution A/C.3/66/L.28 Rev 1, 8th November 2011, paragraph 24; see also UNGA, Resolution A/RES/67/161, 20th December 2012; UNGA, Resolution A/RES/68/156, 18th December 2013.
137. For further detailed discussion including examples of OC and PAVA transfers of potential concern, see Amnesty International and the Omega Research Foundation, European Union: Stopping the Trade in Tools of Torture, POL 34/001/2007, Amnesty International, February 2007; Amnesty International and the Omega Research Foundation (2010) *op.cit.*; Amnesty International and the Omega Research Foundation (June 2012) *op.cit.*; Amnesty International and the Omega Research Foundation (May 2015) *op.cit.*
138. As quoted in European Council General Secretariat, *Implementation of the EU Guidelines on Torture and Other Cruel, Inhuman or Degrading Treatment or Punishment – Stock Taking and New Implementation Measures*, 8407/1/08 REV 1, 18th April 2008.
139. European Parliament Resolution (17th June 2010) *op.cit.*, paragraph 21.
140. United Nations, Charter of the United Nations, 26th June 1945, San Francisco, http://www.un.org/en/documents/charter/index.shtml (accessed 10th August 2015), Article 41, Chapter 7.
141. United Nations (1945) *op.cit.*, Article 39, Chapter 7.
142. United Nations, *UN Security Council Sanctions Committees: An Overview*, http://www.un.org/sc/committees/ (accessed 1st June 2012).
143. Stremlau, J. *Sharpening Economic Sanctions: Toward a Stronger Role for the United Nations*, Report to the Carnegie Commission on Preventing Deadly Conflict, New York: Carnegie Corporation, November 1996.
144. During this period UN embargoes were imposed against non-State actors in Afghanistan, Angola, DRC, Iraq, Lebanon, Liberia, Rwanda, Sierra Leone, Sudan and against Al-Qaeda.
145. Afghanistan (Taliban) [UNSCR 1333(2000)]; Eritrea [UNSCR 1907(2009)]; Eritrea and Ethiopia [UNSCR 1298(2000)]; Haiti [UNSCR 944]; Iraq [UNSCR 661(1990)]; Lebanon (non-governmental forces) [UNSCR 1701(2006)]; Liberia

[UNSCR 788(1992)]; Libya [UNSCR 748 (1992)]; Rwandan rebel groups [UNSCR 918(1994)]; Sierra Leonean Government and rebels [UNSCR 1132(1997), UNSCR 1171(1998)]; Somalia [UNSCR 733(1992)]; Sudan (Darfur) [UNSCR 1556(2004)]; Taliban, Al-Qaida and Osama Bin Laden [UNSCR 1267(1999)].

146. See, for example, UNSCR against Lebanese non-governmental forces: United Nations, UNSCR 1701 Adopted by the Security Council at its 5511th meeting, on 11th August 2006, Article 15.a.
147. Angola (UNITA) [UNSCR 864(1993)]; Cote d'Ivoire [UNSCR 1572(2004)]; DRC [UNSCR 1493(2003)]; Yugoslavia (FRY) [UNSCR 1160(1998)].
148. See, for example, UNSCR 1171 against Sierra Leonean rebel forces, Articles 2 and 3.
149. All EU embargoes – including those implementing UNSCRs – cover, as a minimum, all items in the EU Common Military List, which contains certain ICA weapons, RCAs and related means of delivery.
150. Kirkham, E. and Flew, C. *Strengthening Embargoes and Enhancing Human Security, Biting the Bullet Briefing 17*, Sponsored by the Canadian Department of Foreign Affairs and International Trade, 2003, pp. 10–11.
151. United Nations, UNSCR 687, Adopted by the Security Council at its 2981st meeting, on 3rd April 1991, Articles 8.a and 24.
152. United Nations, UNSCR 1718, Adopted by the Security Council at its 5551st meeting, on 14th October 2006, Article 8.a.(ii).e.
153. Bonn International Center for Conversion (BICC)/Bonn-Berlin Process, Report of Expert Working Group IV, *Monitoring and Enforcing UN Arms Embargoes*, January 2001, p. 5.
154. BICC/Bonn-Berlin Process, Monitoring and Enforcement of Arms Embargoes, in: Brzoska, M. (ed.), *Design and Implementation of Arms Embargoes and Travel and Aviation Related Sanctions*, BICC, Bonn, 2001, p. 114.
155. Report of the panel of experts on violations of Security Council sanctions against UNITA in: United Nations, Security Council, Letter dated 10th March 2000 from the Chairman of the Security Council Committee established pursuant to Resolution 864 (1993) concerning the situation in Angola addressed to the President of the Security Council, S/2000/203, 10th March 2000.
156. Shields, V. *Verifying European Union Arms Embargoes*, VERTIC, 18th April 2005, p. 10.
157. Fruchart, D., Holtom, P., Wezeman, S., Strandow, D. and Wallensteen, P. *United Nations Arms Embargoes: Their Impact on Arms Flows and Target Behaviour*, SIPRI and Uppsala University, September 2007, p. 53.
158. See, for example: UN Sanctions Secretariat, The Experience of the United Nations in Administering Arms Embargoes and Travel Sanctions, an Informal Background Paper prepared by The United Nations Sanctions Secretariat, Department of Political Affairs, for: Smart Sanctions, the Next Step: Arms Embargoes and Travel Sanctions First Expert Seminar, Bonn, 21st–23rd November 1999; Brzoska, M. (ed.), *Design and Implementation of Arms Embargoes and Travel and Aviation Related Sanctions: Results of the 'Bonn–Berlin Process'*, Bonn International Center for Conversion, Bonn, 2001; Sprague, O. *UN Arms Embargoes: An Overview of the Last Ten Years, Control Arms Briefing Note*, Control Arms, 16th March 2006; Fruchart, D. et al. (2007) *op.cit*. See also the reports of the informal UN Working Group on General Issues on Sanctions, established by the UN Security Council on 17th April 2000 to develop general recommendations on how to improve the effectiveness of United Nations sanctions,

reports available at https://www.globalpolicy.org/component/content/article/202/41761.html (accessed 12th July 2015).
159. See, for example, Amnesty International and Human Rights Watch reports highlighting the delayed adoption and weak implementation of the UN arms embargo against Rwandan forces that had perpetrated genocide in that country between April and June 1994. See in particular: Amnesty International, Arming the Perpetrators of Genocide, AI Index: AFR 02/14/95, 13th June 1995; Human Rights Watch, Arming Rwanda: The Arms Trade and Human Rights Abuses in the Rwandan War, 1st January 1994.
160. Kirkham and Flew (2003) *op.cit.*, p. 9.
161. UN, Security Council Sanctions Committees: An overview (undated) *op.cit.*
162. European Commission, Restrictive measures, Spring 2008, http://eeas.europa.eu/cfsp/sanctions/docs/index_en.pdf (accessed 2nd April 2015).
163. In such circumstances these pre-existing EU embargoes can be amended, if necessary, following the adoption of the UNSCR to take into account new embargoed entities, etc.
164. European Union, Council Declaration 7th April 1993; see also Council Common position of 21st October 2002 on the supply of certain equipment into the Democratic Republic of Congo, 2002/829/CFSP.
165. United Nations, Security Council Resolution 1493, 28th July 2003.
166. European Union, Common Position of 19th March 1998 defined by the Council on the basis of Article J.2 of the Treaty on European Union on restrictive measures against the Federal Republic of Yugoslavia, *European Union Journal*, number L 095, 27th March 1998, pp. 1–3.
167. United Nations Security Council, Resolution 1160, adopted by the Security Council at its 3868th session, 31st March 1998.
168. Guinea [CP 2009/788/CFSP]; Indonesia [CP 1999/624/CFSP]; Myanmar [Council Declaration; CP 2004/423/CFSP; CP 2004/730/CFSP]; Nigeria [CP 95/515/CFSP]; Uzbekistan [CP 2005/792/CFSP] and Zimbabwe [CP 2002/145/CFSP]. A further noteworthy embargo introduced just prior to this period was that against China, see Declaration of European Council, Madrid, 27th June 1989.
169. See European Union, Treaty Establishing the European Community, 1992, Article 296; European Union, Treaty on the Functioning of the European Union, Article 346.
170. European Union, European Commission – External Relations, Sanctions or Restrictive Measures, Spring 2008, p. 8.
171. See for example: European Union, Council Common Position 2005/792/CFSP of 14th November 2005 concerning restrictive measures against Uzbekistan.
172. See for example: European Union, Council Regulation (EC) No. 1859/2005 of 14th November 2005 imposing certain restrictive measures in respect of Uzbekistan.
173. European Union, European Commission – External Relations (Spring 2008) *op.cit.*, pp. 7–8.
174. Shields, V. (2005) *op.cit.*, p. 17.
175. Council of the European Union, Guidelines on Implementation and Evaluation of Restrictive Measures (Sanctions) in the Framework of the EU Common Foreign and Security Policy, EU document 15579/03, 3rd December 2003. For the most recent revised text, see Council of the European Union, Guidelines on Implementation and Evaluation of Restrictive Measures (Sanctions) in the Framework of the EU Common Foreign and Security Policy, EU document

17464/09, 15th December 2009. See also Council of the European Union, *Basic Principles on the Use of Restrictive Measures (Sanctions)*, EU document 10198/1/04 REV 1, 7th June 2004, which outlines the Council's view of sanctions, and how and when it would use them.
176. Council of the European Union, EU 15579/03 (2003) *op.cit.*, paragraph 16.
177. *Ibid.*
178. Council of the European Union, EU 15579/03 (2003) *op.cit.*, Annex I, paragraph 23.
179. See for example: European Union, Council Regulation (EC) No. 314/2004 of 19th February 2004 concerning certain restrictive measures in respect of Zimbabwe, Annex I, Article 22.
180. Council of the European Union, EU 17464/09 (2009) *op.cit.*, Annex II.
181. For example, the EU autonomous embargo introduced against Belarus on human rights grounds by Council Decision 2011/357/CFSP and implemented through Council Regulation (EU) No. 588/2011 does not incorporate the relevant text on tear gas or pepper spray within the list of repression equipment.
182. European Union, European Commission – External Relations (Spring 2008) *op.cit.*, p. 9.
183. Shields, V. (2005) *op.cit.*, p. 18.
184. Holtom, P. and Bromley, M. The Limitations of European Union Reports on Arms Exports: The Case of Central Asia, *SIPRI Insights on Peace and Security*, number 2010/5, September 2010, p. 19.
185. Holtom, P. and Bromley, M. (2010) *op.cit.*, p. 19.
186. Council of the European Union, Guidelines on implementation and evaluation of restrictive measures (sanctions) in the framework of the EU Common Foreign and Security Policy, document 15114/05, 2nd December 2005, paragraph 80.
187. Holtom, P. and Bromley, M. (2010) *op.cit.*, p. 5.

11 Application of the United Nations Drug Control Conventions to ICA Weapons

1. League of Nations, International Opium Convention, signed at the Hague on 23rd January 1912, Treaty Series, volume VIII, number 222.
2. Following an initial review, it became clear that there were currently no agreements developed in this context that had any relevancy to the regulation of RCAs or related means of delivery.
3. Single Convention on Narcotic Drugs, 1961, as amended by the Protocol amending the Single Convention on Narcotic Drugs. New York, 8th August 1975.
4. For details of States Parties to the SCND, see UN Treaty Collection Database, available at http://treaties.un.org/Pages/ViewDetails.aspx?src=TREATY&mtdsg_no=VI-18&chapter=6&lang=en (accessed 10th August 2015).
5. Text of SCND available on International Narcotics Control Board website: at https://www.incb.org/documents/Narcotic-Drugs/1961-Convention/convention_1961_en.pdf. (accessed 10th August 2015).
6. UN, SCND (1961) *op.cit*, Article 4.
7. *Ibid.*, Article 3.
8. *Ibid.*, Schedule 1.
9. *Ibid.*, Schedule 4.
10. *Ibid.*, Schedule 1.

11. *Ibid.*, Schedule 1.
12. *Ibid.*, Schedule 1.
13. *Ibid.*, Article 4.
14. *Ibid.*, Article 4.
15. *Ibid.*, Article 1.(1) (w).
16. UN, Commentary on the Single Convention on Narcotic Drugs, 1961, prepared by the Secretary-General in accordance with Paragraph 1 of ECOSOC Resolution 914.D (XXXIV) of 3rd August 1962, published 1973.
17. UN, Commentary on SCND (1973) *op.cit.*, p. 33, paragraph 6.
18. *Ibid.*, paragraph 5.
19. Rabbat, P. [Drug Control Officer with the Narcotics Control and Estimates Section at the International Narcotics Control Board Secretariat Control of substances under the international drug control Conventions, in: International Committee of the Red Cross, Expert Meeting: "Incapacitating Chemical Agents", Law Enforcement, Human Rights Law and Policy Perspectives, Montreux, Switzerland, 24th–26th April 2012, ICRC, Geneva, January 2013, p. 83.
20. Rabbat, P., in: ICRC (2013) *op.cit.*, p. 83.
21. For mandate and functions of the INCB, available at http://www.incb.org/incb/en/about/mandate-functions.html (accessed 2nd April 2015).
22. UN, SCND (1961) *op.cit.*, Articles 19.1 and 19.2.
23. *Ibid.*, Article 20.
24. *Ibid.*, Article 20.3.
25. *Ibid.*, Article 15.
26. An archive containing each Report of the International Narcotics Board from 1980 to 2014 is available at http://www.incb.org/incb/en/publications/annual-reports/annual-report.html (accessed 10th August 2015).
27. INCB, Report of the International Narcotics Control Board for 2003, paragraph 216.
28. UN, SCND (1961) *op.cit.*, Article 14.1.
29. *Ibid.*
30. A review was undertaken of relevant documents publicly available on the INCB website on 10th August 2015.
31. For details of States Parties to the Convention on Psychotropic Substances, see United Nations Treaty Collection Database, available at http://treaties.un.org/Pages/ViewDetails.aspx?src=TREATY&mtdsg_no=VI-16&chapter=6&lang=en (accessed 10th August 2015).
32. See United Nations, Convention on Psychotropic Substances, 1971, available at https://www.incb.org/documents/Psychotropics/conventions/convention_1971_en.pdf (accessed 10th August 2015); Introduction to Convention on INCB website.
33. UN, CPS (1971) *op.cit.*, Article 2.
34. *Ibid.*, Schedule 2.
35. *Ibid.*, Schedule 2.
36. *Ibid.*, Schedule 1.
37. *Ibid.*, Schedule 1.
38. *Ibid.*, Schedule 2.
39. *Ibid.*, Schedule 4.
40. *Ibid.*, Schedule 2.
41. *Ibid.*, Schedule 1.
42. *Ibid.*, Schedule 3.

43. *Ibid.*, Schedule 3.
44. *Ibid.*, Schedule 2.
45. *Ibid.*, Schedule 4.
46. For further discussion of the use of certain incapacitating chemical agents in lethal injection executions and narcoanalysis, see chapters 8 and 10 of this publication.
47. UN, CPS (1971) *op.cit*, Article 5.
48. *Ibid.*, Article 7.a.
49. *Ibid.*, Article 7.b.
50. *Ibid.*, Article 4.a–c.
51. A review was undertaken of relevant documents publicly available on the INCB website on 10th August 2015.
52. UN, CPS (1971) *op.cit.*, Article 16.
53. *Ibid.*, Article 18.
54. Unlike the SCND, there is no provision in the CPS for the INCB to formally receive information from inter-governmental organizations or international NGOs.
55. UN, CPS (1971) *op.cit.*, Article 19.1.a.
56. *Ibid.*
57. UN, CPS (1971) *op.cit.*, Article 19.1.c and Article 19.2
58. A review was undertaken of relevant documents publicly available on the INCB website on 10th August 2015.
59. International Committee of the Red Cross, Toxic Chemicals as Weapons for Law Enforcement, A Threat to Life and International Law?, ICRC, Geneva, September 2012.

12 The Role of Civil Society in Combating the Misuse of Incapacitating Chemical Agents and Riot Control Agents

1. Bohn, L. Rand Corporation Memorandum, 1956, as cited in Rotblat, J. Societal Verification, in Rotblat, J., Steinberger, J. and Udgaonkar, B. (eds), *A Nuclear-Weapon-Free World: Desirable? Feasible?* Boulder, CO: Westview Press, 1993.
2. Melman, S. (ed.), *Inspection for Disarmament*, New York: Columbia University Press, 1958.
3. Clark, G. and Sohn, L. *World Peace through World Law*, 2nd edn, Cambridge, MA: Harvard University Press, 1960, 2000, p. 267.
4. Portnoy, B. Arms Control Procedure: Inspection by the People: A Revaluation and a Proposal, *Cornell International Law Journal*, volume 4, number 2, 1971, pp. 153–165.
5. Deiseroth, D. Societal Verification: Wave of the Future? *2000 Verification Yearbook*, VERTIC, 2000, p. 267.
6. Deiseroth, D. (2000) *op.cit.*; Deiseroth, D. *Societal Verification: Citizen Reporting and Whistleblowing as Integral Elements of the Disarmament Process and of Technological and Other Verification Systems*, 2nd edn, Norderstedt: Books on Demand, 2010.
7. Falter, A. Including Civil Society into Confidence Building: Protecting Whistleblowers and Societal Verification, in Finney, J. and Slaus, I. (eds), *Assessing the Threat of Weapons of Mass Destruction*, NATO Science for Peace and Security Series – E: Human and Societal Dynamics. volume 61, Amsterdam, The Netherlands: IOS Press, 2010, p. 291.

8. Deiseroth, D. (2000) *op.cit.*, p. 265.
9. *Ibid.*
10. Rotblat, J. (1993) *op.cit.*, p. 105
11. *Ibid.*
12. Rotblat, J. (1993) *op.cit.*, p. 105.
13. *Ibid.*
14. *Ibid.*, p. 113.
15. *Ibid.*, p. 113.
16. *Ibid.*, p.108.
17. For further discussion see Crowley, M. and Persbo, A. The Role of Non-Governmental Organizations in the Monitoring and Verification of International Arms Control and Disarmament Agreements, in Borrie, J. and Martin Randin, V. (eds), *Thinking Outside the Box in Multilateral Disarmament and Arms Control Negotiations*. Geneva: United Nations Institute for Disarmament Research (UNIDIR), 2006, pp. 225–252.
18. Other restrictions on open source monitoring arise from limited access to research and primary documents published in certain countries and/or languages. In addition, there is much inaccurate or biased reporting disseminated by both proponents and opponents of "less lethal" weapons.
19. Sunshine Project, Freedom of Information research, available at: http://www.sunshine-project.org/FOIA (last accessed 1st May 2012, subsequently removed).
20. *Ibid.*
21. OPCW, CWC (1993) *op.cit.*, Article IX.
22. Foss, C. Turkey Details 120mm Automatic Mortar, *Jane's Defence Weekly*, 12th November 2003.
23. MKEK was promoting the CS mortar round till at least mid-November 2009 on its website, see http://mkekexport.com/ammunition.html (last accessed 16th November 2009, subsequently removed).
24. *Ibid.*
25. Furkan Defense Industry http://www.furkandefense.com/product_info.php?products_id=147 (last accessed 24th November 2010, subsequently removed); ASCIM Defense Industry http://www.ascim.com.tr/default.asp?inc=urun_detay&b=b&urun_id=78&Urun_kat=180 (last accessed 1st March 2011, subsequently removed).
26. Crowley, M. Regulation of Incapacitants, Riot Control Agents and Their Means of Delivery Under the Chemical Weapons Convention, OPCW, Open Forum Meeting, 29th November 2010.
27. Letter from Ambassador Dogan, Permanent Representative of Turkey to the OPCW, to BNWLRP, ORF and ISS, 25th February 2011 (copy on file with the author).
28. Letter from Mr Utkan, Counsellor, Permanent Representation of Turkey to the OPCW, to BNLWRP, ORF and ISS, 8th July 2011 (copy on file with the author).
29. OPCW, CWC (1993) *op.cit.*, Article III.1.
30. Ambassador Dogan (25th February 2011) *op.cit.*
31. Counsellor Utkan (8th July 2011) *op.cit.*
32. Bahrain Watch, *Leaked Document Shows Massive New Tear Gas Shipment Planned for Bahrain*, 16th October 2013.
33. *Ibid.*
34. *Ibid.*

35. European Parliament News, *Human Rights: Violence against Women in India; Crackdown in Bahrain; Insecurity in Central African Republic*, Press Release, 17th January 2013.
36. US Department of State, Senior Administration Officials on Bahrain, Special Briefing, Senior Administration, 11 May 2012, available at http://www.state.gov/r/pa/prs/ps/2012/05/189810.html (accessed 2nd May 2015).
37. Channel 4, *Bahrain: UK Revoke All Export Licenses to Bahrain*, 18th February 2011.
38. As cited in, and annexed to: Bahrain Watch, *OECD Complaint against Dae-Kwang Chemical Corporation for Possible Violations of the 2011 OECD Guidelines for Multinational Enterprises*, 26th December 2013.
39. Kerr, S. Bahrain Boosts Supplies of Tear Gas as Instability Continues, *The Financial Times*, 21st October 2013.
40. *Ibid*.
41. Bahrain Watch, OECD Complaint (26th December 2013) *op.cit.*, pp.3–4.
42. Bahrain Watch, OECD Complaint (26th December 2013) *op.cit*. The complaint was submitted on behalf of Bahrain Watch and Americans for Democracy and Human Rights in Bahrain.
43. UN Special Rapporteurs on: the Rights to Freedom of Peaceful Assembly and Association; Promotion and Protection of the Right to Freedom of Opinion and Expression; Extrajudicial Killing; Torture and Other Cruel, Inhuman or Degrading Treatment or Punishment.
44. Bahrain Watch, *Shipment Stopped*, Blog, 7th January 2014. https://bahrainwatch.org/blog/2014/01/07/south-korea-halts-massive-tear-gas-shipment-to-bahrain/ (accessed 2nd May 2015).
45. Jung-a, S. and Simeon Kerr, S. South Korea Halts Tear Gas Exports to Bahrain, *The Financial Times*, 9th January 2014.
46. *Ibid*.
47. For further examples see (Bosnia) – Hay, A. Surviving the Impossible: The Long March from Srebrenica. An Investigation of the Possible Use of Chemical Warfare Agents, *Medicine, Conflict and Survival*, volume 14, number 2, 1998, pp. 38–73; (Kurdistan/Iraq) – Physicians for Human Rights, *Iraq: Winds of Death: Iraq's Use of Poison Gas Against Its Kurdish Population. Report of a mission to Turkish Kurdistan*. February 1989; Physicians for Human Rights/Human Rights Watch statement, *Scientific First: Soil Samples Taken from Bomb Craters in Northern Iraq Reveal Nerve Gas – Even Four Years Later*, 29th April 1993, Physicians for Human Rights press release; Physicians for Human Rights, *Iraq: PHR Documentation of Chemical Weapons Attacks against Kurds by Hussein Regime's Anfal Campaign*, 24th August 2006; the Middle East Watch, *Genocide in Iraq: The Anfal Campaign against the Kurds*, 1993; Black R. J., Clarke, R. W. and Reid, M. T. J. Application of Gas Chromatography – Mass Spectrometry and Gas Chromatography – Tandem Mass Spectrometry to the Analysis of Chemical Warfare Samples, Found to Contain Residues of the Nerve Agent Sarin, Sulphur Mustard and Their Degradation Products, *Journal of Chromatography*, A 662, 1994, pp. 301–321, as cited in Perry Robinson, J. Scientists and Chemical Weapons Policies, in Finney, J. and Slaus, I (eds), *Assessing the Threat of Weapons of Mass Destruction*. Amsterdam, The Netherlands: IOS Press, 2010, p. 81.
48. Meselson, M. and Perry Robinson, J. The Yellow Rain Affair: Lessons from a Discredited Allegation, in Clunan, A. L., Lavoy P. R. and Martin, S. B. (eds), *Terrorism,*

War, or Disease? Unravelling the Use of Biological Weapons, Palo Alto, CA: Stanford University Press, 2008, pp. 72–96.
49. Analytichem, Solihull, West Midlands; the Laboratory of the Government Chemist, Teddington, Middlesex.
50. Hay, A., Giacaman, R., Sansur, R. and Rose, S. Skin Injuries Caused by New Riot Control Agent Used Against Civilians on the West Bank, *Medicine, Conflict and Survival*, volume 22, number 4, October–December 2006, pp. 283–291.
51. Hay, A. Giacaman, R. Sansur, R. and Rose, S. (2006) *op.cit.*, p. 287.
52. *Ibid.*, p. 288
53. *Ibid.*, p. 288.
54. Royal Society, Science Policy Centre, *Brain Waves Module 3, Neuroscience, Conflict and Security*, London, February 2012, p. 22.
55. Physicians for Human Rights, *Weaponizing Tear Gas: Bahrain's Unprecedented Use of Toxic Chemical Agents against Civilians*, August 2012.
56. Physicians for Human Rights (August 2012) *op.cit.*, p. 5
57. *Ibid.*, p. 5
58. *Ibid.*, p. 20.
59. *Ibid.*, p. 5 and pp. 25–28.
60. *Ibid.*, p. 5 and pp. 28–31.
61. Physicians for Human Rights, *Tear Gas or Lethal Gas? Bahrain's Death Toll Mounts to 34*, Blog, 16th March 2012, http://physiciansforhumanrights.org/blog/tear-gas-or-lethal-gas.html (accessed 1st March 2015).
62. Physicians for Human Rights (16th March 2012) *op.cit.*
63. Majumdar, S. and Griffin, H. *Bahrain's Continued Weaponizing of Tear Gas*, Physicians for Human Rights, 25th October 2013, http://physiciansforhumanrights.org/blog/bahrains-continued-weaponizing-of-tear-gas.html#sthash.A9H920WF.dpuf (accessed 1st March 2015).
64. Bahrain 2012 Human Rights Report, US State Department, http://www.state.gov/documents/organization/204567.pdf (accessed 1st March 2015), p. 3.
65. BBC World News, *Bahrani Authorities "Weaponising" Tear Gas*, 1st August 2012, http://www.bbc.co.uk/news/world-middle-east-19078659 (accessed 1st March 2015).
66. According to the Legal Information Institute at Cornel University Law School: "Torts are civil wrongs recognized by law as grounds for a lawsuit. These wrongs result in an injury or harm constituting the basis for a claim by the injured party. While some torts are also crimes punishable with imprisonment, the primary aim of tort law is to provide relief for the damages incurred and deter others from committing the same harms." [Legal Information Institute, Cornel University Law School, https://www.law.cornell.edu/wex/tort (accessed 8th June 2015).]
67. United States Code, title 28, part IV, Chapter 85, paragraph 1350.
68. *Filartiga v. Pena-Irala*, 630 F.2d 876 (Second Circuit. 1980).
69. For further discussion, see Hoffman, P. and Stephens, B., International Human Rights Cases Under State Law and in State Courts, UC Irvine Law Review, volume 3, number 1, 2013; Steinhardt, R. Weapons and the Human Rights Responsibilities of Multinational Corporations, in Casey-Maslen, S. (ed.), *Weapons Under International Human Rights Law*. Cambridge University Press, 2014; Stephens, B. Chomsky, J., Green, J., Hoffman, P. and Ratner, M., *International Human Rights Litigation in U.S. Courts*, Martinus Nijhoff Publishers, 2nd ed., 2008.
70. See for example: *Vietnam Association for Victims of Agent Orange v. Dow Chemicals*, 517 f.3d 104 (Second Circuit, 2008); *Romero v. Drummond Co. Inc.*, 552 F.3d 1303 (Eleventh Circuit 2008); *Corrie v. Caterpillar*, 503 F.3d 974 (Ninth Circuit, 2007).

71. A. Fagan (ed.), Human Rights & Peace Law Docket 1945–1993, Meiklejohn Civil Liberties Institute Archives, 2013, available at http://sunsite.berkeley.edu/meiklejohn/meik-peacelaw/meik-peacelaw-16.html#PL-663/34.9 (accessed 8th June 2015).
72. US District Court for the Western District of Pennsylvania, *Abu-Zeineh v. Federal Laboratories, Inc.*, Case No. Civ. A. No. 91-2148, 975 F. Supp. 774, 7 December 1994, available at http://www.leagle.com/decision/19941749975FSupp774_11661.xml/ABU-ZEINEH%20v.%20FEDERAL%20LABORATORIES,%20INC (accessed 8th June 2015).
73. Hoffman, P. and Stephens, B. (2013) *op.cit.*, p. 15.
74. A circuit split is when two or more circuits in the US court of appeals reach opposite interpretations of the law.
75. Bradley, C.A. *International Law and the US Legal System*, Oxford University Press, 2015, p. 220.
76. Supreme Court of the United States, October 2012 term, *Kiobel v. Royal Dutch Petroleum Co.*, 17th April 2013, available at http://www.supremecourt.gov/opinions/12pdf/10-1491_l6gn.pdf (accessed 8th June 2015).
77. Supreme Court of the United States, *Kiobel v. Royal Dutch Petroleum Co.* (17th April 2013) *op.cit.*, Section IV, p. 14.
78. Supreme Court of the United States, *Kiobel v. Royal Dutch Petroleum Co.* (17th April 2013) *op.cit.*, Section IV, p. 14.
79. For further discussion, see Corporate Liability Under Alien Tort Statute in the Second Circuit, Outside Counsel, Expert Analysis, Pepe, D., *New York Law Journal*, volume 253, number 30, 17th February 2015; Guest Post: Is the Alien Tort Statute Headed Back to the Supreme Court?, Opinio Juris, Dodge, W., 13th April 2015; Fourth Circuit's Post-Kiobel Ruling Revives ATS Claims Against U.S. Corporation for Violations Committed Abroad, Human Rights@Harvard Law, Blog posted by Giannini, T. and Farbstein, S., 2nd July 2014.
80. Legal Information Institute (LIC), Cornel University Law School, class action, [available from LIC website].
81. The lawsuit initially also included the Birmingham School Board among the defendants, but this body was subsequently excluded. See U.S. District Court for the Northern District of Alabama, Southern Division, *J.W. ex rel. Williams v. A.C. Roper, et al.*, Third Amended Complaint, 29th July 2011, available at http://www.splcenter.org/sites/default/files/downloads/case/mace_thirdamended110729.PDF (accessed 8th June 2015); See also SPLC, Southern Poverty Law Center Files Federal Lawsuit Targeting Use of Mace on Birmingham Schoolchildren, 1st December 2010; SPLC, Effects of Macing Schoolchildren (undated). [These and all other SPLC documents available from: SPLC website (http://www.splcenter.org/ last accessed 8nd June 2015)].
82. SPLC, case docket, *J.W. ex rel. Williams v. A.C. Roper*, available at http://www.splcenter.org/get-informed/case-docket/jw-et-al-v-birmingham-board-of-education (accessed 8th June 2015).
83. U.S. District Court for the Northern District of Alabama, Southern Division, *J.W. et al., V. Birmingham Board of Education, et al.*, Memorandum Opinion and Order, 31st August 2012, available at http://www.splcenter.org/sites/default/files/downloads/case/Memorandum_ Opinion_and_Order_Class_Cert.pdf (accessed 8th June 2015).
84. SPLC, SPLC goes to trial today to curtail use of pepper spray on children in Birmingham, Ala., public schools, 20th January 2015, available at http://www.splcenter.org/

85. US District Court for the Northern District of Alabama Southern Division, *J.W. et al., v. Birmingham Board of Education, et al.*, Civil Action Number 2:10-cv-03314-AKK, 30 September 2015, available at https://www.splcenter.org/sites/default/files/documents/findings_of_fact_and_conclusions_of_law.pdf (accessed 2 October 2015).
86. US District Court for the Northern District of Alabama Southern Division (30 September 2015) *op.cit.*, p. 3.
87. US District Court for the Northern District of Alabama Southern Division (30 September 2015) *op.cit.*, p. 2.
88. US District Court for the Northern District of Alabama Southern Division (30 September 2015) *op.cit.*, pp. 118–19.
89. For a detailed analysis of the incidents, see UC Davis November 18, 2011 "Pepper Spray Incident" Task Force Report "The Reynoso Task Force Report", March 2012, available at http://reynosoreport.ucdavis.edu/reynoso-report.pdf (accessed 8th June 2015).
90. US District Court for the Eastern District of California, *Baker v. Katehi*, Case No. 2:12-cv-00450, see the report on the University of Michigan Law School's Civil Rights Litigation Clearing House website: http://www.clearinghouse.net/detail.php?id=12478 (accessed 8th June 2015).
91. US District Court for the Eastern District of California, *Baker v. Katehi*, Case No. 2:12-cv-00450, Stipulation for Settlement, 26th September 2012, available at https://www.aclunc.org/sites/default/files/uc_ davis_settlement_agreement.pdf (accessed 3rd June 2015); *Baker v. Katehi*, Closed case, January 10, 2013, American Civil Liberties Union of Northern California, available at https://www.aclunc.org/our-work/legal-docket/baker-v-katehi (accessed 8th June 2015).
92. ICRC, *Preventing Hostile Use of the Life Sciences, from Ethics and Law to Best Practice*, 11th November 2004.
93. *Ibid.*
94. For example, see BTWC Third Review Conference Final Document, Part II, BWC/CONF.III/23, 9th–27th September 1991, p. 3; Report of the Second CWC Review Conference, RC-2/4, 18th April 2008.
95. An education project conducted under the auspices of IUPAC has employed the term "multi-use" rather than "dual use" when discussing the use and misuse of chemical agents. See Mahaffey, P. Multiple Uses of Chemicals, IUPAC, 2007, available on IUPAC website of: Raising Awareness – Multiple Uses of Chemicals and the Chemical Weapons Convention, IUPAC Project 2005-029-1-050, http://multiple.kcvs.ca/ (accessed 2nd April 2015).
96. National Research Council, *Biotechnology Research in an Age of Terrorism* (Fink Report), Washington, DC: National Academies Press, 2004.
97. National Research Council, *Globalisation, Biosecurity and the Future of the Life Sciences* (Lemon Report), Washington, DC: National Academies Press, 2006.
98. National Research Council (2006) *op.cit.*, p. 216
99. For a discussion of such initiatives, see Rappert, B. The Benefits, Risks, and Threats of Biotechnology, *Science & Public Policy*, volume 35, number 1, 2008, pp. 37–44.
100. Rappert, B. Introduction: Education as... in *Education and Ethics in the Life Sciences: Strengthening the Prohibition of Biological Weapons*, Australian National University E-Press, 2010, p. 8.
101. Journal Editors and Authors Group, *Proceedings of the National Academies of Science*, volume 100, number 4, 18th February 2003, p. 1464.

102. Van Aken, J. and Hunger, I. Biosecurity Policies at International Life Science Journals, *Biosecurity and Bioterrorism*, volume 7, number 1, 2009, pp. 61–72.
103. Rappert, B. (2010) *op.cit.*, p. 8. Rappert cited the UK Biotechnology and Biological Sciences Research Council, the UK Medical Research Council, the Wellcome Trust, the Center for Disease Control, and the Southeast Center of Regional Excellence for Emerging Infectious Diseases and Biodefence.
104. Rappert, B. (2010) *op.cit.*, p. 9.
105. *Ibid.*
106. US National Science Advisory Board for Biosecurity, Proposed Framework for the Oversight of Dual Use Life Sciences Research: Strategies for Minimizing the Potential Misuse of Research Information, June 2007.
107. Sunshine Project, *Earth Calling NSABB: Voluntary Compliance Won't Work*, Press Release, 17th April 2007.
108. See for example: Sunshine Project, *Mandate for Failure: The State of IBCs in an Age of Bioweapons Research*, 2004.
109. Supporting correspondence available at http://www.sunshine-project.org/publications/pr/support/deregistries.pdf (last accessed 1st May 2012, subsequently removed).
110. Sunshine Project (17th April 2007) *op.cit.*
111. American Society of Microbiology, Code of Ethics (Revised and approved by ASM Council in 2005). Washington, DC.
112. Royal Society, *The Roles of Codes of Conduct in Preventing the Misuse of Scientific Research*. RS Policy Document 3/05. London: Royal Society, 2005.
113. See, for example, Revil, J. and Dando, M. A Hippocratic Oath for Life Scientists, *EMBO Reports*, volume 7, 2006; Atlas, R. and Somerville, M. Life Sciences and Death Science: Tipping the Balance Towards Life with Ethics, Codes and Laws, in McLeish, C. and Rappert, B. (2007) *op.cit.*, pp. 15–30.
114. Pearson, G. and Mahaffy, P. Education, Outreach, and Codes of Conduct to Further the Norms and Obligations of the Chemical Weapons Convention (IUPAC Technical Report), *Pure and Applied Chemistry*, volume 78, 2006, p. 2186.
115. Pearson, G. and Mahaffy, P. (2006) *op.cit.*, p. 2186.
116. *Ibid.*
117. IUPAC Project: CHEMRAWN XVIII – Ethics, Science and Development, http://www.iupac.org/web/ins/2009-013-1-021 (accessed 2nd April 2015).
118. See Pearson, G., Becker, E. and Sydnes, L. Why Codes of Conduct Matter, *Chemistry International*, volume 33, number 6, November–December 2011, pp. 7–11. See also IUPAC, Statement by Prof. Alastair W. M. Hay and Prof. Graham S. Pearson on Behalf of the International Union of Pure and Applied Chemistry to the Meeting of Experts, BTWC Meeting of Experts, 2008.
119. For further information, see Rappert, B. Biological Weapons and Codes of Conduct, 2010, http://projects.exeter.ac.uk/codesofconduct/Chronology/index.html (accessed 2nd April 2015).
120. Rotblat, J. Letter to Workshop, Report of 2nd Pugwash Workshop Science, Ethics and Society, Ajaccio, Corsica, 10th–12th September 2004, http://web.archive.org/web/20130116075455/http://www.pugwash.org/reports/ees/corsica2004/corsica2004.html (accessed 13th July 2015).
121. Rappert, B. Biological Weapons and the Life Sciences: The Potential for Professional Codes, in *Disarmament Forum*, 2005 volume one, Science Technology and the CBW regimes, Geneva: UNIDIR, 2005, pp. 53–62.
122. See, for example, Bell, C. Pledge by Neuroscientists to Refuse to Participate in the Application of Neuroscience to Violations of Basic Human Rights

or International Law, 2010, available at http://spreadsheets.google.com/viewform?formkey=dEF4RFhhSWZwNktCakYtbTdkd1cxckE6MA (accessed 14th July 2015); Bell, C. Neurons for Peace: Take the Pledge, Brain Scientists, *New Scientist*, Magazine issue 2746, 8th February 2010; Bell. C. Why Neuroscientists Should Take the Pledge: A Collective Approach to the Misuse of Neuroscience, in Giodarno, J. (ed.), *Neurotechnology in National Security and Defense: Practical Considerations, Neuroethical Concerns*. Boca Raton, FL: CRC Press, 2014, pp. 227–238.

123. Perry-Robinson. J. *Near-Term Development of the Governance Regime for Biological and Chemical Weapons*, Science and Technology Policy Research (SPRU), 4th November 2006, p. 16.

124. Corneliussen, F. Adequate Regulation, a Stop-Gap Measure, or Part of a Package? Debates on Codes of Conduct for Scientists Could Be Diverting Attention Away from More Serious Questions, *EMBO Reports*, volume 7 (European Molecular Biology Organization Special Issue), 2006 July, p. 53.

125. Corneliussen, F. (2006) *op.cit.*, p. 54.

126. Rappert, B. (2010) *op.cit.*, p. 14.

127. See, for example: OPCW, Report of the First CWC Review Conference, 28th April–9th May 2003, RC-1/5, 9th May 2003, paragraphs 7.79 and 7.83.d; OPCW, Report of the Second CWC Review Conference, RC-2/4, 7th–18th April 2008, paragraph 9.77.

128. Trapp, R. *OPCW Activities and Perspectives on the Content, Promulgations, and Adoption of Codes of Conduct for Scientists*, Third Meeting of the States Parties of the Biological Weapons Convention, Meeting of Experts, Geneva, 13th–24th June 2005, p. 5.

129. OPCW, Report of the OPCW on the implementation of the CWC in the year 2001, Conference of State Parties, 70th Session C-7/3, 7th–11th October 2002, 10th October 2002, paragraph 3.3 (e).

130. Trapp, R. (2005) *op.cit.*, p. 6.

131. See, for example: OPCW (10th October 2002) *op.cit.*, paragraph 3.3 (e); OPCW, Report of the OPCW on the implementation of the CWC in the year 2002, Conference of State Parties, Eighth Session C-8/5, 20th–24th October 2003, 22nd October 2003, paragraph 3.29.

132. IUPAC, Project: Educational material for raising awareness of the Chemical Weapons Convention and the multiple uses of chemicals, 2005; Hay, A. Multiple Uses of Chemicals: Clear Choices or Dodgy Deals? *Chemistry International*, volume 29, number 6, November–December 2007, pp. 23–25.

133. Trapp, R. (2005) *op.cit.*, p. 6.

134. For example, although activities of the Ethics Project are reported up to 2006 on the OPCW website, no further information appears to be publicly available after this date. See http://www.opcw.org/ (accessed 2nd April 2015).

135. Balali-Mood, M., Steyn, P., Sydnes, L. and Trapp, R. Impact of Scientific Developments on the Chemical Weapons Convention, IUPAC Technical Report, *Pure and Applied Chemistry*, volume 1, 2008, p. 181.

136. See, for example: Updating, Piloting, and Disseminating Educational Material for Raising Awareness of the Multiple Uses of Chemicals and the Chemical Weapons Convention, IUPAC Project No. 2013-020-1-050. For further information on the project see Multiple Uses of Chemicals: IUPAC and OPCW Working Toward Responsible Science, *Chemistry International*, September–October 2014, pp. 9–13.

137. For a discussion of the OPCW activities, see OPCW, Education and Engagement, Final report of the Temporary Working Group on Education and Outreach in Science and Technology Relevant to the Chemical Weapons Convention, November 2014, SAB/REP/2/14.
138. See, for example, educational materials on the "multiple uses of chemicals" website (http://multiple.kcvs.ca/), which was developed by an interdisciplinary team from the Kings Centre for Visualization in Science, working together with scientists and educational specialists from both IUPAC and the OPCW.
139. United Nations, Report of the Meeting of States Parties, BWC/MSP/2008/5, 12th December 2008, paragraph 26.
140. Rappert. B, Chevrier, M. and Dando, M. In-Depth Implementation of the BTWC: Education and Outreach, *Bradford Review Conference Papers*, number 18, November 2006; Online Bradford Project on Strengthening the Biological and Toxin Weapons Convention (BTWC); Whitby, S. and Dando, M. Chapter 10: Biosecurity Awareness-Raising and Education for Life Scientists: What Should Be Done Now?, in *Education and Ethics in the Life Sciences: Strengthening the Prohibition of Biological Weapons*, Australian National University E-Press, 2010.
141. Dando, M. Biologists Napping While Work Militarized, *Nature*, volume 460, issue 7258, 2009, p. 951. Dando's concerns are echoed in an accompanying *Nature* editorial entitled: A Question of Control: Scientists Must Address the Ethics of Using Neuro-Active Compounds to Quash Domestic Crises.
142. Rappert, B., Chevrier, M. and Dando, M. (2006) *op.cit.*, paragraph 18.
143. Whitby, S. and Dando, M. (2010) *op.cit.*, p. 182.
144. For further discussion, see Dando, M. Dual-Use Education for Life Scientists, *Disarmament Forum*, volume 1, 2009, pp. 41–44.
145. This is, in part, due to the very limited resources and restricted mandate of the BTWC ISU (see Chapter 6 for further discussion).
146. For further information concerning the Project on Building a Sustainable Capacity in Dual-Use Bioethics see Bradford University website: http://www.brad.ac.uk/bioethics/About/ (accessed 10th August 2015).
147. Educational Module Resources: see Bradford University website: http://www.brad.ac.uk/bioethics/educationalmoduleresource/ (accessed 10th August 2015).
148. Lecture 17, Weapons Targeted at Nervous System; Lecture 17.b Weapons Targeted at Nervous System [applied version], available from Bradford University Bioethics website, http://www.brad.ac.uk/bioethics/media/ssis/bioethics/emr/englishemr/lecture_ No_17_Applied_Version.pdf (accessed 10th August 2015).
149. All available from Bradford University Bioethics website http://www.brad.ac.uk/bioethics (accessed 10th August 2015).
150. Rappert, B. (2010) *op.cit.*, p. 16.
151. See, for example: Maclean, A. *Historical Survey of the Porton Down Volunteer Programme, Part 4: Human Studies with Incapacitating Agents*, London: Ministry of Defence, June 2006, pp. 109–142; Ketchum, J. S. Chemical Warfare: Secrets Almost Forgotten, self-published, 2006; Gould, C. and Folb, P. *Project Coast: Apartheid's Chemical and Biological Warfare Programme*, UNIDIR, 2002.
152. See, for example: Hess, L. et al. (May 2005) *op.cit.*
153. Gross, M. *Medicalized Weapons and Modern War*, Hastings Center Report 40, number 1, 2010, pp. 34–35.
154. Gross, M. (2010) *op.cit.*, p. 41.

155. Also of relevance are the obligations relating to research and human experimentation established under the Nuremberg Code, requiring voluntary and informed consent of the subjects in such activities. See Nuremberg Code, 1947, Article 1.
156. Adopted by the 29th World Medical Assembly, Tokyo, Japan, October 1975, and editorially revised at the 170th Council Session, Divonne-les-Bains, France, May 2005, and the 173rd Council Session, Divonne-les-Bains, France, May 2006.
157. World Medical Association, Regulations in Times of Armed Conflict, Adopted by the 10th World Medical Assembly, Havana, Cuba, October 1956, last amended by the 35th World Medical Assembly, Tokyo, Japan, 2004 and editorially revised at the 173rd Council Session, Divonne-les-Bains, France, May 2006, paragraph 2.
158. World Medical Association, Declaration on Chemical and Biological Weapons, Adopted by the 42nd World Medical Assembly Rancho Mirage, CA, USA, October 1990 and rescinded at the WMA General Assembly, Santiago 2005.
159. *Ibid.*
160. World Medical Association, Washington Declaration on Biological Weapons, Adopted by the 53rd WMA General Assembly, Washington, DC, USA, October 2002 and editorially revised by the 164th WMA Council Session, Divonne-les-Bains, France, May 2003, and reaffirmed by the 191st WMA Council Session, Prague, Czech Republic, April 2012, paragraphs 18 and 19.
161. For example, the US Army textbook of military ethics has urged medical professionals "to stay in the business of healing and not hurting, which includes not participating in or contributing to weapons research and development". Frisina, M. E. Medical Ethics in Military Biomedical Research, in Beam, T. E. and Sparacino, L. R. (eds), *Military Medical Ethics*, volume 2, Washington, DC: Office of the Surgeon General/Borden Institute, 2003, pp. 533–361, as cited in Gross, M. (2010) *op.cit.*, p. 37.
162. Health Professions Council of South Africa, *Guidelines for Good Practice in the Health Professions: Research, Development and Use of Chemical and Biological Weapons*, 2nd edn, booklet 9, May 2007, Pretoria.
163. HPCSA (2007) *op.cit.*, paragraph 1.2.
164. *Ibid.*, paragraph 1.3.
165. *Ibid.*, paragraph 3.1.
166. *Ibid.*, paragraph 3.2. See paragraphs 3.3 and 3.4 for further elaboration.
167. *Ibid.*, Introduction: The Spirit of Professional Guidelines.
168. *Ibid.*, paragraph 5.
169. HPCSA, Before the Professional Conduct Committee of the Medical and Dental Professions Board, In the matter of Dr W. J. Basson, accused, Charge sheet and annexures, undated, available at http://www.hpcsa.co.za/downloads/inquiries/basson_charge_sheet.pdf (accessed 14th July 2015).
170. *SABC News*, Basson's Application to Bar HPCSA Inquiry Dismissed, 10th May 2010.
171. *HSPCA Press Release*, HSPCA Applauds High Court Ruling on Basson, 10th May 2010.
172. Smith, J. Wouter Basson to Face South Africa Misconduct Hearing, *The Guardian*, 27th January 2012.
173. HPCSA, Dr W. J. Basson, accused, Charge sheet and annexures (undated) *op.cit.*, charge 2.2.
174. *Ibid.*, charge 4.1.
175. *Ibid.*, charge 4.2.

176. *Ibid.*, charge 5.
177. HPCSA, Professional Conduct Committee: Concerning Dr W. Basson, Reasons of the Committee, Issued by the HPCSA, 18th December 2013.
178. HPCSA, PCC (18th December 2013) *op.cit.*
179. *Ibid.*
180. *Ibid.*
181. HPCSA, Dr Wouter Basson Found Guilty of Unprofessional Conduct, *the e-bulletin*, http://web.archive.org/web/20140311012114/http://www.hpcsa-blogs.co.za/dr-wouter-basson-found-guilty-of-unprofessional-conduct (accessed 15th July 2015).
182. *Ibid.*
183. See, for example: World Medical Association, Declaration of Helsinki, *Adopted by the 18th WMA General Assembly, Helsinki, Finland, June 1964 and last amended at the 59th WMA General Assembly, Seoul, October 2008*; Nuremberg Code, as detailed in *Trials of War Criminals Before the Nuremberg Military Tribunals Under Control Council Law No. 10*, volume 2, Washington, DC: U.S. Government Printing Office, 1949, pp. 181–182.
184. Nathanson, V. Ethical Issues for Health Professionals, in *Incapacitating Chemical Agents: Implications for International Law*, ICRC, Expert meeting, Montreux, Switzerland, 24–26th March 2010, pp. 29–30.
185. British Medical Association, *The Use of Drugs as Weapons: The Concerns and Responsibilities of Healthcare Professionals*, London: BMA, 2007.
186. British Medical Association (2007) *op.cit.*, p. 20.
187. *Ibid.*, p. 24.
188. A review was undertaken of all relevant publicly available WMA documentation from January 1990 to August 2015.
189. Rotblat, J. Remember Your Humanity, Acceptance and Nobel Lecture, Oslo, 1995, available at http://www.nobelprize.org/nobel_prizes/peace/laureates/1995/rotblat-lecture.html (accessed 14th July 2015).
190. Deiseroth, D. (2000) *op.cit.*, p. 266.
191. Falter. A., (2010) *op.cit.*, p. 289.
192. International Committee of the Red Cross, *Preventing Hostile Use of the Life Sciences* (11th November 2004) *op.cit.*
193. International Committee of the Red Cross, *Preventing Hostile Use of the Life Sciences* (11th November 2004) *op.cit.*
194. South African Protected Disclosures Act 26 of 2000, available at http://www.justice.gov.za/legislation/acts/2000-026.pdf (accessed 14th July 2015).
195. UK Government, Public Interest Disclosure Act 1998 (PIDA).
196. US Federal Whistleblower Protection Act (5 USC sec. 1201), 9th July 1989.
197. Martin, B. Whistle-Blowing: Risks and Skills, in Feakes, D., Rappert, B. and McLeish, C. (eds), *Web of Prevention* (2007) *op.cit.*, p. 6.
198. For example, the AAAS Science and Human Rights Program (SHRP), which works with scientists to "advance science and serve society" through human rights, has successfully campaigned on behalf of a number of scientific whistle-blowers, further information is available at http://www.aaas.org/program/scientific-responsibility-human-rights-law (accessed 2nd April 2015).
199. Dando. M., Pearson. G., Rozsa. L., Robinson. J. and Wheelis, M. Analysis and Implications, in Wheelis, M., Rozsa. L. and Dando. M. (eds), *Deadly Cultures*, Cambridge, MA: Harvard University Press, 2006, p. 373.

200. Perry Robinson, J. P. Scientists and Chemical Weapons Policies, in Finney, J. and Slaus, I. (eds), *Assessing the Threat of Weapons of Mass Destruction*. Amsterdam, Netherlands: IOS Press, 2010, p. 89.
201. *Statement of the International Committee of the Red Cross, First Special Session of the Conference of the States Parties to Review the Operation of the Chemical Weapons Convention*, The Hague, 28th April–9th May 2003.
202. Ruppe, D., CWC: Red Cross Says It Was Muzzled Over Stand on Incapacitating Weapons, *Global Security Newswire*, 30th April 2003.
203. Harvard Sussex Program, *Open Forum on the Chemical Weapons Convention: Challenges to the Chemical Weapons Ban*, 1st May 2003; OPCW, *Open Forum: Chemical Weapons Convention – Recent Experience and Future Prospects*, 9th April 2008.
204. Panel discussion: 'The Chemical Weapons Ban and the Use of Incapacitants in Warfare and Law Enforcement', Harvard Sussex Program (2003) *op.cit.*, pp. 23–39.
205. Wheelis, M. *Toxic Chemicals and Law Enforcement*, Washington, DC: Scientists' Working Group on Biological and Chemical Weapons, Center for Arms Control and Nonproliferation, 2008.
206. Editorial, *CBW Conventions Bulletin 60*, Harvard Sussex Program (June 2003) *op.cit.*, pp. 4–5.
207. See for example, Crowley, M. Dangerous Ambiguities: Regulation of Incapacitants and Riot Control Agents Under the Chemical Weapons Convention, 2nd December 2009, Open Forum, CSP-14, OPCW, The Hague, Netherlands; Crowley, M. Regulation of Incapacitants, Riot Control Agents and Their Means of Delivery Under the Chemical Weapons Convention, OPCW, Open Forum Meeting, 29th November 2010.
208. Chemical Weapons Convention Coalition (CWCC): Founding Document, agreed by consensus by the participants at the meeting to establish the CWCC, 2nd–3rd December 2009 in The Hague, Netherlands.
209. For example, the Second Annual Meeting of the CWCC was hosted by the OPCW and was addressed by the Director General. The full details of the meeting, together with a video of the Director General's interventions, are available on the OPCW website. See OPCW Hosts Second General Meeting of the Chemical Weapons Convention Coalition, 15th April 2011, http://www.opcw.org/news/article/opcw-hosts-2nd-general-meeting-of-chemical-weapons-convention-coalition/ (accessed 1st September 2012) and http://www.opcw.org/fileadmin/OPCW/video/DG-CWCC_01.mp4 (accessed 1st September 2012).
210. Available on OPCW website, http://www.opcw.org/our-work/readings/ (accessed 14th July 2015).
211. Discussion with Technical Secretariat official, 28th November 2011.
212. Schneidmiller, C., New Coalition Aims to Promote Chemical Weapons Disarmament, Nonproliferation, *Global Security Newswire*, 22nd January 2010, available at http://www.nti.org/gsn/article/new-coalition-aims-to-promote-chemical-weapons-disarmament-nonproliferation/ (accessed 14th July 2015).
213. Schneidmiller, C. (2010) *op.cit.*
214. OPCW, Report of the Third Review Conference, Part B (19th April 2013) *op.cit.*, paragraph 1.6. Rule 33 of the Rules of Procedure was amended to read: "Representatives of non-governmental organisations may attend the plenary sessions of the Conference, and participate in the activities of the review conferences, in accordance with such rules or guidelines as the Conference has approved."

215. Guthrie, R. CWC Review Conference Report, number 4, Completion of the General Debate and the Start of the Thematic Review, 11th April 2013.
216. Statement on behalf of Bradford Non-Lethal Weapons Project and the Omega Research Foundation to be delivered to the CWC Review Conference. Michael Crowley, University of Bradford and Joe Farha, Omega Research Foundation, 11th April 2013.
217. Guthrie, R. CWC Review Conference Report, number 1, The Third CWC Review Conference: Setting the Scene, 8th April 2013.
218. Details of all Statements and National Papers delivered by States Parties at the Third CWC Review Conference are available on the OPCW website at: https://www.opcw.org/documents-reports/conference-states-parties/third-review-conference/national-statements/ (accessed 2nd May 2015).
219. OPCW, Report of the Third Review Conference, Part B (19th April 2013) *op.cit.*, paragraph 9.144.
220. *Ibid.*, paragraph 9.155 (n).

13 Conclusions

1. Organisation for the Prohibition of Chemical Weapons (OPCW), Convention on the Prohibition of the Development, Production, Stockpiling and Use of Chemical Weapons and on Their Destruction (Chemical Weapons Convention [CWC]), 1993, Article VIII.22.
2. The utility of such a model for addressing CWC-related issues requiring clarification has previously been proposed. See Mathews, R. Convergence of Biology and Chemistry: Implications for the Verification Regime of the Convention, in Mashhadi, H., Paturej, K., Runn, P. and Trapp, R. (eds), *Seminar on the OPCW's Contribution to Security and the Non-Proliferation of Chemical Weapons.* 11th–12th April 2011, pp. 178–179.
3. Such mechanisms could be facilitated by independent, expert and respected bodies such as the ICRC or the Pugwash Conferences on Science and World Affairs. As they would be outside the OPCW, such processes could address constraints imposed upon ICA weapons and RCA development and use under all relevant international law (e.g. BTWC, IHL, IHRL, UN Drugs Conventions) not just the CWC, and could present their findings to all relevant control regimes.
4. OPCW, Switzerland, *Riot Control and Incapacitating Agents Under the Chemical Weapons Convention*, RC-2/NAT.12, 9th April 2008, p. 5.
5. And potentially Article I.5 of the CWC, if the means of delivery contains an RCA.
6. OPCW, CWC (1993) *op.cit.*, Article III.1.(e).
7. The permissibility of developing, stockpiling, transferring and using chemical agents other than RCAs for law enforcement purposes (such as ICAs) is currently contested and will remain so until States Parties establish their status under the Convention.
8. Ekeus panel, Report of the advisory panel on future priorities of the Organisation for the Prohibition of Chemical Weapons, OPCW Director General, S/951/2011, 25th July 2011, p. 18, paragraph 71.
9. OPCW, Note by the Technical Secretariat, The OPCW in 2025: Ensuring A World Free of Chemical Weapons, Office of Strategy and Policy, S/1252/2015, 6th March 2015, paragraphs 21 and 23.

10. For example, such constraints could include a prohibition on the use of RCAs in confined spaces or against restrained individuals.
11. See for example: ICRC, Toxic Chemicals as Weapons for Law Enforcement: A Threat to Life and International Law? Synthesis document, September 2012.

Index

Note: locators followed by n refer to notes.

Advanced Riot Control Agent Technology (ARCAT), 12
Alien Tort Statute (ATS), 239–40
  Case of *Abu-Zeineh vs. Federal Laboratories Inc*, 239
  Case of *Kiobel v. Royal Dutch Petroleum Co*, 239
Amnesty International (AI), 69, 71, 76, 77, 78, 79–80, 91, 128, 211
Armament Research, Development and Engineering Center (ARDEC), 102
arms embargoes, European Union, 218–20, 275
  effectiveness, 217, 218
  items covered, 220–1
  monitoring/verification, 217–18
arms embargoes, United Nations, 215–18, 275
  effectiveness, 218
  items covered, 216–17
  monitoring/verification, 217–18
Arms Trade Treaty (ATT), 205, 222
Australia, contemporary position on ICA weapons, 131
Australia Group (AG), 199, 202–4, 222
  items controlled, 203
  transfer control criteria, 203–4

Bahrain, misuse of RCAs/means of delivery by law enforcement officials, 48, 49, 76–7, 237–8
Bahrain Watch, 234–5
Bangladesh, misuse of RCAs by law enforcement officials, 77
Basson, Wouter, 195, 252–4
biochemical threat spectrum, 9–10
Biological and Toxin Weapons Convention (BTWC), 109, 136–42, 150, 152, 196–7, 201–2, 222, 241–2, 244, 246, 247, 255, 258, 260, 265–6, 268, 269, 272
  activities regulated, 140–1
  agents covered, 135–6
  biological weapons, defined, 136–7
  effectiveness of, 141–2
  ICA weapons and, 136–7, 138–9
  object and purpose, 136–7
  RCAs and, 139–40
  Review Conferences, 136, 137, 138, 139, 142, 202, 247, 257, 260
  transfer controls, 201–2
bioregulators, 2, 36, 137, 140
Bradford Non-Lethal Weapons Research Project (BNLWRP), 27, 233, 258
Brazil, misuse of RCAs by law enforcement officials, 70–1
British Medical Association (BMA), 254–5
B'Tselem, 45, 71–2

Centre for Non-Proliferation Studies, 83
Chemical Weapons Convention Coalition (CWCC), 258–9
Chemical Weapons Convention (CWC), 1, 13, 31, 32, 37, 44, 80, 81, 95, 102, 107–33, 135, 138, 142, 145–6, 149, 151, 158, 171, 187, 193, 196–7, 199–200, 202, 203, 206, 207, 233, 234, 237, 244, 246, 248, 255, 257–9, 268–9, 276
  chemical weapons, definition of, 109–10
  definitional lacunae, 110–11
  general obligations, 107
  general purpose criterion, 109–10
  ICA weapons; implications of use by Russian Federation, 126–9; prohibition of use in armed conflict, 124; under CWC, 123–32; use in law enforcement, 123–32
  malodorants, 113–14
  means of delivery, definition of, 119–20
  operative provisions, 112

purposes not prohibited, 109–10
RCAs, definition of, 113; means of delivery, 119–23; prohibition as method of warfare, 115; regulation of use in law enforcement, 115–18; reporting and transparency measures, 14, 118–19
scheduled chemicals, 109, 200–1
toxic chemical, definition of, 108–9
transfer controls, 199–200, 202
China
development of "wide area" RCA means of delivery, 95, 101
misuse of RCAs by law enforcement officials, 54
possession and promotion of ICA weapons, 20–1
research relating to ICAs, 21–2
China Ordnance Institute, 100
Civil law class action, 240–2
Case of *Baker v. Katehi*, 241
Case of *J.W. ex rel. Williams v. A.C. Roper*, 240–1
civil society, 229–62, 275–7
codes of conduct/ethical standards, 244–6, 249–55, 275–6
health professionals, 249–55
life and chemical scientists, 244–6
Combined Systems, Inc, 97
Cote d'Ivorie, misuse of RCAs by law enforcement officials, 55
Czech Republic
prohibition on use of ICA weapons for law enforcement, 131
research relating to ICAs, 22–5

Dae-Kwang Chemical Company Ltd, 235
Defence Research and Development Organisation (DRDO), 25–7
Defence science and technology laboratory (DSTL), 15, 30, 31
Department of Defense (DOD), 33
Desert Wolf, 104
"Dual-use" research, 242–4, 275
Fink report, 242
Lemon report, 242
regulation of, 241–4

Economic Community of West African States (ECOWAS), 218
The Economist, 46
EC Regulation [1236/2005], 211–15
measures to strengthen the regulation, 214
range of goods covered, 212
Edgewood Chemical Biological Center (ECBC), 31, 32, 41, 45, 103
education and awareness raising, 247–9
amongst chemical science community, 246–7
amongst life science community, 247–8
Egypt, misuse of RCAs by law enforcement officials, 74–5
Equatorial Guinea, misuse of RCAs by law enforcement officials, 79–80
Ettlingen Non-Lethal Weapons Symposium, 22, 23, 24, 29
EU Common Position 2008/944/CFSP, 208–11, 275
range of items covered, 210
transfer control criteria, 208–10
European Committee for the Prevention of Torture and Inhuman or Degrading Treatment or Punishment (CPT), 174, 182, 183
European Convention on Human Rights (ECHR), 170, 171, 186
European Court of Human Rights (EctHR), 183
Case of *Abdullah Yaşa and Others v. Turkey*, 185–6
Case of *Ali Güneş v. Turkey*, 178, 185–6
Case of *Ataykaya v. Turkey*, 186
Case of *Finogenov and others v. Russia*, 169–71
Executive Order 11850 (US), 115, 135

Federal Laboratories, 50
The Financial Times, 235

General Dynamics, 102–3
Geneva Conventions, 152, 154, 156, 157, 159, 163, 164, 181, 187, 195, 273
Additional Protocol I, 152, 154–5, 156, 157, 158, 159–64, 187
Additional Protocol II, 152, 155

Geneva Protocol, 81, 134–6, 137, 139, 143, 145, 151, 152, 191, 193, 202
  effectiveness of, 136
  scope of coverage, 135–6
Germany, contemporary position on ICA weapons, 130–1
The Guardian, 75, 103

Harvard Sussex Draft Convention, 196–8
Harvard Sussex Program (HSP), 196, 197
Health Professions Council of South Africa (HPCSA), 251–4
holistic arms control (HAC), 4–7, 263–77
Honduras, misuse of RCAs by law enforcement officials, 73–4
Human Rights Foundation of Turkey (HRFT), 90
Human Rights Watch (HRW), 75, 76, 78–9, 85–9, 90, 94
Hurriyet Daily News, 94, 146

Iman Hossein University, 27
incapacitating chemical agent (ICA) weapons, 9–38
  concerns regarding, 12–14
  definition of, 10–11
  delivery systems, 20–1, 23, 34
  description of, 2–3
  lethality of, 16–17
  perceived operational utility of, 12
  see also incapacitating chemical agents
incapacitating chemical agents (ICAs), 9–38
  description of, 9–11
  dual-use research into, 19–35
  (potential) ICAs; alpha-2 adrenoceptor agonists, 19; anaesthetics, 11, 16, 16, 19, 20, 21, 22, 24, 25, 131, 212, 213; analgesics, 11, 22, 31, 32, 128; anticholinergics, 18; antidepressants, 11, 32; antipsychotics, 11, 32; benzodiazepines, 19; BZ (3-quinuclidinyl benzilate), 18, 96, 147, 203, 207, 253; carfentanil, 15; dexmedetomidine, 22, 23; fentanyl, 12, 15, 20, 22, 23, 25–7, 28, 29, 30, 127, 129, 207, 224, 226; glycollates, 18; ketamine, 23, 24; lysergic acid diethylamide (LSD), 18, 226; medetomidine, 22, 23, 27; midazolam, 22, 23, 226; naphylmedetomidine, 24; opioids, 11, 19, 23, 25, 30, 32; peptides, 138, 140; phencyclidine, 226; remifentanil, 15; sedatives, 11, 16, 19, 22, 26, 32, 131
  use in interrogation, 180
  use in "lethal injection" execution, 171–3
  use in torture or other cruel, inhuman or degrading treatment or punishment, 95, 117, 176–8
India
  Prohibition on forced narcoanalysis, 180
  research relating to ICAs, 25–7
  use of ICAs for narcoanalysis, 180
Inter-American Commission on Human Rights, 73
Inter-American Convention to Prevent and Punish Torture, 181
Inter-American Court of Human Rights, 274
International Code of Conduct for Private Security Service Providers (ICoC), 86
International Committee of the Red Cross (ICRC), 13, 14, 20, 36, 128, 153, 154, 155, 156, 157, 158–63, 168, 170, 188, 228, 241, 254, 256, 257–8
International Court of Justice (ICJ), 152, 156, 166
International Covenant on Civil and Political Rights (ICCPR), 167, 172, 175, 179
International Covenant on Economic, Social and Cultural Rights (ICESCR), 181
International Criminal Court (ICC) [and Rome Statute], 153, 177, 190–3, 198
International criminal law, 190–8
  internationalised hybrid/domestic courts and tribunals, 193–4
  national courts, 194–6
International Criminal Tribunal for the former Yugoslavia (ICTY), 153, 193
International Criminal Tribunal for Rwanda (ICTR), 193

International Federation of Human Rights (FIDH), 55
international humanitarian law (IHL), 152–65, 273
  Article 36 reviews of "new" weapons, 159–64
  constraints upon medical personnel, 155
  customary law relating to specific weapons, 158
  investigative and enforcement measures, 157–8
  principles of humanity and the dictates of public conscience ("Martens clause"), 157
  prohibition against indiscriminate attacks or deliberate attacks on civilians, 156
  protection of persons *hors de combat*, 154
  requirement to respect and ensure respect of IHL, 156–7
  scope of IHL, 153
  superfluous injury or unnecessary suffering (SIRUS), 155
international human rights law, (IHRL)
  considerations for RCA means of delivery, 184–7
  freedom of assembly and association, right to, 174
  health, right to, 176–81
  IHRL considerations in Article 36 reviews, 160–1
  liberty and security, right to, 175–6
  life, right to, (and restrictions on use of force), 167–74
  obligations to review and monitor all "new" weapons, 187–8
  opinion and expression, right to, 175
  protection from torture and other cruel, inhuman or degrading treatment or punishment, 176–81
  scope of human rights law, 166
International Narcotics Control Board (INCB), 225
International Union of Pure and Applied Chemistry (IUPAC), 19, 244–5, 247

Iran
  misuse of RCA/means of delivery by law enforcement officials, 49, 74, 92
  prohibition on use of ICA weapons for law enforcement, 131
  research relating to ICAs, 27–8
Iraq
  development of large calibre RCA munitions, 96
  misuse of RCAs by US private military company, 84
  UN investigation of chemical weapons use, 143
  use of chemical weapons by, 194, 195, 196
Israel
  development of malodorants, 45
  ICAs, research relating to, 17, 28–9
  ICA weapon, use, 28–9, 35
  misuse of malodorants, 71–2
  misuse of RCAs/means of delivery by law enforcement officials, 48, 49, 50, 69–70, 237–8
Israel Institute for Biological Research (IIBR), 28–9

Joint Non-Lethal Weapons Directorate (JNLWD), 103
Joint Non-Lethal Weapons Program (JNLWP), 44
Judge Advocate General (JAG) [Office of], 125, 139

law enforcement, meaning of, 315 n
"less-lethal" weapons, 14, 16, 34, 45, 71, 79, 90, 97, 98, 101, 105, 106, 135, 150, 156, 159, 174, 182, 187–8, 203, 205, 212, 249, 254, 255, 264, 274

Maldives, misuse of RCAs by law enforcement officials, 71
malodorants
  and the CWC, 44, 113–14
  development, 44–6
  Skunk, 45–6, 71–2
  use, 70

means of delivery and dispersal, for RCAs, 88–106
  automatic grenade launchers, 99–100
  cluster munitions, 104
  constraints under CWC, 113–14
  constraints under human rights law, 184
  fixed installation devices, 92
  hand held sprayers, 88–9
  hand thrown canisters and grenades, 89–91
  heliborne munition dispensers, 103–4
  high capacity RCA smoke dispersal systems, 96–7
  large back-back and tank sprayers, 95–6
  large calibre aerial munitions, 103–4
  mortar munitions, 101–2, 123, 232–4
  multiple munition launchers, 98
  rocket propelled grenades, 100
  unmanned aerial vehicles (UAVs), 104
  unmanned ground vehicles (UGVs), 97, 104
  water cannon, 93–4
  weapons-launched single projectiles, 89–90
  "wide area", 93–106
medical ethics, 180, 251–4, 277
military operations other than war (MOOTW), 141
military operations in urban terrain (MOUT), 97, 120, 121
Mistral, Inc., 45
MKEK, 102, 123, 232–3
MSI Delivery Systems, 96–7

National Institute of Justice (NIJ), 32
National Institutes of Health (NIH), 243
National Research Council (NRC), 20, 32, 36, 37
National Science Advisory Board for Biosecurity (NSABB), 243
NATO Research Technology Organisation (NATO RTO), 122, 126
The New York Times, 85
Nonlethal Technologies, Inc., 98
Norinco, 21
North Atlantic Treaty Organisation (NATO), 9, 23, 24, 83, 176

Omega Research Foundation (ORF), 47, 92, 93, 211, 232
OPCW fact-finding mission (FFM) in Syria, 149
OPCW/UN investigation in Syria, 146–9
Organisation of American States (OAS), 217
Organisation for Economic Co-operation and Development (OECD), 235
Organisation for the Prohibition of Chemical Weapons (OPCW), 1–2, 25, 27, 30, 31, 35, 40, 41, 42, 44, 101, 111, 112, 114, 117, 119, 122, 123, 128, 129, 130, 131, 132, 141, 144, 145, 146, 187, 201, 206, 232, 233, 246, 247, 248, 257–62, 262, 265, 269
  Conference of States Parties, 1, 112, 122, 131, 258–9, 268, 269, 271, 272
  Director General, 1–2, 11, 107, 113, 127, 232, 244
  Executive Council, 43, 112, 122, 132, 269
  Review Conferences of, 130–1
  Scientific Advisory Board, 20, 43, 113, 122, 123, 131, 269
  Technical Secretariat, 25, 107, 112, 113, 119, 122, 128, 232, 233, 258, 259, 266, 271

Papua New Guinea, misuse of RCAs by private security company, 85–6
Pennsylvania State University, research on ICAs, 11–12, 32
Pepperball Technologies, Inc, 236
Physicians for Human Rights (PHR), 48, 48, 237–8
Police Scientific Development Branch, 288 n
Project Coast, 195, 251–4
Pugwash, 230

QinetiQ, 105

riot control agents (RCAs), 3–4, 40–87
  CN (2-Chloroacetophenone), 40, 88, 89, 206

CR (dibenz (b, f)-1, 4-oxazepine, 42, 88, 89, 2069
CS (2-chlorobenzalmalononitrile), 40–2, 48, 49, 75, 81, 82, 84, 85, 89, 89, 93, 206
   definition of, 39
   DM (adamsite), obsolete as, 43–4, 206
   Executive Order 11850 and, 115
   health and safety concerns, 46–50
   misuse of RCAs/means of delivery by law enforcement officials, 50–70
   OC (oleoresin capsicum), 33, 42–3, 46, 47, 48, 88, 89, 93, 206
   PAVA (pelargonic acid vanillyamide), 43, 88, 89, 90, 206
   pepper spray, 70, 71, 77, 182, 240
   PS (chloropicrin), obsolete as, 44
   use in armed conflict, 80–2
   use by armed opposition groups, 84
   use in caves, 80–1
   use in "collective punishment", 71–2
   use in enclosed space, 70
   use with excessive force or firearms, 79
   use in law enforcement, 50
   use by non-State actors, 83–4
   use of out-of-date RCA products, 49–50
   use in peace-keeping operations, 82–3
   use by private military or security companies, 84–6
   use in suppression of freedom of expression and assembly, 76–9
   use in torture or cruel, inhuman, or degrading treatment or punishment, 50, 70–2
Royal Society, 10, 17, 19, 154–5, 156–7, 179, 237
Russian Federation
   development and promotion of "wide area" RCA means of delivery, 97–8, 103
   research relating to ICAs, 29–31
   use of ICA weapon, 15–16

SIRUS project, 155
societal monitoring and verification, 229–38
South Africa, development and promotion of "wide area" RCA means of delivery, 104
South Korea, use of RCAs by law enforcement officials, 49
Spain, development and promotion of "wide area" RCA means of delivery, 104–5
Spiez laboratory, 10
Sri Lanka, misuse of RCAs by armed opposition group, 84
staphylococcal enterotoxin B (SEB), 18–19
State 9616 Plant, 21
Sunshine Project, 232, 243–4
Switzerland, contemporary position on ICA weapons, 130–1
Syria, alleged use of ICA weapons, 146–9

tear gas, *see* riot control agents
Technorobot, 104
toxins, 109
transfer control mechanisms, 199–222, 275
Turkey
   destruction of "wide area" RCA mortar munition, 234
   development and promotion of "wide area" RCA means of delivery, 101, 233
   misuse of RCAs/means of delivery by law enforcement officials, 77–7, 90–1
   use of RCAs by military in armed conflict, 80

UN (authorised) peace operations, misuse of RCAs
   Stabilization Force (SFOR), 83
   UN Mission in Liberia (UNMIL), 83
   UN Stabilization Mission in Haiti (MINUSTAH), 83
UN Basic Principles on the Use of Force and Firearms by Law Enforcement Officials (UNBP), 167–8, 175, 187, 188
UN Body of Principles for the Protection of All Persons under Any Form of Detention or Imprisonment, 177, 180

UN Code of Conduct for Law
    Enforcement Officials (UNCoC),
    167, 176
UN Committee Against Torture (CAT),
    189
UN Convention on Psychotropic
    Substances (CSP), 226–8
UN Convention against Torture (CAT),
    176
UN General Assembly (UNGA), 205
UN High Commissioner for Human
    Rights, 77
UN Human Rights Committee, 128, 166,
    167, 169, 189
UN Human Rights Council, 174, 187,
    188
United Kingdom (UK)
    contemporary position on ICA
        weapons, 130–1
    research relating to ICAs, 31
United States (US)
    contemporary position on ICA
        weapons, 130–1
    development and promotion of "wide
        area" RCA means of delivery,
        94–6, 97, 99–100, 101
    misuse of RCAs by law enforcement
        officials, 50
    research and/or development of ICA
        weapons (1950s-1970s), 17–19
    research relating to ICAs, 31–5
    use of RCAs by military, 80–2, 83
UN Principles of Medical Ethics, 180
UN Secretary Generals Investigation
    Mechanism, 142–9
    employment in Syria, 146–9
    limitations, 144–5
    scope of coverage, 145–6, 149
UN Secretary General (UNSG), 79, 185
UN Security Council Resolution
    (UNSCR) 1540, 149–51
    effectiveness, 150–1
    scope of coverage, 150
UN Security Council (UNSC), 205
UN Single Convention on Narcotic
    Drugs (SCND), 223–4
UN Special Rapporteurs, on
    extra-judicial, summary or arbitrary
        executions, 74, 77, 91, 128,
        169, 189

promotion and protection of the right
    to freedom of opinion and
    expression, 76
the rights to freedom of peaceful
    assembly and of association, 77,
    174
rights to physical and mental health,
    183–4
the situation of human rights on
    Palestinian Territories occupied
    since 1967, 185
torture and other cruel, inhuman or
    degrading treatment or
    punishment, 77, 170–4, 176–7,
    182, 189, 201, 215
UN Subcommittee on the prevention of
    torture (SPT), 70, 71
UN Vienna Convention on the Law of
    Treaties, 101, 129–30
US Air-force Office for Scientific
    Research, 33

Vietnam War
    proposed use of BZ in, 18
    use of RCAs/means of delivery during,
        43, 81, 96, 135

Wassenaar Arrangement (WA),
    206–8
    range of items controlled,
        206–8
    transfer control criteria, 208
whistle-blowing, 255–7
World Health Organisation (WHO), 147,
    148, 224
World Medical Association (WMA), 250,
    251, 255, 276
World Organisation Against Torture
    (OMCT), 299 n

Zimbabwe
    misuse of RCAs by law enforcement
        officials, 74
    misuse of RCAs/means of delivery by
        military, 94
Zimbabwe Lawyers for Human
    Rights, 74

Printed and bound by CPI Group (UK) Ltd, Croydon, CR0 4YY